Current Topics in
Developmental Biology
Volume 38

Current Topics in Developmental Biology

Volume 38

Edited by

Roger A. Pedersen
Reproductive Genetics Division
Department of Obstetrics, Gynecology,
and Reproductive Sciences
University of California
San Francisco, California

Gerald P. Schatten
Departments of Obstetrics–Gynecology
and Cell and Developmental Biology
Oregon Regional Primate Research Center
Oregon Health Sciences University
Beaverton, Oregon

Academic Press
San Diego   London   Boston   New York   Sydney   Tokyo   Toronto

Cover photograph: Male and hermaphrodite flowers of an inflorescences of
Rumex acetosa. (Courtesy Charles Ainsworth, Wye College, University of London.)

This book is printed on acid-free paper. ∞

Academic Press
a division of Harcourt Brace & Company
525 B Street, Suite 1900, San Diego, California 92101-4495, USA
http://www.apnet.com

Academic Press Limited
24-28 Oval Road, London NW1 7DX, UK
http://www.hbuk.co.uk/ap/

International Standard Book Number: 0-12-153138-4

PRINTED IN THE UNITED STATES OF AMERICA
97 98 99 00 01 02 EB 9 8 7 6 5 4 3 2 1

Contents

1

Paternal Effects in *Drosophila*: Implications for Mechanisms of Early Development
Karen R. Fitch, Glenn K. Yasuda, Kelly N. Owens, and Barbara T. Wakimoto

2

Drosophila* Myogenesis and Insights into the Role of *nautilus
Susan M. Abmayr and Cheryl A. Keller

3

Hydrozoa Metamorphosis and Pattern Formation
Stefan Berking

4

Primate Embryonic Stem Cells
James A. Thomson and Vivienne S. Marshall

5

Sex Determination in Plants
Charles Ainsworth, John Parker, and Vicky Buchanan-Wollaston

6

Somitogenesis
Achim Gossler and Martin Hrabě de Angelis

Contributors

Numbers in parentheses indicate the pages on which the authors' contributions begin.

Susan M. Abmayr (35) Department of Biochemistry and Molecular Biology and Center for Gene Regulation, The Pennsylvania State University, University Park, Pennsylvania 16802

Charles Ainsworth (167) Plant Molecular Biology Laboratory, Wye College, University of London, Kent TN25 5AH, United Kingdom

Stefan Berking (81) Zoological Institute, University of Cologne, D-50923 Cologne, Germany

Vicky Buchanan-Wollaston (167) Plant Molecular Biology Laboratory, Wye College, University of London, Kent TN25 5AH, United Kingdom

Karen R. Fitch (1) Department of Genetics, University of Washington, Seattle, Washington 98195

Achim Gossler (225) The Jackson Laboratory, Bar Harbor, Maine 04609

Martin Hrabě de Angelis (225) Institute for Mammalian Genetics, 85758 Neuherberg, Germany

Cheryl A. Keller (35) Department of Biochemistry and Molecular Biology and Center for Gene Regulation, The Pennsylvania State University, University Park, Pennsylvania 16802

Vivienne S. Marshall (133) Wisconsin Regional Primate Research Center, University of Wisconsin, Madison, Wisconsin 53715-1299

Kelly N. Owens (1) Department of Genetics, University of Washington, Seattle, Washington 98195

John Parker (167) University Botanic Garden, Cambridge University, Cambridge CB2 1JF, United Kingdom

James A. Thomson (133) Wisconsin Regional Primate Research Center, University of Wisconsin, Madison, Wisconsin 53715-1299

Barbara T. Wakimoto (1) Departments of Genetics and Zoology, University of Washington, Seattle, Washington 98195

Glenn K. Yasuda (1) Department of Biology, Seattle University, Seattle, Washington 98122

Preface

This volume addresses developmental mechanisms in a variety of experimental systems, including plants, hydrozoans, *Drosophila,* and mammals, and focuses on several developmental processes, including sex determination, paternal effects, myogenesis, somitogenesis, and developmental potency of stem cells. The conceptual sequence of topics begins with paternal effects in *Drosophila,* continues with myogenesis in this system, then turns to metamorphosis in *Hydra,* derivation and differentiation of primate embryonic stem cells, and sex determination in plants, and ends with vertebrate somitogenesis. More specifically, the chapter by Fitch and co-authors combines classic studies of chromosome imprinting in insects that leads to paternal chromosome loss with recent genetic studies on the role of sperm-supplied products required for fertilization and initiation of embryogenesis. Abmayr and Keller summarize current knowledge about the molecular control of myogenesis in *Drosophila,* with particular emphasis on the role of the myogenic regulatory factor *nautilus.* The chapter by Berking provides a definitive review of metamorphosis and pattern formation in the model marine hydrozoan *Hydra.* By contrast, Thomson and Marshall concentrate on the derivation, differentiation, and potential clinical uses of primate embryonic stem cells, including the use of these cells as models for human embryogenesis. Ainsworth and co-authors examine the mechanisms of sex determination in flowering plants, asking the question, "Are there analogies with sex determination in animals?" Finally, the chapter by Gossler and Hrabě de Angelis eloquently reviews the processes of segmentation involved in vertebrate somitogenesis.

Together with other volumes in this series, this volume provides a comprehensive survey of major issues at the forefront of modern developmental biology. These chapters should be valuable to researchers in the fields of both animal and plant development, as well as to students and other professionals who want an introduction to current topics in cellular and molecular approaches to developmental biology. This volume in particular will be essential reading for anyone interested in the role of paternal factors in fertilization and early development, the use of primate embryonic stem cells to understand human development, or the mechanisms of somitogenesis, myogenesis, sex determination, and metamorphosis.

This volume has benefited from the ongoing cooperation of a team of participants who are jointly responsible for the content and quality of its material. The authors deserve the full credit for their success in covering their subjects in depth yet with clarity, and for challenging the reader to think about these topics in new ways. We thank the members of the Editorial Board for their suggestions of topics and authors. We thank Diana Myers in Madison for her exemplary administrative expertise, and Craig Panner and Melanie Gross at Academic Press in San Diego for their unwavering editorial support. We are also indebted to the researchers who took time from their busy academic lives to share their knowledge in these chapters, and to their funding agencies for enabling their research, and thus their insights. Research and teaching are said to be related in the same way as are sin and confession: if you have not done any of the former, you will not have much to say in the latter.

Gerald P. Schatten
Roger A. Pedersen

1

Paternal Effects in *Drosophila:* Implications for Mechanisms of Early Development

Karen R. Fitch,[1] Glenn K. Yasuda,[2] Kelly N. Owens,[1] and Barbara T. Wakimoto[1,3]
Departments of Genetics[1] and Zoology[3]
University of Washington
Seattle, Washington 98195

Department of Biology[2]
Seattle University
Seattle, Washington 98122

The study of paternal effects on development provides a means to identify sperm-supplied products required for fertilization and the initiation of embryogenesis. This review describes paternal effects on animal development and discusses their implications for the role of the sperm in egg activation, centrosome activity, and biparental inheritance in different animal species. Paternal effects observed in *Caenorhabditis elegans* and in mammals are briefly reviewed. Emphasis is placed on paternal effects in *Drosophila melanogaster*. Genetic and cytologic evidence for paternal imprinting on chromosome behavior and gene expression in *Drosophila* are summarized. These effects are compared to chromosome imprinting that leads to paternal chromosome loss in sciarid and coccid insects and mammalian gametic imprinting that results in differential expression of paternal and maternal loci. The phenotypes caused by several early-acting maternal effect mutations identify specific maternal factors that affect the behavior of paternal components during fertilization and the early embryonic mitotic divi-

Current Topics in Developmental Biology, Vol. 38

sions. In addition, maternal effect defects suggest that two types of regulatory mechanisms coordinate parental components and synchronize their progression through mitosis. Some activities are coordinated by independent responses of parental components to shared regulatory factors, while others require communication between paternal and maternal components. Analyses of the paternal effects mutations *sneaky, K81, paternal loss,* and *Horka* have identified paternal products that play a role in mediating the initial response of the sperm to the egg cytoplasm, participation of the male pronucleus in the first mitosis, and stable inheritance of the paternal chromosomes in the early embryo. Copyright © 1998 by Academic Press.

I. Introduction

The sperm and the egg are highly specialized cells that differ greatly in their developmental histories, morphologies, and contributions to embryogenesis. The interaction between sperm and egg during fertilization initiates a series of rapid changes in the nuclear, cytoskeletal, and cytoplasmic architecture of both cells. These changes are required to coordinate activities of the gametes, ensure the inheritance of maternal and paternal components, and activate the embryonic program of cell division and gene expression.

Our understanding of the mechanisms regulating the flurry of events that constitute fertilization and allow embryogenesis to begin has been based largely on cytologic and biochemical studies. Cytologic studies have provided a wealth of information on the sequence of morphologic transitions that occur in the sperm and egg in a variety of different species (see Longo, 1973; Schatten and Schatten, 1987). Progress has also been made on the biochemistry of fertilization in a few model organisms, most notably in sea urchins, frogs, and mice. These animals offer one or more of the following advantages: production of large quantities of gametes, external fertilization, and availability of *in vitro* or cell-free systems that permit investigators to experimentally alter sperm components or egg extracts (see Poccia and Collas, 1996). From biochemical studies, it has been possible to identify signal transduction pathways (Whitaker and Swann, 1993) and cell surface receptors (Wassarman and Litscher, 1995; Ohlendieck and Lennarz, 1996) that mediate sperm–egg interactions and to study the nuclear and cytoplasmic conditions that support development of pronuclei (Poccia and Collas, 1996; Poccia and Green, 1996), nuclear assembly (Almouzani and Wolffe, 1993), and mitotic cycling *in vitro* (Murray and Kirschner, 1989).

As noted by the authors of earlier reviews (Poccia and Collas, 1996; Karr, 1996), a genetic approach to the study of fertilization has received little attention in the past. However, in recent years, an increasing number of mutations that disrupt sperm–egg interactions and the earliest stages of embryogenesis in the nematode *Caenorhabditis elegans* or the fruit fly *Drosophila melanogaster* have been subject to in-depth analyses. Consequently, the potential of a genetic analysis of fertilization is just beginning to be realized. In this review, we will summarize the progress made in these genetic studies, with an emphasis on those that address the function of the sperm. As has been elegantly demonstrated for other processes, such as sex determination and pattern formation, a developmental genetic approach provides a powerful means to define functionally important molecules that may be impossible to identify using strictly cytologic or biochemical means. The molecular insights gained from studies of genetically tractable organisms such as *C. elegans* and *D. melanogaster* should complement the cytologic and biochemical studies of other organisms and should also provide information on species-specific aspects of fertilization.

We expect mutations that affect sperm–egg interactions and the activation of development to show parental effects, since these events occur before transcription of the embryonic genome. Some of these mutations may result in male or female sterility by preventing sperm from entering the egg. Others may act as maternal or paternal effect mutations, causing developmental arrest after sperm entry. In this review, we focus on paternal effects as an avenue for understanding the behavior of the sperm during fertilization and the role of paternal contributions to embryogenesis. We first describe the few known cases of paternal effects in *C. elegans* and mammals, then concentrate on studies of *Drosophila*. We review cytologic data which show that *Drosophila* fertilization has much in common with that of other organisms but also has peculiar features. We also present cytologic and genetic evidence for paternal imprinting effects on chromosome behavior and gene expression in *Drosophila* and other insects. In order to evaluate the extent to which maternal factors regulate the activities of sperm components, we examine the phenotypes caused by known maternal effect mutations that affect early development. Finally, we describe studies of paternal effect mutations and discuss the implications of these findings for understanding the mechanisms controlling early development.

II. Paternal Effects in *Caenorhabditis elegans* and Mammals

Organisms differ in the extent to which the sperm is required for early embryogenesis. In parthenogenetic organisms, embryos develop from unfertilized eggs, demonstrating that sperm are completely dispensable for reproduction in some animals. However, in other organisms, sperm may be required to block polyspermy, stimulate the oocyte to complete meiosis, activate protein synthesis and DNA replication, induce mitotic cycling, and/or contribute a centrosome for the first embryonic division. By fusing with the female pronucleus, the sperm pronucleus also functions to restore the diploid state. In many organisms, the female and male genomic contributions can be considered functionally equivalent, since normal development ensues if diploidy is restored by natural or experimental means, with the fusion of either two paternal or two maternal haploid nuclei. However, this is not the case in mammals. Functional differences exist in parental genomes of mammals due to gametic imprinting (see Gold and Pedersen, 1994).

Table I.   Known Paternal Effect Mutations in Animals

Organism	Gene(s)	Paternal Effect Phenotype	References
C. elegans	*spe-11*	Incomplete egg activation, spindle, and cytokinesis defects in the first mitosis	Hill *et al.,* 1989; Browning and Strome, 1996
Axolotl	*ts-1*	Developmental arrest at gastrula stage	Malacinski and Barone, 1985
Bovine	—	Variation in sperm aster formation	Navara *et al.,* 1996
Human	—	Centrosomal and microtubule defects	Simerly *et al.,* 1995
Mouse, Human	Imprinted genes	Differential expression of paternal versus maternal genes	Reviewed by Gold and Pederson, 1994
Sciarid insects	—	Paternal X chromosome elimination	Reviewed by Gerbi, 1986; de Saint Phalle and Sullivan, 1996
Coccid insects	—	Paternal genome elimination or heterochromatization	Reviewed by Nur, 1990
D. melanogaster	—	Paternal effects on gene expression	This review
	ms(1)7	Arrest after sperm entry	Dybas *et al.,* 1981
	snky	Failure in sperm decondensation and centrosome formation	Fitch *et al.,* submitted
	K81	Haploid development or anaphase bridge formation	Fuyama, 1984, 1986a; Yasuda *et al.,* 1995
	pal	Paternal chromosome loss	Baker, 1975; Tomkiel, 1990; Owens, 1996; Wilson and Wakimoto, unpublished observations
	Horka	Paternal chromosome loss	Szabad *et al.,* 1995

As a result, mammalian embryos have the additional requirement of transcription from both paternal and maternal genomes during development. In theory, the number and types of mutations that induce paternal effects will differ among organisms and will reflect these variations in sperm function. In practice, only a few paternal effect mutations have been isolated (Table I).

Mutations causing paternal effects in *C. elegans* have been identified from two collections. At least two maternal effect mutations are known to also induce paternal effects on embryogenesis (Wood *et al.,* 1980; D. Shakes, unpublished observations). The most extensively studied paternal effect mutation, *spe-11* (Hill *et al.,* 1989; Browning and Strome, 1996), was recovered in a screen for mutations causing male sterility (L'Hernault *et al.,* 1988). This mutation acts as a strict paternal effect lethal mutation. When sperm from *spe-11* mutants fertilize normal oocytes, development is abnormal, but the reciprocal interaction of normal sperm and *spe-11* oocytes results in viable offspring (Hill *et al.*, 1989). The analysis of *spe-11* has provided intriguing clues about the role of sperm-supplied products in egg activation. Oocytes fertilized by *spe-11* sperm show a natural block to polyspermy, normal centrosomal activity, pronucleus formation, and migration. However, these oocytes fail to complete meiosis, form weak eggshells, misorient the first spindle, and fail in cytokinesis. This phenotype suggests that *spe-11* may function to coordinate a subset of the events required to activate the egg and initiate embryogenesis (Browning and Strome, 1996). Molecular cloning has shown that SPE-11 is a novel, sperm-specific protein that is synthesized during spermatogenesis and delivered to the egg on fertilization. Remarkably, the protein can carry out its function when it is ectopically expressed and loaded into the egg during oogenesis. Further studies are required to determine whether SPE-11 is indeed an activation factor and to elucidate its mechanism of action and targets. It will also be interesting to know whether full activation of the egg is dependent on other paternally provided factors. As noted by Browning and Strome (1996), an attractive biologic rationale for having at least one essential activation factor delivered by the sperm is to prevent precocious egg activation and parthenogenesis.

To our knowledge, only one paternal effect mutation has been genetically identified in vertebrates. Malacinski and Barone (1985) described phenotypic studies of the *ts-1* mutation of the axolotl. Sperm produced by homozygous *ts-1* males fertilize eggs, but the resulting embryos fail to develop past the early gastrula stage. Further studies of this mutation were hampered by the difficulties of genetic and molecular analyses in the axolotl, so it is not known whether *ts-1* is a loss-of-function allele or if it results in the production of an aberrant gene product. However, in either case, the interesting property of this paternal effect mutation is that its effects are manifested at a rather late stage of embryogenesis.

Paternal effects have been observed in two different aspects of mammalian development, centrosome function and gene expression. Navara *et al.* (1996) showed that the ability of bull sperm to organize an aster after *in vitro* fertilization varied among individual bulls. The success of aster formation was positively correlated with reproductive success of each male, suggesting that variation in centrosome activity may influence the *in vivo* developmental success of their offspring. Simerly *et al.* (1995) surveyed human couples who had experienced repeated failures with *in vitro* fertilization for possible paternal effects. Nearly half of the *in vitro*–fertilized oocytes obtained from these couples arrested development shortly after sperm entry. Some of the defects could be attributed to paternal effects on centrosome or microtubule function. These data provided evidence that the human centrosome, unlike that of rodent species, is paternally provided. In addition to documenting specific paternal effect defects during early development, the results of this cytologic study have practical implications for diagnosing whether certain types of human infertility are treatable by *in vitro* approaches.

The effects of gametic imprinting on gene expression in mice and humans are by far the most extensively studied parental effects. Studies of imprinting in the mouse have led to the identification of several genes that are differentially imprinted during spermatogenesis and oogenesis and differentially expressed during development. In addition, strain variation in the patterns of imprinting has permitted some investigators to map candidate loci that may confer the imprint or otherwise influence its inheritance (e.g., Latham, 1994). We refer the reader to a review by Gold and Pedersen (1994) for a detailed discussion of current models of the mechanisms of gametic imprinting and the implications for mammalian embryogenesis.

III. Paternal Effects on Chromosome Behavior in Insects

Studies with insects provided the first evidence for the existence of paternal imprinting effects on chromosome behavior in the embryo. The studies of the dipteran *Sciara* describe a fascinating example in which the paternal and maternal X chromosomes exhibit different behaviors during embryogenesis (see Gerbi, 1986; de Saint Phalle and Sullivan, 1996). In *Sciara coprophila*, all embryos receive one X chromosome from their mother and two X chromosomes from their father. The maternal X is stably inherited throughout development. However, embryos destined to become females eliminate one of the paternal X chromosomes from their soma during the ninth mitotic cycle, while embryos destined to become males eliminate both paternal X chromosomes simultaneously or sequentially during the seventh through ninth cycles. The mechanism of elimination is failure of the imprinted X chromosome to complete sister chromatid separation during anaphase (de Saint Phalle and Sullivan, 1996). Loss of one paternal X also occurs in the germ line of both sexes, but at a later stage of embryogenesis and by a different mechanism (see Gerbi, 1986). Crouse (1960) first coined the term "imprinting" to describe the parental effect in *Sciara* that resulted in differential behavior of homologous chromosomes in the embryo. The molecular nature of this imprint remains unknown. At minimum, the recognition of the imprint during embryogenesis requires an interaction between the paternal X and maternal factors. A *cis*-acting region, required for elimination, has been localized on the paternal X chromosome (Crouse, 1960) and has been shown to act at a distance to interfere with sister chromatid separation (de Saint Phalle and Sullivan, 1996). Furthermore, the choice between eliminating one versus two paternal X chromosomes and the timing of elimination depend on the type of egg laid by the mother (Gerbi, 1986). de Saint Phalle and Sullivan (1996) proposed an attractive model to explain how mothers may determine the timing of chromosome loss in their offspring. These authors suggested that mothers provide their eggs with a factor that is essential for the paternal X to undergo normal sister chromatid separation. The mother of sons produce a reduced quantity of this factor, compared with mothers of daughters. The titration of this factor by the paternal X chromosomes at each mitotic division eventually results in chromosome elimination, but at different cell cycles in the two sexes.

Several species of coccids (scale insects) show even more extreme chromosome loss than that observed in *Sciara*. A variety of different mechanisms underlie chromosome loss among different coccid species, but in the best-studied cases the entire paternal genome is either eliminated or heterochromatized during mid-cleavage (see Nur, 1990). In both *Scira* and coccids, paternal chromosome loss occurs as a part of the normal developmental process and as a mechanism for sex determination. The event responsible for the differences observed is most commonly believed to be gametic imprinting, as proposed by Crouse (1960). However, the possibility remains that the imprinting event occurs in the egg cytoplasm after sperm entry but before parental genomes have joined together. For instance, Nur (1990) proposed for that some species of coccids genomes may be imprinted according to their position within the egg. The *Sciarid* and coccid insects serve to illustrate the potential in insect species for the differential behavior of parental genomes during embryogenesis. As discussed later, particularly in the section on the *paternal loss* (*pal*) gene, these findings are also relevant for considering paternal imprinting effects on gene expression and chromosome behavior in *Drosophila*.

IV. Paternal Effects on Gene Expression in *Drosophila*

Genetic evidence exists for paternal effects on gene expression in *D. melanogaster*. Several investigators have shown that the level of expression of a variegating gene on a chromosome rearrangement can vary dramatically depending on whether the gene is transmitted through the father or the mother (Spofford, 1976; Dorn, Krauss, *et al.*, 1993; Dorn, Szidonya, *et al.* 1993; Bishop and Jackson, 1996). Remarkably, the variegating genes monitored in these studies are not expressed in the progeny until the late larval or pupal stages of development. Hence, the parental effect can persist through many cell divisions before its consequences are manifested. In some instances, genetic crosses show that the paternal effect can be attributed to the presence of the Y chromosome in males (Dorn, Krauss, *et al.*, 1993). *Trans*-acting modifier mutations have been identified that strongly influence the paternal effect of the Y, and several of these mutations are known to reside in genes that encode chromosomal proteins (Dorn, Krauss, *et al.*, 1993; R. Dorn and G. Reuter, unpublished observations). Further analyses of these proteins may provide insight into the molecular nature of this imprinting phenomenon. A paternal effect was also described by Kuhn and Packert (1988), who reported that paternal transmission enhanced the homeotic transformation caused by a *bithorax* complex mutation associated with an inversion. Current hypotheses propose that these examples reflect differential gametic imprinting of genes at the level of chromatin structure (Dorn, Krauss, *et al.*, 1993; Bishop and Jackson, 1996). It is likely to be significant that the genes for which parental effects have been observed are present either on chromosome rearrangements or in mutant backgrounds. These conditions may impose a greater imprinting effect than occurs in normal backgrounds, or they may simply increase the sensitivity of detecting such effects. In either case, it appears that imprinting effects on gene expression can be revealed in *Drosophila*, although they can be quite subtle and highly susceptible to variation in genetic backgrounds (Bishop and Jackson, 1996).

These observations raise the question as to whether imprinting in *Drosophila*, like that in mammals, results in functional differences in parental genomes that are essential for survival of the embryo. Some *Drosophila* species are capable of parthenogenesis, demonstrating that in these species the sperm is dispensable and development can occur with two maternal complements (Templeton, 1983). Although extensive searches have been carried out for parthenogenetic strains of *D. melanogaster*, viable parthenogenotes have not been recovered from natural populations, nor have they been obtained from laboratory strains (Stalker, 1954). However, it has been possible to use mutations that interrupt normal meiosis or fertilization to produce gynogenetic and androgenetic individuals, which develop from fertilized eggs but contain nuclei derived only from maternal or paternal genomes, respectively. Fuyama (1984; 1986b) showed that diploid gynogenetic offspring are produced at extremely low frequencies from crosses of wild-type females to males mutant for the *ms(3)K81* gene. A higher frequency of gynogenetic offspring result when the mothers from the *gyn-F9* strain are mated to *K81* males (Fuyama, 1986a). As discussed in more detail later, *K81* is a paternal effect mutation that often prevents the male pronucleus from participating in development and *gyn-F9* increases the frequency with which two female meiotic products fuse to form a diploid nucleus. Komma and Endow (1995) showed that diploid androgenetic offspring can be produced by females mutant for the α-*TUB67C* gene. These females produce eggs lacking an α-tubulin used specifically during oogenesis and early embryogenesis. As a result, a functional female pronucleus is not formed (Matthews *et al.*, 1993). Fusion or nondisjunction of the haploid mitotic products of the sperm is the proposed mechanism leading to the diploid androgenetic nucleus (Komma and Endow, 1995).

The viability of the gynogenetic and androgenetic diploids demonstrates the capacity of two maternal and two paternal genomes to support normal development in *D. melanogaster*. Since true parthenogenesis, the development of an unfertilized egg into a viable adult, has not been demonstrated in this species, the sperm must provide an essential factor required for normal development. Cytologic and genetic evidence indicates that, at minimum, the sperm contributes a centrosome that allows mitotic cycling to begin (see Ripoll *et al.*, 1992). Additional information on the unique contributions of the sperm in *D. melanogaster* has been obtained from cytologic analyses of fertilization and the study of paternal effect mutants.

V. Cytologic Aspects of *Drosophila* Fertilization and Early Embryogenesis

In *Drosophila*, as in other insects, fertilization occurs within the reproductive tract of the female. Fertilization takes place extremely rapidly, and the first mitosis is completed by approximately 20 min after sperm entry (Rabinowitz, 1944). Most often, the first few nuclear cycles are completed before the egg is laid by the female. Because of the rapid progression of the early stages and the lack of a suitable *in vitro* system for studying sperm–egg interactions, most of our knowledge of *Drosophila* fertilization has been based on cytologic analyses of fixed material. These descriptive data permit us to order some of the morphologic transitions that occur, but the precise timing of other events, such as the breakdown of the sperm nuclear envelope and the initiation and completion of pronuclear DNA replication, have not been resolved. In the following discussion, we consider the events depicted in Fig. 1 as they relate to sperm entry, formation of the pronuclei, pronuclear migration, and formation of the gonomeric spindle. Our emphasis is on aspects useful for comparing *Drosophila* fertilization with other model systems and for interpreting the defects caused by mutations. For an in-depth discussion of aspects relevant to cell cycle control, we refer the reader to the excellent review by Foe *et al.* (1993).

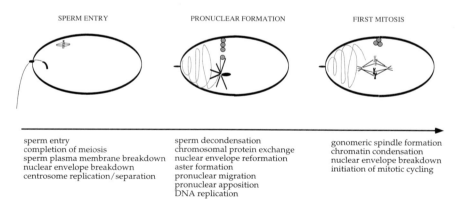

SPERM ENTRY	PRONUCLEAR FORMATION	FIRST MITOSIS
sperm entry	sperm decondensation	gonomeric spindle formation
completion of meiosis	chromosomal protein exchange	chromatin condensation
sperm plasma membrane breakdown	nuclear envelope reformation	nuclear envelope breakdown
nuclear envelope breakdown	aster formation	initiation of mitotic cycling
centrosome replication/separation	pronuclear migration	
	pronuclear apposition	
	DNA replication	

Fig. 1   Overview of fertilization in *Drosophila melanogaster.* The stages of fertilization, from sperm entry to formation of the first mitotic spindle, are represented, with an emphasis on behavior of sperm components. Landmark events are listed below each diagram. The micropyle, located at the left of each egg, marks the anterior end. Female meiotic nuclei are located at the dorsal surface of the egg. After sperm entry, the sperm tail forms a coiled structure at the anterior end of the egg.

A. Sperm Entry

During mating, *Drosophila* males transfer sperm and a complex mixture of accessory proteins to females. Several of these proteins elicit interesting behavioral responses in females, such as an increase in egg laying and a resistance to remating, and may also facilitate storage of sperm in the female's sperm storage organs (Clark *et al.*, 1995; see Wolfner, 1997). The number of sperm that are transferred during mating is estimated to be at most only a few thousand (Lefevre and Jonsson, 1962), a small number compared with the prodigious quantities characteristic of most animals. *Drosophila* females use sperm in a highly efficient manner, by somehow carefully controlling the release of sperm from the sperm storage organs. Sperm enter the anterior end of the egg through the micropyle, a channel-like extension of the vitelline envelope. It is believed that the sperm acrosome reaction occurs as the sperm contacts the vitelline envelope. Perotti and Pasini (1995) showed that glycoconjugates, as detected by concanavalin A or β-*N*-acetylglucosamine binding sites, are enriched on the sperm plasma membrane that lies above the acrosome and on the surface of the micropyle. Whether these molecules are used for sperm–egg recognition in *Drosophila*, as they are in echinoderms (Foltz and Lennarz, 1993) and mammals (Wassarman and Litscher, 1995), is not known. Suggestive evidence has been provided by Perotti *et al.* (1994), who showed that two different male sterile mutants produce sperm that are defective in fertilization and lack the acrosomal carbohydrate-binding sites.

An unusual aspect of sperm entry in *Drosophila* is that the entire sperm, surrounded by its plasma membrane, enters the egg. This was first observed by Perotti (1975) in her ultrastructural analysis of fertilization. The absence of a fusion event between sperm and egg plasma membranes during fertilization is unprecedented. The mechanism that allows the *Drosophila* sperm to traverse the plasma membrane of the egg remains a mystery. One possibility is that the sperm activates, then transits through, a large channel that resides in the egg plasma membrane beneath the micropyle. Clearly the most extraordinary feature of the *Drosophila* sperm is its length, which can exceed 50 mm in some species (Karr and Pitnick, 1996). The sperm tail can enter the egg, as it does in some mammals (Simerly *et al.*, 1993), but the degree of tail incorporation varies substantially among different *Drosophila* species (Counce, 1963; Karr and Pitnick, 1996). The entire 1.8-mm sperm enters the egg in *D. melanogaster*, with the tail forming a coiled structure that occupies the anterior third of the egg (Karr, 1991). The axoneme and portions of sperm plasma membrane persist as intact structures until late in embryogenesis (Perotti, 1975). Karr (1996) speculated that these persistent structures may play a provisional or morphogenetic role for the developing embryo. However, comparative studies show that if a postfertilization role for the sperm tail exists, it is not conserved among *Drosophila* species (Counce, 1963; Pitnick *et al.*, 1995). The sperm also introduces an elaborate mitochondrial derivative into the egg. Kondo *et al.* (1990) showed that a low level of paternal transmission of mitochondrial DNA can be detected in *Drosophila simulans*. Thus, mitochondria contributed by sperm are capable of replicating and regaining function in the embryo.

In many organisms, sperm entry is required to activate a metabolically quiescent egg (Whitaker and Swann, 1993) and to induce the blocks to polyspermy (Schultz and Kopf, 1995). In *Drosophila*, the sperm is not required for egg activation. The completion of meiosis and the rise in protein synthesis occur in unfertilized eggs on egg deposition or after *in vitro* hypotonic treatments of mature ovarian eggs (Doane, 1960; Mahowald *et al.*, 1983). It is unclear whether there are any postfertilization mechanisms to prevent polyspermy in *Drosophila*, or whether supernumerary sperm even pose a threat to the developing embryo. Polyspermy is a regular occurrence, with rates varying from 1% to as high as 9.5% in different strains (Hildreth and Lucchesi, 1963). Since the early mitotic divisions occur in a syncytium, extra centrosomes are not expected to be as detrimental to the insect embryo as they are in animals with complete cleavage. Indeed, haploid spindles, nucleated by accessory sperm, can occasionally be cytologically detected in early embryos (Callaini and Riparbelli, 1996). The majority of these spindles, along with the sperm chromatin, eventually disintegrate. Rarely, accessory sperm retain mitotic activity and give rise to large patches of haploid tissue (Bridges, 1925; Tomkiel, 1990). Indeed, there may not be a need for a rapid block to polyspermy in *Drosophila* since the number of sperm approaching the egg micropyle is low (Lefevre and Jonsson, 1962). The long sperm tail may provide a transient mechanical block within the micropyle, perhaps limiting the rate of polyspermy to less than 10% (Sander, 1985).

B. Formation of Pronuclei

The mature spermatozoan of *Drosophila* is a prototypical sperm in that it has a nucleus that is transcriptionally inactive and highly condensed. In a variety of different organisms condensation during spermiogenesis involves either the modification of histones or the replacement of histones with highly basic, sperm-specific chromosomal proteins (see Poccia, 1986; Risley, 1990). This change in chromosomal proteins is reversed during fertilization and is required to transform the condensed sperm nucleus into a functional pronucleus. The steps involved in this transformation have been studied in several model organisms and have been reviewed by Poccia and Collas (1996). Nuclear envelope breakdown is required to expose the sperm chromatin to cytoplasmic factors that mediate decondensation and the exchange of chromosomal proteins (Yamashita *et al.*, 1990; Longo and Kunkle, 1978). A well-studied example of a factor that facilitates the exchange of sperm-specific chromosomal proteins for embryonic histones is nucleoplasmin, a protein abundant in amphibian oocytes (Philpott *et al.*, 1991). The exchange of chromosomal proteins in sperm nuclei is followed by reformation of the nuclear envelope and DNA replication. The result is the formation of a functional male pronucleus.

Cytochemical studies by Das *et al.* (1964a, 1964b) provided the first evidence that *Drosophila* sperm undergo an exchange of chromosomal proteins. Using an alkaline-fast green staining procedure, these authors showed that lysine-rich nuclear proteins were replaced by arginine-rich proteins during the final stages of spermiogenesis. Several genes have since been identified in *Drosophila* that encode highly basic, sperm-specific proteins. Two loci were identified by Russell and Kaiser (1993) in a molecular screen for regions containing male-specific transcripts. The *mst77F* and *mst35* genes were found to encode small proteins that are similar to the protamines of mammals in their enrichment in basic residues (Russell and Kaiser, 1993; S. R. H. Russell, unpublished data). The function of the *mst77F* protein is not yet known. However, when the *mst35* site, which contains two related genes, is deleted, sperm heads fail to condense, proving that this protein is required for nuclear condensation during spermiogenesis (S. R. H. Russell, unpublished data).

The cytochemical studies of Das *et al.* (1964b) suggested that pronuclei and early cleavage nuclei contain arginine-rich chromosomal proteins that are either modified or replaced no later than the tenth nuclear division by proteins that show typical somatic histone staining patterns. The results of more recent studies are consistent with this idea. The histone H1 epitope is lost in the male germ line during spermatogenesis (Kremer *et al.*, 1986), and H1 cannot be detected in embryonic chromatin until cycle 7 (Ner and Travers, 1994). However, HMG-D, the *Drosophila* HMG1 homologue, is abundant during the earliest stages of embryogenesis. HMG-D levels in embryonic nuclei decrease gradually with each mitotic division, while the H1 levels increase. These embryo-specific changes in chromosomal proteins may be required to allow the rapid cycling that occurs during the early syncytial divisions and before the onset of transcription of the embryonic genome (Ner and Travers, 1994).

Although no direct experimental data exist for *Drosophila* sperm, *in vitro* studies of other animals (Poccia and Collas, 1996) predict that *Drosophila* sperm require exposure to the egg cytoplasm to accomplish decondensation. As noted previously, ultrastructural studies showed that the *Drosophila* sperm enters the egg surrounded by its plasma membrane (Perotti, 1975). Therefore, two membrane types, the plasma membrane and the nuclear envelope, must be broken down after the sperm entry into the egg (Fig. 2). Perotti (1975) noted that shortly after sperm entry the plasma membrane surrounding the head and tail vesiculates and, subsequently, tail components are dispersed into the surrounding cytoplasm. Evidence for nuclear envelope vesiculation has not been reported. Nuclear envelopes are known to be present in mature sperm (Perotti, 1969; Tokuyasu, 1974), although sperm are believed to lack lamins (Liu *et al.*, 1997). Since mitotic nuclear envelope disassembly may require lamin phosphorylation (see Nigg, 1992; Poccia and Collas, 1996), it is unclear how the sperm nuclear envelope

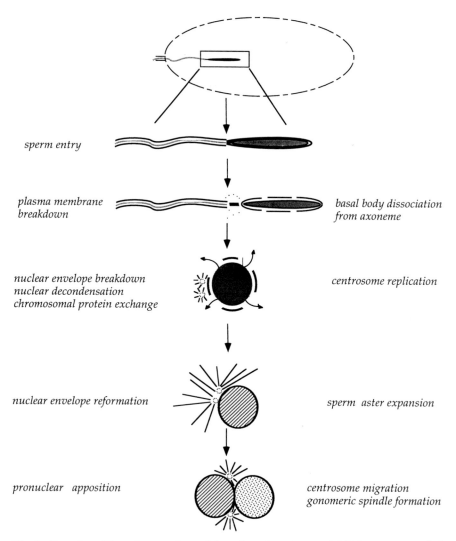

Fig. 2 Formation of the male pronucleus and the embryonic centrosome. Multiple steps are required to transform the condensed sperm into a functional pronucleus and to reconstitute a functional centrosome. Both processes are likely to share the initial requirement for plasma membrane breakdown. The breakdown and reformation of the sperm nuclear envelope has not been directly observed in *Drosophila* but is deduced to occur on the basis of observations in other animals. Curved arrows depict exchange of sperm-specific chromosomal proteins for maternally provided proteins. The timing of centrosome duplication and migration is based on the cytologic studies of Callaini and Riparbelli (1996). Aster formation is required to bring the male and female pronuclei, shown in the last figure, into close proximity.

dissociates from the chromatin. In addition, nuclear envelopes fragment, rather than vesiculate, during the embryonic mitoses in *Drosophila* (Stafstrom and Staehelin, 1993), and this may also be true of the sperm nuclear envelope. Additional studies are needed to clarify the requirements for membrane breakdown and decondensation. Ideally, an *in vitro* system for sperm decondensation could provide much-needed insights. However, this has not yet been possible due to the difficulties in isolating large quantities of *Drosophila* sperm. Several groups have successfully prepared cytoplasmic extracts from *Drosophila* embryos to analyze maternal components (Ulitzur and Gruenbaum, 1989; Berrios and Avilion, 1990; Crevel and Cotterill, 1991), but these investigators have used *Xenopus* or chicken sperm as donor nuclei. With the use of these extracts, two cytoplasmic proteins were identified that promote decondensation of frog sperm (Kawasaki *et al.*, 1994; Crevel and Cotterill, 1995). These proteins are immunologically distinct from nucleoplasmin, but their activities suggest functional similarities in the mechanisms promoting remodeling of sperm chromatin.

C. Pronuclear Migration and Formation of the Gonomeric Spindle

In *Drosophila*, as in the majority of organisms studied, the sperm provides the centrosomal material required for pronuclear migration. A single centriole is present in the mature *Drosophila* sperm (Perotti, 1969; Tokuyasu, 1975) and serves as the basal body of the sperm tail and, later, as the microtubule organizing center for the sperm aster. Maternally provided centrosomal components, such as the CP190 protein (Callaini and Riparbelli, 1996) and centrosomin (CNN) (Heuer *et al.*, 1995), associate with the sperm centrosome immediately after sperm entry. In order for the centriolar material of the sperm to join with the maternal centrosomal proteins, it is thought that the basal body must dissociate from the sperm tail axoneme. Presumably, this dissociation is triggered by products in the maternal cytoplasm, such as centrin, as described for other organisms (Uzawa *et al.*, 1995; Schatten, 1994).

Many of the nuclear events leading to pronuclear migration and the formation of the gonomeric spindle were described in the early histologic studies of Huettner (1924) and Sonnenblick (1950) and have been reviewed by Foe *et al.* (1993). The description here incorporates findings by Callaini and Riparbelli (1996), which provide increased resolution of the timing of nuclear events relative to centrosomal activity and nuclear envelope dynamics. The sperm head decondenses rapidly after sperm entry, while the female nuclei resume meiosis. The male pronucleus is associated with a small aster at these early stages. However, when the female nuclei are in telophase II, the centrosome associated with the male pronucleus appears to duplicate, giving rise to two distinct foci, each organizing an expanding array of microtubules (Callaini and Riparbelli, 1996, Fig. 2). As the sperm asters mature, microtubules elongate and extend to the periphery of the egg (Foe *et al.*, 1993). The female haploid nucleus located closest to the interior migrates along the aster microtubules to the center of the egg to meet the male pronucleus. The male and female pronuclei become closely apposed but do not fuse. As the chromosomes condense, the sperm centrosomes become widely separated (Callaini and Riparbelli, 1996). The first mitotic spindle is formed and a single centrosome is present at each pole (Sonnenblick, 1950; Callaini and Riparbelli, 1996). This first spindle is a gonomeric spindle, since maternal and paternal chromosomes remain as separate groups of chromosomes, on separate bundles of microtubules, as they congress at the metaphase plate (Huettner, 1924). The two genomes also retain separate nuclear envelopes until late metaphase (Callaini and Riparbelli, 1996). At the end of metaphase, a single envelope appears to encompass both genomes, but nuclear envelope fragments are observed between parental genomes, presumably maintaining their separation. The intermixing of the paternal and maternal genomes does not occur until chromatids approach the spindle poles at late anaphase of the first division (Huettner, 1924; Callaini and Riparbelli, 1996).

VI. Coordination between Maternal and Paternal Contributions in *Drosophila*

One of the most challenging aspects of studying fertilization is to understand how the activities of the gametes are coordinated to ensure that paternal and maternal components are competent to enter mitosis at the same time. Since the sperm and egg components share a common cytoplasm, coordination could simply result from parallel but independent responses to the same regulatory factors. Alternatively, active signaling and feedback mechanisms may exist to synchronize the behavior of sperm and egg counterparts. We consider these possibilities in the following discussion of a set of maternal effect mutations causing early developmental arrest (Fig. 3). Several of these mutations have provided useful material for examining the general issue of how nuclear and cytoplasmic events are regulated during the cell cycle (Foe *et al.*, 1993; Ripoll *et al.*, 1992). We concentrate here on the phenotypes relevant for addressing the following questions: What aspects of sperm behavior are controlled by maternal cytoplasmic factors? Do paternal and maternal nuclei respond in similar fashion to these maternal factors? To what extent are the changes that occur in paternal and maternal nuclei interdependent?

Experimental studies of fertilization in marine invertebrates and mammals have shown that the egg must be at certain stage of maturation to promote sperm aster formation and decondensation and replication of the sperm nucleus (Longo *et al.*, 1991; Maleszewski, 1992; see Poccia and Collas, 1996). A large number of *Drosophila* mutations cause eggs to arrest without initiating development (*e.g.*, see Schupbach and and Wieschaus, 1989; Szabad *et al.*, 1989; Erdelyi and

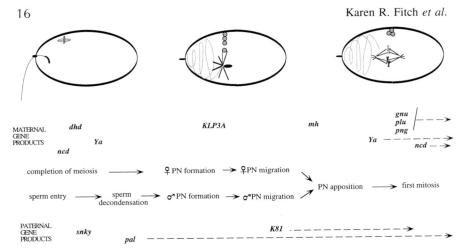

Fig. 3 Genetic control of fertilization in *Drosophila melanogaster*. The phenotypes associated with maternal and paternal effect mutations disrupting meiosis, fertilization, or early embryogenesis are described in the text. The relative order of the *Ya* and *gnu/plu/png* mutations has been determined from epistasis studies (Lin and Wolfner, 1991; Shamanski and Orr-Weaver, 1991). The other mutations are placed according to the stage at which defects are first observed. Several mutations also induce defects at later stages of embryogenesis. PN, pronucleus.

Szabad, 1989; Leiberfarb *et al.*, 1996), but so far only a few have been characterized with respect to behavior of the sperm nucleus (Erdelyi and Szabad, 1989; Salz *et al.*, 1994). One of these mutations is *deadhead* (*dhd*), which is a maternal effect mutation with high penetrance. More than 90% of the eggs produced by *dhd* mothers fail to complete meiosis I but allow sperm to enter (Salz *et al.*, 1994). Karr (1996) has reported that the sperm do not undergo nuclear decondensation. This phenotype may reflect a direct requirement of the sperm for the maternally provided DHD protein. Alternatively, it is possible that the completion of meiosis is a prerequisite for sperm decondensation. This latter hypothesis is consistent with the sequence of events reported by Callaini and Riparbelli (1996). Their study showed that the sperm does not fully decondense, nor does the sperm aster mature, until after meiosis is completed.

Recent studies on the kinesin-like protein KLP3A provide a particularly interesting example of a maternal effect mutation affecting pronuclear behavior (Williams *et al.*, 1997). KLP3A mutant females produce eggs that complete meiosis and allow sperm entry. However, all of the meiotic products remain at the cortex of the egg, and a female pronucleus fails to form and migrate to the male pronucleus. Williams *et al.* (1997) proposed that the KLP3A microtubule motor activity may be required to keep the meiotic products from precociously fusing with one another after meiosis is complete. This role is consistent with the localization of the protein to the equatorial region of spindles. The male pronucleus proceeds to migrate into the interior of the KLP3A-deficient egg, where it forms a bipolar spindle. Remarkably, this haploid spindle arrests with the paternal chromosomes in a metaphase-like configuration. As noted by Williams *et al.* (1997), it is puzzling why this spindle remains arrested, since previous studies showed that KLP3A mutations permit meiotic and mitotic spindles to progress through anaphase (Williams *et al.*, 1995). It is possible that progression through the first cell cycle requires an interaction between pronuclei or factors associated with the pronuclei in close proximity. The survival of androgenetic and gynogenetic diploids would seem at first to argue against this possibility (Williams *et al.*, 1997). However, these exceptional individuals develop from fertilized eggs. Therefore, interaction between parental components may have occurred before subsequent defects resulted in the elimination of either the maternal or the paternal genome. The possibility that efficient progression through mitosis requires an interaction between maternal and paternal pronuclear factors could also explain why the accessory sperm frequently degenerate and so rarely give rise to haploid tissues, in spite of the 1–10% occurrence of polyspermy in some strains. Ripoll *et al.* (1992) suggested a related hypothesis to account for the phenotype of fertilized eggs produced by females mutant for the *abnormal spindle* (*asp*) gene. These females show high levels of meiotic nondisjunction, and a proportion of their eggs show no detectable DNA. Although these eggs can be fertilized, there are no apparent signs of division of the sperm nucleus. This observation led Ripoll *et al.* (1992) to suggest that the female nucleus, or a factor associated with it, is required to activate nuclear divisions in fertilized embryos.

Zalokar (1975) characterized several maternal effect mutations that resulted in haploid development. The best-studied of these is the *maternal haploid* (*mh*) mutation, which results in the production of progeny carrying only the maternal genome. These haploid embryos die before hatching. Edgar *et al.* (1986) proposed that haploidy is the consequence of the failure of the maternal and paternal pronuclei to become closely apposed. Karr (1996) reported abnormalities in the pattern of sperm tail coiling, suggesting that sperm migration is abnormal in these eggs. Whatever the mechanism is that prevents the male pronucleus from participating in development, the centrosome or perhaps other factors supplied by the sperm must come into close enough proximity with the female pronucleus to activate its mitotic cycling.

Several maternal effect mutations affect the maternal meiotic products and the male and female pronuclei in similar ways, suggesting that these nuclei respond to common factors present in the egg cytoplasm. Eggs produced by *fs(1)Young arrest* (*Ya*) mutant females are characterized by asynchrony in the behavior of the polar bodies and arrest of the paternal and maternal nuclei at the pronuclear apposition stage (Lin and Wolfner, 1991; Liu *et al.*, 1995). Genetic and molecular studies have shown that the YA protein is a specialized nuclear lamina protein that is required for female meiosis and progression through the early mitotic divisions (Lopez *et al.*, 1994; Lin and Wolfner, 1991). The incorporation of YA into both pronuclei may play a role in synchronizing their activities before their entry into the first mitosis. The nuclear lamina localization suggests the possibility that YA may fulfill this function by interacting with chromatin, perhaps by regulating chromatin condensation or DNA replication (Liu *et al.*, 1995). The products of three other genes, *giant nuclei* (*gnu*), *plutonium* (*plu*), and *pan gu* (*png*), are required for the normal regulation of DNA synthesis before and after fertilization (Freeman and Glover, 1987; Shamanski and Orr-Weaver, 1991). Whereas unfertilized eggs are normally incapable of DNA synthesis, the onset of DNA synthesis occurs precociously in the unfertilized eggs of mothers mutant for any one of these genes. In addition, DNA synthesis and nuclear divisions become uncoupled, resulting in the production of giant polyploid nuclei. All four female meiotic nuclei and the sperm nucleus form giant nuclei, showing that they share a requirement for the wild-type GNU, PLU, and PNG proteins.

On the basis of the maternal effect phenotypes described, we suggest that coordination between the maternal and paternal components may be achieved by two means. The parental genomes respond independently to common regulatory factors to allow some activities, such as DNA replication or the coupling of replication with nuclear division, to occur in the nuclei within the same time frame. The coordination of other activities may depend on communication between maternal and paternal counterparts. Three possible interactions are proposed. First, the completion of meiosis may be required to allow the sperm to undergo nuclear decondensation and full aster expansion. Second, although the sperm nucleus reaches the interior of the egg before meiosis is complete, its competency to complete the first mitosis may be facilitated by an interaction with factors supplied by the female pronucleus. Third, the initiation of mitotic cycling of the female pronucleus requires its association with a sperm-supplied component, most likely the centrosome.

The *non-claret disjunctional* (*ncd*) mutation is remarkable in that its phenotype demonstrates a clear difference in the behavior of maternal and paternal chromosomes during the first mitotic cycles. Mutations in *ncd* result in high levels of nondisjunction and chromosome loss during female meiosis. In addition, *ncd* mutant females also produce progeny that lose chromosomes during the early mitotic divisions (see Davis, 1969). Chromosome loss during embryogenesis is a feature *ncd* shares with other female meiotic mutants, and this phenotype has been attributed to the disruptive effect of an aberrant meiotic division on the early mitotic divisions, rather than a direct effect of a deficiency of *ncd* on mitosis (Hawley *et al.*, 1993). However, NCD is known to be a minus-end directed microtubule motor protein (McDonald *et al.*, 1990; Walker *et al.*, 1990) that localizes to the spindle microtubules of the meiotic spindle as well as the mitotic spindles and the centrosome of the early embryo (Endow and Komma, 1996). Hence, chromosome loss during embryogenesis may reflect a direct role for NCD in the first few embryonic mitoses (Endow and Komma, 1996). Moreover, Davis (1969) provided genetic data to unlink the meiotic and mitotic phenotypes by showing that the probability of a chromosome's suffering loss during mitosis does not correlate with its disjunctional history during meiosis. The remarkable aspect of *ncd*-induced mitotic chromosome loss is that maternal and paternal chromosomes differ in the frequency and timing of loss during development. Nelson and Szauter (1992) analyzed the proportion of tissue showing either maternal or paternal X in gynandromorph progeny of *ncd* mothers. Their data show that the maternal X chromosomes are lost almost exclusively at the first division. Paternal X chromosomes are lost during the second and later mitoses and at a 10-fold lower frequency. These differences in parental chromosome loss have been attributed to the unusual structure of the gonomeric spindle, with *ncd* mutations resulting in differential effects on centrosomes or perhaps other structures that may organize the maternal versus the paternal halves of the spindle (Nelson and Szauter, 1992). However, cytologic studies reveal that a single centrosome organizes each spindle pole (Sonnenblick, 1950; Callaini and Riparbelli, 1996). Since embryos of *ncd* mutant mothers show centrosomal and spindle defects during preblastoderm mitoses (Endow and Komma, 1996), it is not clear why maternal X chromosome loss is restricted to the earliest mitoses. We suggest that the differences observed in the timing of maternal versus paternal chromosome loss reveal differences in the composition of parental genomes, specifically in the chromosomal proteins that interact with NCD or other spindle components. The narrow time intervals during which chromosome loss occurs indicate not only that parental genomes show a distinction at the first mitosis but that dynamic changes in the composition of both parental genomes take place during subsequent mitoses. The persistence of sperm-specific proteins on the paternal chromosomes may render them relatively resistant to the effects of the *ncd* deficiency at the first division. Subsequent remodeling of the paternal chromosomes could account for the increased sensitivity to *ncd*-induced loss at later stages. Further evidence for the differential behavior of maternal and paternal chromosomes during early mitotic cycles and the potential importance of compositional differences between parental genomes is provided in the discussion of paternal effect mutations.

VII. Paternal Effect Mutations of *Drosophila*

A. General Considerations

Drosophila offers several key advantages for genetic studies of fertilization and paternal effects on development. It is one of the few genetically tractable organisms currently being used to study fertilization. Unlike *C. elegans*, which produces amoeboid sperm, *Drosophila* produces flagellated sperm that are morphologically similar to those of most other animals. As described previously, a large number of early-acting maternal effect mutations exist, and these provide a rich source of material to identify maternal factors affecting sperm behavior.

Finally, large-scale screens for *Drosophila* paternal effect mutations are feasible, and the results should provide information on the number and types of genes that cause paternal effect defects when mutated.

We are currently screening for paternal effect mutations in our laboratory. Potential candidates are first identified as mutant males that produce progeny showing paternal chromosome loss during embryogenesis or as sterile males that produce motile sperm capable of inseminating eggs. This latter criterion eliminates the largest class of male sterile mutants, those in which spermatogenesis is affected (Hackstein, 1991). A limitation of the screen is that it selects against potentially interesting candidates with disruption of functions required for both gametogenesis and embryogenesis. Available data suggest that paternal effect mutations are rare. From a collection of 400 male sterile mutations induced by ethyl methane sulfate (EMS), Hackstein (1991) identified 10 that produced motile sperm. Two of these mutations were defective in sperm–egg recognition (Perotti *et al.*, 1994). The number of paternal effect mutations among the remaining eight mutations was not determined (J. H. P. Hackstein, unpublished data). Of the known paternal effect genes, only one, the *ms(3)sneaky (snky)* gene, was recovered in a search for paternal effects among a collection of 89 male sterile mutants (Fitch *et al.,* submitted). Other known paternal effect mutations were recovered fortuitously in studies of processes. The *ms(1)7* and *ms(3)K81 (K81)* mutations were recovered as male sterile mutations (Dybas *et al.*, 1981; Fuyama, 1984). The first *pal* allele was isolated from a screen for male meiotic mutants (Sandler, 1971). The *ms(1)7* mutant has been lost, and available data allow only a brief description of the mutation as causing a postfertilization defect (Dybas *et al.*, 1981) (see Table I). The characteristics of the four remaining paternal effect mutations and their implications for paternal contributions to development are discussed in the following sections.

B. The *ms(3)sneaky* Gene

The paternal effect mutation, *snky* results in the earliest developmental arrest phenotype so far described (Fitch *et al.,* submitted). Males that are homozygous for this mutation produce sperm that are able to enter the egg, but 99% of these eggs fail to initiate development. Cytologic studies of *snky*-fertilized eggs reveal that the behavior of the female meiotic products is typical of that seen in unfertilized eggs. Meiosis is completed, and the resulting haploid products frequently fuse and then condense to form a single star-like configuration. Sperm entry is efficient, with the entire sperm tail entering the egg. However, rather than showing the regularly spaced spiral configuration of wild-type sperm, the tail forms a more disorganized structure. In addition, the sperm head is located within the anterior portion of the egg, frequently occupies a cortical position, and remains tightly condensed.

Several observations suggest that *snky* sperm fail to interact with maternally provided products. For example, immunostaining with the use of antibodies against centrosomin failed to reveal the association of this protein with the sperm head. Similarly, using antitubulin antibodies we did not detect microtubule organizing capacity characteristic of a normal sperm centrosome. We have also tested whether the sperm nucleus responds to an absence of the maternally provided GNU protein. The results showed that eggs that are produced by *gnu*-mutant mothers and fertilized by *snky* fathers form giant polyploid egg nuclei, but the *snky* sperm head remains tightly condensed. Thus, the sperm does not respond to the *gnu* maternal effect defect that promotes repeated rounds of DNA replication in wild-type sperm and in egg nuclei (Figure 3).

Why do *snky* sperm fail to show signs of activity once inside the egg? A formal possibility is that the mutant defects result from the abnormal placement of the sperm within the egg. The SNKY protein could be required, perhaps as a component of the tail, to place the sperm in the interior of the egg and allow it to interact with assymetrically distributed cytoplasmic factors that promote decondensation and centrosome formation. Cytologic observations of dispermic but otherwise wild-type embryos argue against this possibility (Callaini and Riparbelli, 1996). Occasionally, when two sperm enter an egg, one sperm forms a small haploid spindle at the anterior end of the egg. This observation suggests that factors required for the transformation from a condensed sperm to a pronucleus with centrosomal activity are indeed present in the anterior regions of the egg.

We favor the hypothesis that the plasma membrane surrounding *snky* sperm fails to break down after the sperm enters the egg. As a consequence, the sperm chromatin and basal body do not have access to cytoplasmic factors that are required for nuclear decondensation and centrosomal activity. In the absence of these interactions, the sperm aster fails to form. Studies of sea urchin sperm suggest that the aster is the motive force that carries the sperm head into the interior of the egg (Schatten and Schatten, 1987). If *Drosophila* uses a similar mechanism for sperm placement, the lack of sperm aster formation could account for the cortical location of *snky* sperm and the abnormal bundling of its tail. This interpretation is also consistent with studies of other organisms which have shown that sperm must be demembranated in order to achieve decondensation of the sperm head (see Poccia and Collas, 1996).

Ultrastructural characterization of the mutant phenotype may provide the resolution required to determine whether the plasma membrane remains intact in *snky* sperm after its entry into the egg. It may also be possible to monitor the activity of *snky* sperm, with and without detergent treatment to remove the plasma membrane, after microinjection of sperm nuclei into activated eggs. The *snky* gene has not yet been cloned, but it will be interesting to determine the function and subcellular localization of its product. The gene may encode a sperm plasma membrane protein, a finding that would support a direct role for the protein, perhaps as a receptor, in the initial response of the sperm to maternal cytoplasmic cues. If this is the case, then studies of the SNKY protein should provide insight into the puzzle of why the *Drosophila* sperm retain the plasma membrane during sperm entry. Alternatively, SNKY may mediate a downstream event, perhaps by acting as a signaling molecule that coordinates the activities of the sperm membrane and nuclear and centrosomal components.

C. The *ms(3)K81* Gene

Studies of the *K81* gene have been instructive for addressing several different aspects of the paternal contributions to the embryo. In addition to defining the role of the *K81* gene product, studies of the mutant phenotype have also suggested that coordination between the paternal and maternal pronuclei is controlled, at least in part, by paternal products. Finally, as described previously, *K81* mutations provide genetic evidence that the sperm contributes the centrosome for mitotic divisions in the embryo.

The *K81* gene was initially described by Fuyama (1984), who isolated the first allele of this intriguing male sterile mutation. Fuyama discovered that the embryos of wild-type mothers and *K81¹* fathers arrest at one of two different developmental stages. Approximately 90% of the embryos arrest during the early cleavage divisions, while 10% survive until just prior to hatching. These late embryos are haploid individuals that contain only the maternal chromosomal complement. Anaphase bridges are observed in *K81¹*-derived embryos as early as the first mitotic division (Fuyama *et al.*, 1988; Yasuda *et al.*, 1995). However, spermatogenesis in mutant males appears to be normal at the ultrastructural level (Fuyama *et al.*, 1988).

We recovered new *K81* alleles to determine whether the unusual dual–lethal phase phenotype was an allele-specific property (Yasuda *et al.*, 1995). Five new mutations were isolated and tested in all possible hemizygous and heteroallelic combinations. In all cases, the mutations resulted in the early and late arrest phenotypes, with frequencies similar to that of *K81¹*. These results, and the association of several alleles with molecular deficiencies, strongly suggest that the dual–lethal phase phenotype is caused by complete lack of *K81* product in the male. We also found that paternal effect lethality was the only defect common to the six mutations, establishing that *K81* is a strict paternal effect gene.

Why does loss of *K81* function in males result in two different phases of developmental arrest? Cytologic studies suggest that in the majority of the embryos of *K81* fathers, the paternal chromosomes are incorporated into the gonomeric spindle (Yasuda *et al.*, 1995). This invariably results in the formation of chromosome bridges and breaks during anaphase. Chromosome breakage continues to occur at a high frequency in subsequent divisions. Embryos showing these defects arrest with fewer than 64 heteropyknotic nuclei but with a variable number of degrading nuclei that does not match that predicted for the normal doubling during each cleavage cycle (Yasuda *et al.*, 1995). The broad range of this early-arrest phenotype may reflect unsynchronized attempts at division by chromosomes that were damaged in the very first divisions. This proposal for the genesis of the *K81* early-arrest class proposes that the *K81* pronucleus essentially acts as a "poison product" for the first cleavage division. If our hypothesis to account for the early-arrest class is correct, then the haploid embryos that arrest late in embryogenesis must result from either a failure in pronuclear apposition or the removal of all of the paternal chromosomes during the first division. We have observed embryos with phenotypes consistent with the former explanation. These embryos contained two haploid spindles, one in a position typical of the gonomeric spindle and a second in a slightly more anterior position (Yasuda *et al.*, 1995).

Further support for the "poison product" hypothesis and the proposed mechanism for haploid development is provided by Fuyama's studies of the interaction of *K81* with *gyn-F9* (Fuyama, 1986a). A substantial number of viable progeny are produced from *gyn-F9* mothers and *K81* fathers, all of which are diploid females with the genetic constitution of their mother. The *gyn-F9* stock has two defects that allow it to rescue the *K81*-induced lethality. The first defect is an increase in the production of diploid nuclei by the fusion of two female pronuclei. Significantly, the second defect is a decrease in the frequency of pronuclear apposition and the incorporation of the male pronucleus into the gonomeric spindle (Fuyama, 1986a). If the *K81*-derived paternal chromosomes act to disrupt division, then diploids could arise in those embryos in which pronuclear apposition has failed. In this context, a defect in pronuclear apposition would be a necessary attribute for the *gyn-F9* stock to produce viable diploids when crossed to *K81*.

A proposal for mechanism of action of the K81 protein must explain both loss-of-function phenotypes. The most parsimonious explanation proposes that a single defect is responsible for both phenotypes. We favor the hypothesis that the *K81* defect lies in the remodeling of the sperm chromatin that is required to transform a condensed sperm into a functional male pronucleus (Yasuda *et al.*, 1995). If the decondensation or the remodeling of sperm chromatin is abnormal or delayed, defects in migration or alignment of the sperm pronucleus may occur. The frequency of haploid embryos would suggest that this defect occurs in 10% of the *K81*-fertilized eggs. Most frequently, the paternal chromosomes are incorporated into the gonomeric spindle. However, defects in chromatin remodeling may result in incomplete replication or abnormal condensation such that the paternal chromosomes are poorly prepared for mitosis and unable to segregate normally during anaphase. The result is the formation of anaphase bridges in the first division and, ultimately, the early-arrest phenotype. We propose that the K81 protein is normally delivered by the sperm and functions in the fertilized embryo either in the removal of sperm-specific chromosomal proteins or in the repackaging of sperm chromatin with maternally provided proteins. We are currently testing this proposal by characterizing the K81 protein. We have recently rescued the *K81* mutant by germ line transformation and have identified a transcript encoding a novel protein. Further analysis of this protein is required to determine its subcellular localization and function.

Our interpretation of the *K81* mutant phenotypes provides additional insight into the mechanisms of early development. The observations underscore the fact that the sperm centrosome is required for mitotic activity, whether the embryonic nuclei are normal diploid nuclei of biparental origin, diploid monoparental nuclei such as those produced by the *gyn-F9* or *α-TUB67C* mutants, or haploid nuclei with either parental genome. The observations on *K81* haploid and diploid development clearly demonstrate that the normal functioning of the sperm centrosome in the embryo does not depend on the normal functioning of the sperm pronucleus.

D. The *paternal loss* Gene

Unlike *snky* and *K81* mutations, mutations in the *pal* gene do not result in a strong effect on male sterility. Instead, *pal* mutant males are fertile, but a proportion of their progeny show paternal chromosome loss. Baker (1972, 1975) characterized the phenotype caused by the *pal¹* mutation in a series of elegant genetic analyses. He found that homozygous *pal¹* males produce nulloexceptional progeny that lack one or more paternal chromosomes and mosaic progeny that lose paternal chromosomes during the early mitoses. These loss events appeared to be caused by the loss of entire chromosomes, rather than chromosome fragments, since *pal¹* did not appear to induce chromosome breakage as assayed by genetic criteria (Baker, 1972). Based on the average proportion of tissue showing chromosome loss in adult mosaic progeny of mutant males, Baker (1975) estimated that the majority of loss occurs during the first three cleavage divisions. Furthermore, all of the paternal chromosomes are susceptible to loss, but chromosome 4, the smallest chromosome of the complement, is lost at the highest rate ($\sim$17%), whereas the X and Y are lost at lower frequencies (3.4% and 0.7%, respectively). Two X chromosomes showing different loss rates were used to map the region conferring sensitivity to *pal* by recombination. The sensitive region maps within the proximal half of the X chromosome, the region that contains the heterochromatin and the centromere. Based on these data, Baker considered *pal* to be a male-specific meiotic mutant and proposed that the *pal+* gene specifies a component of the centromere. He suggested that paternal chromosome loss during embryogenesis was the consequence of inheriting chromosomes with defective centromeres.

Studies in our lab have provided new insights into how the *pal+* product functions. Tomkiel (1990) used a series of Y chromosome deficiencies to analyze the *cis*-requirement for chromosome loss. Since these chromosomes originated

Entire Y Chromosome (0.34)

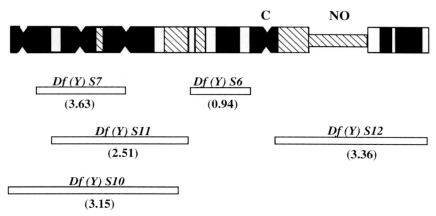

Fig. 4 Frequency of *pal*-induced Y chromosome loss. The cytogenetic map depicting chromosome bands on the entire Y chromosome is shown at the top of the figure, with the location of the centromere (C) and nucleolar organizer (NO) indicated. The regions deleted in different deficiencies of the Y chromosome, *Df(Y)*, are shown by the clear rectangles. The frequency of *pal*-induced loss of these chromosomes was monitored in the adult progeny of homozygous *pal¹* fathers. The frequency is shown in parentheses, as the number of loss events per 1000 progeny.

from the same parent chromosome, they shared a common centromere and differed only in location and size of the deletion. All five deficiencies were lost at a higher frequency (3–10-fold) when compared with the entire Y chromosome (Fig. 4). We conclude that susceptibility to loss is not conferred by a localized region but instead maps throughout the Y chromosome. It should be noted that the Y is entirely heterochromatic, so it is possible that heterochromatin plays an important role in mediating *pal+* function. Additional studies are required to more critically evaluate this relation and to examine the relation between chromosome size and loss rates.

All existing data are consistent with the classification of the *pal* gene as a strict paternal effect gene. Complete deletions of the gene result only in the paternal effect defect (Owens, 1996). The frequency of paternal chromosome loss caused by *pal* deficiencies is similar to that characteristic of *pal¹* and never exceeds a total frequency of 13% in our hands. Thus, while the wild-type gene is required to maintain the paternal chromosome with high fidelity, the incomplete penetrance of the deletion phenotype suggests that its function may be partially compensated by the actions of other products. Either paternal or maternal products could fulfill this role. In fact, the rates of chromosome loss induced by *pal* mutations are highly susceptible to both paternal and maternal background effects (Baker, 1975; Owens, 1996).

Cytologic analyses of the embryos of *pal*-mutant fathers suggest that chromosome loss can occur at different phases of the cell cycle (K. Wilson and B. Wakimoto, unpublished data). The earliest defect we have observed is the presence of micronuclei adjacent to the male pronucleus. This is the first evidence for premitotic loss of paternal chromosomes in the embryo and provides one explanation for the production of nullosomic progeny. We also observed mitotic chromosome loss, as evidenced by the presence of chromosomes that lie outside of the spindle at metaphase and chromosomes that lag at anaphase. Chromosomes that are lost during mitosis can form micronuclei in the following interphase. Micronuclei are observed in embryos during the early mitotic divisions and, rarely, in stages as late as cycle 14. In the later stages, the micronuclei are located at the surface of the embryo and lie adjacent to larger nuclei. These superficial micronuclei are likely to represent loss events that occur after cycle 10 (the stage at which nuclei complete their migration to the periphery), rather than loss events that occur before cycle 10 and are followed by the migration of individual micronuclei. This is an important distinction, since the occurrence of late loss has profound implications for models of *pal* function. The observation that chromosomes inherited from *pal* fathers may be maintained through as many as 10–14 mitotic divisions before the loss occurs suggests that a distinction between paternal and maternal chromosomes may persist through many mitoses and be recognized as late as middle to late cleavage. This feature suggests similarities between the chromosome loss induced by *pal* mutations and the normal processes of chromosome imprinting and elimination that occur in sciarid and coccid insects.

How can the absence of the *pal+* gene in males result in premitotic and mitotic defects observed in their progeny? Molecular characterization of the gene (Owens, 1996) permits us to suggest new models. Our studies have shown that the gene encodes a transcript that is detectable only in spermatocytes, thus accounting for the sex-specific effect of the mutations. Sequence analysis shows that the transcript encodes a small protein of 121 amino acids with a predicted pI of 10.96. Although the protein lacks significant amino acid homologies with known proteins, it is similar to other sperm-specific proteins in its small size and its enrichment in basic residues. Preliminary immunolocalization studies are consistent with the proposal that the PAL protein is a chromosomal protein. It is detected in the nuclei of elongating spermatids and in mature sperm. The protein is also detected in the sperm head immediately after its entry into the egg. Although we have not ruled out a role for PAL in sperm condensation, the fact that we have not observed defects during spermiogenesis has led us to suggest another possibility. We favor the idea that PAL is a sperm-specific chromosomal protein that binds at multiple sites, if not entirely along the paternal chromosomes, and functions in nuclear decondensation. As suggested from studies of other organisms, decondensation requires multiple steps: breakdown of the sperm nuclear envelope, exchange of sperm-specific proteins for maternal chromosomal proteins, and reformation of the nuclear envelope (Longo and Kunkle, 1978; Poccia and Collins, 1996). These events occur rapidly as the aster expands and the sperm migrates through the egg cytoplasm. PAL could be necessary for one or more of these steps. PAL may be a general chromosomal protein whose replacement facilitates the assembly of the maternal chromosomal proteins onto the sperm chromatin. These changes in chromatin state may serve to hold the paternal chromosomes together during nuclear envelope breakdown and reformation, by promoting interchromosomal interactions or interactions of the chromosomes with nuclear envelope fragments or with the sperm aster. PAL-deficient chromosomes that are improperly repackaged or tethered may be lost premitotically during pronuclear formation and form micronuclei. Mitotic instability could reflect a second role for PAL in chromosome segregation during mitosis, or it could result from the failure of poorly condensed chromosomes to interact with spindle components. Studies of HMG-D and H1 show that the exchange for maternal proteins occurs progressively during the early mitoses and may not be completed until middle or late cleavage cycles (Ner and Travers, 1994). If chromosomes deficient in PAL also fail in this exchange, then chromosome losses may occur after cycle 10. In the past, sperm-specific chromosomal proteins have been implicated mainly in condensation or gene repression activities during spermatogenesis (Bloch, 1969). Our hypotheses suggest a role for one small, highly basic sperm protein in decondensation and chromosome maintenance in the embryo. Genetic and molecular studies are currently being pursued to determine how long the PAL protein persists in the embryo, to further define its chromosomal localization, and to identify other proteins with which it interacts.

E. The *Horka* Gene

The *Horka* mutation, which was originally isolated as a dominant female sterile mutation, also induces a dominant paternal effect on development (Szabad et al., 1995). This mutation is similar to *pal* mutations in that it induces mitotic loss of paternal chromosomes in the progeny of mutant males. However, in contrast to *pal*, the *Horka* mutation does not induce Y chromosome loss, and *Horka* males show abnormal chromosome segregation during meiosis. Reversion studies have shown that the paternal effect defect results from a dominant, gain-of-function mutation, rather than a loss-of-function mutation. In addition, Szabad et al. (1995) report that this unusual mutation is an allele of *mus309*, a gene that was initially defined by recessive mutations that induce larval lethality (Gausz et al., 1981) or confer hypersensitivity to killing by chemical mutagens (Boyd et al., 1981). The *mus309* gene encodes a protein that binds to the inverted repeats of the transposable P-element and has homology to Ku p70, a mammalian DNA-binding protein (Beall *et al.*, 1994). Studies have demonstrated that *mus309* mutants show a greatly reduced ability to repair double-strand DNA breaks induced by P-element excision (Beall and Rio, 1996). Thus, the *Horka/mus309* gene differs from *pal* since it does not encode a paternal-specific product but rather a general cellular function. The *Horka* mutant phenotype demonstrates that abnormal expression of this function can interfere with the normal transmission of paternal chromosomes through meiosis and embryogenesis.

VIII. Summary and Perspectives

Numerous steps and complex interactions are required to transform the sperm into a functional pronucleus and ensure the inheritance of the paternal contributions to the embryo. Analyses of paternal effects on development have provided new insights into the regulation that exists to ensure biparental inheritance. The paternal effect mutations of *D. melanogaster* identify three specific products required for the sperm to participate in the initial stages of embryogenesis. Specifically, the *snky*+ gene product is required to form a functional male pronucleus, and both the *K81*+ and *pal*+ products are needed to maintain the paternal chromosomes during the mitotic divisions of the embryo (Figure 3).

The defects caused by several early-acting maternal effect mutations provide evidence that specific maternal factors play a role in coordinating the activities of the maternal and paternal components during and after fertilization. It is clear that some coordination results from independent responses of sperm and egg components to common cytoplasmic factors. The extent to which coordination relies on communication or feedback mechanisms between paternal and maternal components is less clear, and this is an issue that deserves further attention. We look forward to more extensive analyses of early-acting maternal effect mutations for their effects on sperm behavior and more comprehensive searches for new paternal effect mutations to address this and other issues. Areas that we feel are particularly interesting and in need of molecular inroads are the assembly and replication of the embryonic centrosome, the mechanisms of chromosome imprinting that cause differential behavior of paternal and maternal chromosomes, and the nature of signaling pathways that regulate interactions between parental components. It is hoped that new molecular tools will permit higher-resolution cytologic and molecular studies and facilitate the development of more biochemical and *in vitro* approaches for the study of *Drosophila* fertilization. These molecular tools should also allow us to identify the molecular mechanisms of fertilization that are the most highly conserved among animals. Finally, as suggested by Simerly *et al.* (1995), genetic studies of fertilization in model organisms such as *D. melanogaster* and *C. elegans* provide paradigms that may also be useful for diagnosing the possible causes of infertility in humans.

Acknowledgments

Studies of paternal effect mutations in our laboratory were supported by National Science Foundation Research Grant DCB9506927, by National Institutes of Health Genetics Training Grant T32 GM07735 for support of K.R.F. and K.N.O., and by a National Science Foundation Visiting Professor Fellowship to B.T.W. We thank Kathleen Wilson and Hak Chang for their contributions to our studies of *pal* and *K81*. We are grateful to colleagues, especially Steve Russell, Byron Williams, and Michael Goldberg, for allowing us to cite their unpublished data.

References

Almouzani, G., and Wolffe, A. P. (1993). Nuclear assembly, structure and function: The use of *Xenopus in vitro* systems. *Exp. Cell Res.* **205**, 1–15.

Baker, B. S. (1972). Tests for chromosome breakage in the meiotic mutant *paternal loss*. *Dros. Inform. Serv.* **49**, 55.

Baker, B. S. (1975). *Paternal loss (pal):* A meiotic mutant in *Drosophila melanogaster* causing paternal chromosome loss. *Genetics* **80**, 267–296.

Beall, E. L., Admon, A., and Rio, D. C. (1994). A *Drosophila* protein homologous to the human p70 Ku autoimmune antigen interacts with the P transposable element inverted repeats. *Proc. Natl. Acad. Sci. U.S.A.* **91**, 12681–12685.

Beall, E. L., and Rio, D. C. (1996). Drosophila IRBP/Ku p70 corresponds to the mutagen-sensitive *mus309* gene and is involved in P-element excision *in vivo*. *Genes Dev.* **10**, 921–933.

Berrios, M., and Avilion, A. A. (1990). Nuclear formation in a *Drosophila* cell-free system. *Exp. Cell Res.* **191**, 64–70.

Bishop, C. P., and Jackson, C. M. (1996). Genomic imprinting of chromatin in *Drosophila melanogaster. Genetica* **97**, 33–37.

Bloch, D. P. (1969). A catalog of sperm histones. *Genetics* **61** (Suppl), 93–111.

Boyd, J. B., Golino, M. D., Shaw, K. E. S., Osgood, C. J., and Green, M. M. (1981). Third-chromosome mutagen-sensitive mutants of *Drosophila melanogaster. Genetics* **97**, 607–623.

Bridges, C. B. (1925). Haploidy in *Drosophila melanogaster. Genetics* **11**, 706–710.

Browning, H., and Strome, S. (1996). A sperm-supplied factor required for embryogenesis in C. *elegans. Development* **122**, 391–404.

Callaini, G., and Riparbelli, M. G. (1996). Fertilization in *Drosophila melanogaster:* Centrosome inheritence and organization of the first mitotic spindle. *Dev. Biol.* **176**, 199–208.

Clark, A. G., Aguade, M., Prout, T., Harshman, L. G., and Langley, C. H. (1995). Variation in sperm displacement and its association with accessory gland protein loci in *Drosophila melanogaster. Genetics* **139**, 189–201.

Counce, S. (1963). Fate of the sperm tails within the *Drosophila* egg. *Dros. Inform. Serv.* **37**, 71.

Crevel, G., and Cotterill, S. (1991). DNA replication in cell-free extracts from *Drosophila melanogaster. EMBO J.* **10**, 4361–4369.

Crevel, G., and Cotterill, S. (1995). DF 31, a sperm decondensation factor from *Drosophila melanogaster:* Purification and characterization. *EMBO J.* **14**, 1711–1717.

Crouse, H. V. (1960). The controlling element in sex chromosome behavior in *Sciara. Genetics* **45**, 1429–1443.

Das, C. C., Kaufmann, B. P., and Gay, H. (1964a). Histone-protein transition in *Drosophila melanogaster.* I. Changes during spermatogenesis. *Exp. Cell Res.* **35**, 507–514.

Das, C. C., Kaufmann, B. P., and Gay, H. (1964b). Histone protein transition in *Drosophila melanogaster.* II. Changes during early embryonic development. *J. Cell Biol.* **23**, 423–430.

Davis, D. G. (1969). Chromosome behavior under the influence of *claret-nondisjunctional* in *Drosophila melanogaster. Genetics* **61**, 577–594.

de Saint Phalle, B., and Sullivan, W. (1996). Incomplete sister chromatid separation is the mechanism of programmed chromosome elimination during early *Sciara coprophila* embryogenesis. *Development* **122**, 3775–3784.

Doane, W. (1960). Completion of meiosis in uninseminated eggs of *Drosophila melanogaster. Science* **132**, 677–678.

Dorn, R., Krauss, V., Reuter, G., and Saumweber, H. (1993). The enhancer of position-effect variegation of *Drosophila, E(var)3-93D*, codes for chromatin protein containing a conserved domain common to several transcriptional regulators. *Proc. Natl. Acad. Sci. U.S.A.* **90**, 11376–11380.

Dorn, R., Szidonya, J., Korge, G., Sehnert, M., Taubert, H., Archoukieh, E., Tschiersch, B., Morawietz, H., Wustmann, G., Hoffmann, G., and Reuter, G. (1993). P transposase induced dominant enhancer mutations of position effect variegation of *Drosophila melanogaster. Genetics* **133**, 279–290.

Dybas, L. K., Tyl, B. T., and Geer, B. W. (1981). Aberrant spermiogenesis in X-linked male-sterile mutants of *Drosophila melanogaster. J. Exp. Zool.* **216**, 299–310.

Edgar, B. A., Kiehle, C. P., and Schubiger, G. (1986). Cell cycle control by the nucleocytoplasmic ratio in early *Drosophila* development. *Cell* **44**, 365–372.

Endow, S. A., and Komma, D. J. (1996). Centrosome and spindle function of the *Drosophila* ncd microtubule motor visualized in live embryos using Ncd-GFP fusion proteins. *J. Cell Sci.* **109**, 2429–2442.

Erdelyi, M., and Szabad, J. (1989). Isolation and characterization of dominant female sterile mutations of *Drosophila melanogaster.* I. Mutations on the third chromosome. *Genetics* **122**, 111–127.

Fitch, K. R., Reith, M. A., and Wakimoto, B. W. The paternal effect gene *ms(3)sneaky* is required for sperm function and the initiation of embryogenesis in *Drosophila melanogaster,* submitted.

Foe, V. E., Odell, G. M., and Edgar, B. A. (1993). Mitosis and morphogenesis in the *Drosophila* embryo: Point and counterpoint. *In* "The Development of *Drosophila melanogaster,*" vol. 1 (M. Bate and A. Martinez Arias, Eds.), pp. 149–300. Cold Spring Harbor Laboratory Press, Cold Spring Harbor, New York.

Foltz, K. R., and Lennarz, W. J. (1993). The molecular basis of sea urchin gamete interactions at the egg plasma membrane. *Dev. Biol.* **158**, 46–61.

Freeman, M., and Glover, D. M. (1987). The *gnu* mutation of *Drosophila* causes inappropriate DNA synthesis in unfertilized and fertilized eggs. *Genes Dev.* **1**, 924–930.

Fuyama, Y. (1984). Gynogenesis in *Drosophila melanogaster. Jap. J. Genet.* **59**, 91–96.

Fuyama, Y. (1986a). Genetics of parthenogenesis in *Drosophila melanogaster.* II. Characterization of a gynogenetically reproducing strain. *Genetics* **114**, 4995–5009.

Fuyama, Y. (1986b). Genetics of parthenogenesis in *Drosophila melanogaster.* I. The modes of diploidization in the gynogenesis induced by male-sterile mutant, *ms(3)K81. Genetics* **112**, 237–248.

Fuyama, Y., Hardy, R. W., and Lindsley, D. L. (1988). Parthenogenesis in *Drosophila melanogaster:* The mechanisms of gynogenesis induced by *ms(3)K81* sperm. *Jap. J. Genet.* **63**, 553–554.

Gausz, J., Gyurkovics, H., Bencze, G., Awad, A. A. M., Holden, J. J., and Ish-Horowicz, D. (1981). Genetic characterization of the region between 86F1,2 and 87B15 on chromosome 3 of *Drosophila melanogaster. Genetics* **98**, 775–789.

Gerbi, S. A. (1986). Unusual chromosome movements in sciarid flies. *In* "Results and Problems in Cell Differentiation: Germ Line–Soma Differentiation" (W. Hennig, Ed.), pp. 71–104. Springer-Verlag, Berlin.

Gold, J. D., and Pedersen, R. A. (1994). Mechanisms of genomic imprinting in mammals. *Curr. Top. Dev. Biol.* **29**, 227–280.

Hackstein, J. H. P. (1991). Spermatogenesis in *Drosophila:* A genetic approach to cellular and subcellular differentiation. *Eur. J. Cell Biol.* **56**, 151–169.

Hawley, R. S., McKim, K. S., and Arbel, T. (1993). Meiotic segregation in *Drosophila melanogaster* females: Molecules, mechanisms and myths. *Annu. Rev. Genet.* **27,** 281–317.

Heuer, J. G., Li, K., and Kaufman, T. C. (1995). The *Drosophila* homeotic target gene, *centrosomin (cnn)* encodes a novel centrosomal protein with leucine zippers and maps to a genomic region required for gut morphogenesis. *Development* **121,** 3861–3876.

Hildreth, P. E., and Lucchesi, J. C. (1963). Fertilization in *Drosophila*. I. Evidence for the regular occurrence of monospermy. *Dev. Biol.* **6,** 262–278.

Hill, D. P., Shakes, D. C., Ward, S., and Strome, S. (1989). A sperm-supplied product essential for initiation of normal embryogenesis in *C. elegans* is encoded by the paternal effect lethal gene, *spe11. Dev. Biol.* **136,** 154–166.

Huettner, A. F. (1924). Maturation and fertilization in *Drosophila melanogaster. J. Morphol.* **39,** 249–265.

Karr, T. (1991). Intracellular sperm/egg interactions in *Drosophila*: A three-dimensional structural analysis of a paternal product in the developing egg. *Mech. Dev.* **34,** 101–112.

Karr, T. (1996). Paternal investment and intracellular sperm-egg interactions during and following fertilization in *Drosophila. Curr. Top. Dev. Biol.* **34,** 89–115.

Karr, T., and Pitnick, S. (1996). The ins and outs of fertilization. *Nature* **379,** 405–406.

Kawasaki, K., Philpott, A., Avilion, A. A., Berrios, M., and Fisher, P. A. (1994). Chromatin decondensation in *Drosophila* embryo extracts. *J. Biol. Chem.* **269,** 10169–10176.

Komma, D. J., and Endow, S. A. (1995). Haploidy and androgenesis in *Drosophila. Proc. Natl. Acad. Sci. U.S.A.* **92,** 11884–11888.

Kondo, R., Satta, Y., Matsuura, E. T., Ishiwa, H., Takahata, N., and Chigusa, S. I. (1990). Incomplete maternal transmission of mitochondrial DNA in *Drosophila. Genetics* **126,** 657–663.

Kremer, H., Hennig, W., and Dijkhof, R. (1986). Chromatin organization in the male germ line of *Drosophila hydei. Chromosoma* **94,** 147–161.

Kuhn, D. T., and Packert, G. (1988). Paternal imprinting of inversion *Uab1* causes homeotic transformations in *Drosophila. Genetics* **118,** 103–107.

Latham, K. (1994). Strain-specific differences in mouse oocytes and their contributions to epigenetic inheritance. *Development* **120,** 3419–3426.

Lefevre, G., and Jonsson, U. B. (1962). Sperm transfer, storage, displacement, and utilization in *Drosophila melanogaster. Genetics* **47,** 1719–1736.

L'Hernault, S. W., Shakes, D. C., and Ward, S. (1988). Developmental genetics of chromosome I spermatogenesis-defective mutants in the nematode, *Caenorhabditis elegans. Genetics* **120,** 435–452.

Lieberfarb, M. E., Chu, T., Wreden, C., Theurkauf, W., Gergen, J. P., and Strickland, S. (1996). Mutations that perturb polyA dependent mRNA activation block the initiation of development. *Development* **122,** 579–588.

Lin, H., and Wolfner, M. F. (1991). The *Drosophila* maternal-effect gene *fs(1)Ya* encodes a cell cycle–dependent nuclear envelope component required for embryonic mitosis. *Cell* **64,** 49–62.

Liu, J., Lin, H., Lopez, J. M., and Wolfner, M. F. (1997). Formation of the male pronuclear lamina in *Drosophila melanogaster. Dev. Biol.* **184,** 187–196.

Liu, J., Song, K., and Wolfner, M. (1995). Mutational analyses of *fs(1)Ya,* an essential, developmentally regulated, nuclear envelope protein in *Drosophila. Genetics* **141,** 1473–1481.

Longo, F. J. (1973). Fertilization: A comparative ultrastructural review. *Biol. Reprod.* **9,** 149–215.

Longo, F. J., and Kunkle, M. (1978). Transformations of sperm nuclei upon insemination. *Curr. Top. Dev. Biol.* **12,** 149–184.

Longo, F. J., Cook, S., and Mathews, L. (1991). Pronuclear formation in starfish eggs inseminated at different stages of meiotic maturation: Correlation on sperm nuclear transformations and activity of the maternal chromatin. *Dev. Biol.* **147,** 62–72.

Lopez, J. M., Song, K., Hirshfeld, A. B., Lin, H., and Wolfner, M. (1994). The *Drosophila fs(1)Ya* protein, which is needed for the first mitotic division, is in the nuclear lamina and in the envelopes of cleavage nuclei, pronuclei and nonmitotic nuclei. *Dev. Biol.* **163,** 202–211.

Mahowald, A. P., Goralski, T. J., and Caulton, J. H. (1983). *In vitro* activation of *Drosophila* eggs. *Dev. Biol.* **98,** 437–445.

Malacinski, G. M., and Barone, D. (1985). Towards understanding paternal extragenic contributions to early amphibian pattern specification: The axolotl *ts-1* gene as a model system. *J. Emb. Exp. Morph.* **89,** 53–68.

Maleszewski, M. (1992). Behavior of sperm nuclei incorporated into parthenogenetic mouse eggs prior to the first cleavage division. *Mol. Reprod. Dev.* **33,** 215–221.

Matthews, K. A., Rees, D., and Kaufman, T. C. (1993). A functionally specialized a-tubulin is required for oocyte meiosis and cleavage mitoses in *Drosophila. Development* **117,** 977–991.

McDonald, H. B., Stewart, R. J., and Goldstein, L.S.B. (1990). The kinesin-like *ncd* protein of *Drosophila* is a minus end-directed microtubule motor. *Cell* **63,** 1159–1165.

Murray, A. W., and Kirschner, M. W. (1989). Cyclin synthesis drives the early embryonic cell cycle. *Nature* **339,** 275–280.

Navara, C. S., First, N. L., and Schatten, G. (1996). Phenotypic variation among paternal centrosomes expressed within the zygote as disparate microtubule lengths and sperm aster organization: Correlations between centrosome activity and developmental success. *Proc. Natl. Acad. Sci. U.S.A.* **93,** 5384–5388.

Nelson, C. R., and Szauter, P. (1992). Timing of mitotic chromosome loss caused by the *ncd* mutation of *Drosophila melanogaster. Cell Motility and Cytoskeleton* **23,** 34–44.

Ner, S. S., and Travers, A. A. (1994). HMG-D, the *Drosophila melanogaster* homologue of HMG1 protein is associated with early embryonic chromatin in the absence of histone H1. *EMBO J.* **13,** 1817–1822.

Nigg, E. A. (1992). Assembly and cell cycle dynamics of the nuclear lamina. *Semin. Cell Biol.* **3,** 245–253.

Nur, U. (1990). Heterochromatization and euchromatization of whole genomes in scale insects (Coccoidea: Homoptera). *Development* (Suppl), 29–34.

Ohlendieck, K., and Lennarz, W. J. (1996). Molecular mechanisms of gamete recognition in sea urchin fertilization. *Curr. Top. Dev. Biol.* **32,** 39–55.

Owens, K. N. (1996). Genetic, molecular and cytological characterization of the paternal effect gene, *paternal loss* of *Drosophila melanogaster.* Ph.D. thesis, Department of Genetics, University of Washington, Seattle, Washington.

Perotti, M-E. (1969). Ultrastructure of the mature sperm of *Drosophila melanogaster* Meig. *J. Submicr. Cytol.* **1,** 171–196.

Perotti, M-E. (1975). Ultrastructural aspects of fertilization in *Drosophila. In* "The Functional Anatomy of the Spermatozoan," Proceedings of the Second International Symposium (B. A. Afzelins, Ed.), pp. 57–68. Pergamon Press, Oxford.

Perotti, M-E., and Pasini, M. E. (1995). Glycoconjugates of the surface of the spermatozoa of *Drosophila melanogaster:* A qualitative and quantitative study. *J. Exp. Zool.* **271,** 311–318.

Perotti, M-E., Pasini, M. E., and Hackstein, J. H. P. (1994). Detection of carbohydrate binding sites on *Drosophila* sperm surface. *In* "Electron Microscopy: Applications in Biological Sciences" (B. Jouffrey and C. Colliex, Eds.), pp. 317–318. Les Ulis Cedex, France.

Philpott, A., Leno, G. H., and Laskey, R. A. (1991). Sperm decondensation in *Xenopus* egg cytoplasm is mediated by nucleoplasmin. *Cell* **65,** 569–578.

Pitnick, S., Spicer, G. S., and Markow, T. A. (1995). How long is a giant sperm? *Nature* **375,** 109.

Poccia, D. (1986). Remodeling of nucleoproteins during gametogenesis, fertilization, and early development. *Int. Rev. Cytol.* **105,** 1–54.

Poccia, D., and Collas, P. (1996). Transforming sperm nuclei into male pronuclei *in vivo* and *in vitro. Curr. Top. Dev. Biol.* **34,** 25–88.

Poccia, D. L., and Green, G. R. (1996). Packaging and unpackaging the sea urchin sperm genome. *Trends Biol. Sci.* **17,** 223–227.

Rabinowitz, M. (1944). Studies on the cytology and early embryology of the egg of *Drosophila melanogaster. J. Morphol.* **69,** 1–49.

Ripoll, P., Carmena, M., and Molina, I. (1992). Genetic analysis of cell division in *Drosophila*. *Curr. Top. Dev. Biol.* **27**, 275–307.

Risley, M. S. (1990). Chromatin organization in sperm. *In* "Chromosomes: Eukaryotic, Prokaryotic and Viral II" (S. Adolph, Ed.), pp. 61–78. CRC Press, New York.

Russell, S. H., and Kaiser, K. (1993). *Drosophila melanogaster* male germ line specific transcripts with autosomal and Y-linked genes. *Genetics* **134**, 293–308.

Salz, H. K., Flickinger, T. W., Mittendorf, E., Pellicena-Palle, A., Petschek, J. P., and Albrecht, E. B. (1994). The *Drosophila* maternal effect locus *deadhead* encodes a thioredoxin homolog required for female meiosis and early embryonic development. *Genetics* **136**, 1075–1086.

Sander, K. (1985). Fertilization and egg cell activation in insects. *In* "Biology of Fertilization" (C. Metz and A. Monroy, Eds.), pp. 409–430. Academic Press, New York.

Sandler, L. (1971). Induction of autosomal meiotic mutants by EMS in *D. melanogaster*. *Dros. Inform. Serv.* **47**, 68.

Schatten, G. (1994). The centrosome and its mode of inheritance: The reduction of the centrosome during gametogenesis and its restoration during fertilization. *Dev. Biol.* **165**, 299–335.

Schatten, G., and Schatten, H. (1987). Fertilization: Motility, the cytoskeleton and the nuclear architecture. *Oxford Reviews of Reproductive Biology* **9**, 322–378.

Schultz, R. M., and Kopf, G. S. (1995). Molecular basis of mammalian egg activation. *Curr. Top. Dev. Biol.* **30**, 21–62.

Schupbach, T., and Wieschaus, E. (1989). Female sterile mutations on the second chromosome of *Drosophila melanogaster*. I. Maternal effect mutations. *Genetics* **121**, 101–117.

Shamanski, F. L., and Orr-Weaver, T. L. (1991). The *Drosophila plutonium* and *pan gu* genes regulate entry into S phase at fertilization. *Cell* **66**, 1289–1300.

Simerly, C., Wu, G-J., Zoran, S., Ord, T., Rawlins, R., Jones, J., Navara, C., Gerrity, M., Rinehart, J., Binor, Z., Asch, R., and Schatten, G. (1995). The paternal inheritance of the centrosome, the cell's microtubule-organizing center, in humans, and the implications for infertility. *Nat. Med.* **1**, 47–52.

Simerly, C. R., Hecht, N. B., Goldberg, E., and Schatten, G. (1993). Tracing the incorporation of the sperm tail in the mouse zygote and early embryo using an anti-testicular a-tubulin antibody. *Dev. Biol.* **158**, 536–548.

Sonnenblick, B. P. (1950). The early embryology of *Drosophila melanogaster*. *In* "Biology of Drosophila" (M. Demerec, Ed.), pp. 62–167. Wiley, New York.

Spofford, J. B. (1976). Position-effect variegation in *Drosophila*. *In* "The Genetics and Biology of Drosophila," vol. 1C (M. Ashburner and E. Novitski, Eds.), pp. 955–1018. Academic Press, New York.

Stafstrom, J. P., and Staehelin, L. A. (1993). Dynamics of the nuclear envelope and of nuclear pore complexes during mitosis in the *Drosophila* embryo. *Eur. J. Cell Biol.* **34**, 179–189.

Stalker, H. D. (1954). Parthenogenesis in *Drosophila*. *Genetics* **39**, 4–33.

Szabad, J., Erdelyi, M., Hoffman, G., Szidonya, J., and Wright, T. R. F. (1989). Isolation and characterization of dominant female sterile mutations of *Drosophila melanogaster*. II. Mutations of the second chromosome. *Genetics* **122**, 823–835.

Szabad, J., Mathe, E., and Puro, J. (1995). *Horka*, a dominant mutation of *Drosophila*, induces nondisjunction and, through paternal effect, chromosome loss and genetic mosaics. *Genetics* **139**, 1585–1599.

Templeton, A. R. (1983). Natural and experimental parthenogenesis. *In* "The Genetics and Biology of Drosophila," vol. 3A (M. Ashburner, H. L. Carson, and J. N. Thompson, Eds.), pp. 343–398. Academic Press, London.

Tokuyasu, K. T. (1974). Dynamics of spermiogenesis in *Drosophila melanogaster*. IV. Nuclear transformation. *J. Ultrastr. Res.* **48**, 284–303.

Tokuyasu, K. T. (1975). Dynamics of spermiogenesis in *Drosophila melanogaster*. V. Head-tail alignment. *J. Ultrastr. Res.* **50**, 117–129.

Tomkiel, J. T. (1990). Genetic studies of the interaction of the *abnormal oocyte* mutation with heterochromatin in *Drosophila melanogaster.* Ph.D. thesis, Department of Genetics, University of Washington, Seattle, Washington.

Ulitzur, N., and Gruenbaum, Y. (1989). Nuclear envelope assembly around sperm chromatin in cell-free preparations from *Drosophila* embryos. *FEBS Lett.* **259,** 113–116.

Uzawa, M., Grams, J., Madden, B., Toft, D., and Salisbury, J. L. (1995). Identification of a complex between centrin and heat shock proteins in CSF-arrested *Xenopus* oocytes and dissociation of the complex following oocyte activation. *Dev. Biol.* **171,** 51–59.

Walker, R. A., Salmon, E. D., and Endow, S. A. (1990). The *Drosophila claret* segregation protein is a minus-end directed motor molecule. *Nature* **347,** 780–782.

Wassarman, P. M., and Litscher, E. S. (1995). Sperm-egg recognition mechanisms in mammals. *Curr. Top. Dev. Biol.* **30,** 1–19.

Whitaker, M., and Swann, K. (1993). Lighting the fuse at fertilization. *Development* **117,** 1–12.

Williams, B. C., Dernberg, A. F., Puro, J., Nokkala, S., and Goldberg, M. L. (1997). The *Drosophila* kinesin-like protein KLP3A is required for proper behavior of male and female pronuclei at fertilization. *Development* **124,** 2365–2376.

Williams, B. C., Riedy, M. F., Williams, E. V., Gatti, M., and Goldberg, M. L. (1995). The *Drosophila* kinesin-like protein KLP3A is a midbody component required for central spindle assembly and initiation of cytokinesis. *J. Cell Biol.* **129,** 709–723.

Wolfner, M. (1997). Tokens of love: Function and regulation of *Drosophila* male accessory gland products. *Insect Biochem. Molec. Biol.* **27,** 179–192.

Wood, W. B., Hecht, R., Carr, S., Vanderslice, R., Wolf, N., and Hirsh, D. (1980). Parental effects and phenotypic characterization of mutations that affect early development in *Caenorhabditis elegans. Dev. Biol.* **74,** 446–469.

Yamashita, M., Onozato, H., Nakanishi, T., and Nagahama, Y. (1990). Breakdown of the sperm nuclear envelope is a prerequisite for male pronucleus formation: Direct evidence from the gynogenetic crucian carp *Carassius auratus* langsdorfii. *Dev. Biol.* **13,** 155–160.

Yasuda, G. K., Schubiger, G., and Wakimoto, B. T. (1995). Genetic characterization of the *ms(3)K81,* a paternal effect gene of *Drosophila melanogaster. Genetics* **140,** 219–229.

Zalokar, M., Audit, C., and Erk, I. (1975). Developmental defects of female-sterile mutants of *Drosophila melanogaster. Dev. Biol.* **47,** 419–432.

2

Drosophila Myogenesis and Insights into the Role of nautilus

Susan M. Abmayr and Cheryl A. Keller
Department of Biochemistry and Molecular Biology and Center for Gene Regulation
The Pennsylvania State University
University Park, Pennsylvania 16802

I. Introduction

Myogenesis is one of the first developmental processes of which our understanding has been aided by the identification of genes that are direct regulators of a specific program of differentiation. The family of transcription factors encoded by these genes, the so-called myogenic regulatory factors (MRFs), have provided a molecular entry point from which both upstream and downstream events can be examined. These proteins have had tremendous impact as molecular markers for events in myogenesis, in part as a consequence of their high degree of conservation in both vertebrate and invertebrate species. In addition, fundamental aspects of their biochemical, and perhaps biological functions extend to many different organisms. The conservation of this gene family therefore provides biochemical, genetic, and cell biological approaches to their function, both in vivo and in vitro. While all aspects of myogenesis are unlikely to be conserved in organisms as diverse as fruitflies and mammals, the experimental strengths of each system may provide insights that are applicable to other organisms.

Current Topics in Developmental Biology, Vol. 38

II. *Drosophila* Myogenesis

A. Early Events

The somatic muscles of a *Drosophila* larva are derived from cells that become committed to the mesodermal lineage approximately 2 hr after fertilization (Beer *et al.*, 1987). Early specification of this mesodermal germ layer begins with establishment of the dorsoventral axis, a process that requires the maternally-provided Dorsal protein. A morphogenetic gradient is generated by the selective localization of Dorsal protein only to nuclei on the ventral side of the embryo (reviewed in Govind and Steward, 1991; Bate, 1993). Dorsal, which has homology to the transcription factor NfκB, positively regulates transcription of mesoderm-specific genes such as *twist* and *snail*. These genes, in turn, specify and define the mesodermal analgen. The *twist* gene encodes a basic Helix-Loop-Helix (bHLH)–containing transcription factor, TWI, that is essential for formation of the mesodermal layer (Thisse *et al.*, 1987, 1988) and is initially expressed in mesodermal cells at stage 5, approximately 2.5 hr after egg laying (AEL) (Thisse *et al.*, 1988; Leptin, 1991). At this stage, transplantation experiments have suggested that the ventrally-located cells are fated to be mesoderm, but they do not yet exhibit features reminiscent of any particular mesodermal derivative (Beer *et al.*, 1987). Expression of *twist* declines during stage 11 but persists in a smaller number of cells that appear to be larval myoblasts (Thisse *et al.*, 1988; Bate, 1993), suggesting the possibility of a later role for *twist* in larval myogenesis (see later discussion). Additionally, *twist* expression continues in cells that will contribute to the adult musculature (Bate *et al.*, 1991). By comparison to *twist*, the *snail*-encoded protein contains a zinc-finger DNA-binding domain that is also detected during stage 5 (Kosman *et al.*, 1991; Leptin, 1991) and appears to define the mesodermal borders.

Coincident with the development of the dorsoventral axis and the specification of the mesodermal anlagen is the process of cell division. In *Drosophila*, nuclei undergo 13 synchronous divisions without cytokinesis prior to stage 5. During cellularization, the nuclei of the syncitical blastoderm embryo migrate to the periphery and are surrounded by invaginating plasma membrane. This process is followed by three less synchronous waves of cell division. The third post-blastoderm mitosis, which appears to be the final mitotic cycle for many cell types, occurs at approximately 5.5 hr AEL in mesodermal cells (Beer *et al.*, 1987). It precedes subdivision of the mesodermal epithelium into precursors that include the visceral, somatic, and cardiac mesoderm. Many mesodermal cells undergo a fourth cell division, including at least a subset of the myoblasts (Bate, 1993; Carmena *et al.*, 1995; Rushton *et al.*, 1995; Azpiazu *et al.*, 1996). Indeed, this cell division may be critical for the specification of founder cells for some muscle fibers (see later discussion and Carmena *et al.*, 1995; Azpiazu *et al.*, 1996). However, significant proliferation beyond this stage has not been reported. Thus, by the time that myogenesis begins to be evident (see later), the number of cells necessary to form the somatic musculature is largely present. Significant myoblast proliferation analogous to that of vertebrates (Bischoff and Holtzer, 1969; Holtzer, Rubinstein, *et al.*, 1975; Holtzer, Strahs, *et al.*, 1975; Bischoff, 1978) is not apparent in *Drosophila*.

B. Subdivision of the Mesoderm

At approximately 6 hr AEL, subsequent to the third postblastoderm cell division, the mesoderm becomes subdivided into progenitors of tissues that include cells of the dorsal vessel, visceral and somatic musculature, and fat body. Historically, the mesodermal subdivisions have been termed the splanchnopleura and the somatopleura (Crossley, 1978; Campos-Ortega and Hartenstein, 1985). The splanchnopleura, the inner layer, was suggested as the primary source of cells that form the visceral musculature, while the somatopleura, which is in close proximity to the ectoderm, was thought to contribute largely to the somatic musculature (Crossley, 1978; Campos-Ortega and Hartenstein, 1985).

More recent studies have suggested that subdivision of the mesoderm occurs along the dorsoventral axis early in development such that dorsolateral meso- derm gives rise to visceral musculature and cardiac lineages (Bodmer *et al.*, 1990; Bodmer, 1993; Azpiazu *et al.*, 1996), while more ventrally located meso- dermal cells give rise to somatic muscle (Leptin, 1991; Bate and Rushton, 1993; Azpiazu *et al.*, 1996). Specification of dorsolateral mesoderm occurs in response to secretion of the *decapentaplegic* (*dpp*) gene product, a *Drosophila* homologue of transforming growth factor-β (TGFβ) (Padgett *et al.*, 1987), from cells of the dorsal ectoderm (Staehling-Hampton *et al.*, 1994; Frasch, 1995; Maggert *et al.*, 1995). DPP induces synthesis of the autonomously acting, homeobox-containing gene *tinman* (Frasch, 1995), which is essential for the formation of the heart and visceral musculature (Bodmer *et al.*, 1990; Azpiazu and Frasch, 1993; Bodmer, 1993). TIN subsequently activates *bagpipe*, a homeobox-containing gene that is expressed exclusively in the visceral mesoderm and is critical for differentiation of the visceral musculature (Azpiazu and Frasch, 1993). By contrast to the induction of dorsolateral mesoderm, and its derivatives, DPP represses *pox-meso*, a gene expressed in ventrally located mesodermal cells that contribute to the somatic musculature (Bopp *et al.*, 1989; Staehling-Hampton *et al.*, 1994). Addi- tional support for the importance of ectodermal signaling in the specification of cells of the somatic musculature comes from studies showing induction of *nau- tilus* (*nau*), a marker that distinguishes subsets of somatic muscle progenitors (see later discussion), by ventral ectoderm (Baker and Schubiger, 1995). There- fore, as with the visceral and cardiac mesoderm, the somatic mesoderm appears to be specified by molecules secreted from the ectoderm that differ along the dorsoventral axis such that ventrally located mesodermal cells become specified primarily as progenitors of the somatic musculature.

In addition to the dorsoventral axis, studies have established that the mesoderm is also subdivided into parasegmental units along the anteroposterior axis (Lawrence, 1992; Lawrence et al., 1995). The regional location of cells within this germ layer reflects the tissue type that they will ultimately generate: cells located in the anterior region of each parasegment give rise to visceral musculature and fat body, while those located in the posterior portion give rise to the heart and somatic musculature (Dunin-Borkowski et al., 1995; Azpiazu et al., 1996). As in the dorsoventral axis, subdivision along the anteroposterior axis is controlled, at least in part, by signaling molecules. For example, the formation of even-skipped– and tinman-expressing heart precursors is dependent on the presence of the product of the wingless (wg) gene (Lawrence et al., 1995; Wu et al., 1995; Park et al., 1996), while the bagpipe-expressing progenitors of the visceral musculature are expanded in wingless mutant embryos (Azpiazu et al., 1996). By comparison, overexpression of wingless and other molecules in the wingless pathway induce hyperplasia of the heart precursors (Lawrence et al., 1995; Park et al., 1996), reportedly at the expense of the visceral mesoderm (Park et al., 1996). Results suggest that wingless and hedgehog influence development of the posterior and anterior portions, respectively, of each parasegment (Azpiazu et al., 1996). These studies demonstrate that hedgehog is required for formation of the visceral musculature and fat body through the induction of bagpipe and serpent expression. Conversely, wingless antagonizes the action of hedgehog in the posterior region of the parasegment, preventing bagpipe and serpent expression and promoting development of the heart and somatic musculature (Azpiazu et al., 1996). The wingless gene is also essential for the specification of a subset of muscle precursors because mutations in wingless result in the loss of particular founder cell clusters (Bate and Rushton, 1993; Volk and VijayRaghavan, 1994; Baylies et al., 1995; Ranganayakulu et al., 1996). Unlike its action in heart precursors, however, overexpression of wingless does not appear to result in ectopic clusters of founder cells or muscle precursors (Ranganayakulu et al., 1996). Finally, it should be noted that the signaling molecules responsible for these inductive events can originate in overlying ectodermal cells as well as within the mesoderm itself (Lawrence et al., 1994, 1995; Baylies et al., 1995; Azpiazu et al., 1996; Ranganayakulu et al., 1996), leading to the suggestion that they may play a permissive rather than an instructive role (Lawrence et al., 1995).

Studies have established that, after domains of muscle progenitors are set aside, autonomously acting molecules within the mesoderm are critical for their subsequent development. In particular, studies examining the embryonic pattern of TWI, a bHLH-containing transcription factor described previously, demonstrated domains of high and low expression that correspond to progenitors of different mesodermal derivatives (Dunin-Borkowski et al., 1995; Baylies and Bate, 1996). These studies further showed that domains with high levels of endogenous TWI correspond to somatic mesoderm (Dunin-Borkowski et al.,

1995; Baylies and Bate, 1996), and that high levels of TWI can propel cells into somatic myogenesis when provided artificially (Baylies and Bate, 1996). By comparison, high TWI levels suppress development of the visceral musculature and heart. Of note, domains of TWI-expressing cells correspond closely with mesodermal cell clusters that express the proneural gene *lethal of scute* (*l'sc*) (Carmena *et al.*, 1995), which appears to be essential for cell fate decisions within the somatic musculature and is discussed further in the section on muscle patterning.

In comparison to the roles of *bagpipe* and *tinman* in the visceral musculature and the heart, a single gene has not yet been identified that has an analogous role in development of the precursors to the somatic musculature. The paired-domain–containing protein encoded by the *pox-meso* gene is one candidate for such a regulatory molecule. *pox-meso* transcripts are observed in the somatic mesoderm in a segmentally repeated pattern during stages 10 and 11 (Bopp *et al.*, 1989). As described earlier, DPP induces expression of genes associated with the visceral musculature but represses this ventrally-located *pox-meso* expression (Bopp *et al.*, 1989; Staehling-Hampton *et al.*, 1994). Nevertheless, it remains to be shown whether *pox-meso* is actually necessary for development of all somatic muscles. Another protein that clearly functions in this pathway is TWI, as already described (Dunin-Borkowski *et al.*, 1995; Baylies and Bate, 1996). However, understanding of its role in the specification of all somatic muscles through genetic loss-of-function mutations is hampered by the earlier requirement for TWI in mesoderm formation (Leptin, 1991). Therefore, direct loss-of-function genetic studies have not yet revealed molecules that serve functions in the somatic mesoderm similar to those of *tinman* and *bagpipe* in other mesodermal derivatives.

Several additional genes appear to be expressed in broad domains in the early mesoderm and may play a role in the specification and/or subdivision of the mesoderm. Examples include *DFR1* (Shishido *et al.*, 1993), *Drac1* (Luo *et al.*, 1994), *zfh-1* (Lai *et al.*, 1991), *mef2* (Lilly *et al.*, 1994; Nguyen *et al.*, 1994), and the P2 transcript of *pointed* (Klambt, 1993). Embryos mutant for *Drosophila* fibroblast growth factor receptor *DFR1* (renamed *heartless*), exhibit abnormalities in several mesodermal derivatives as a consequence of defects in cell migration and spreading (Bieman *et al.*, 1996; Gisselbrecht *et al.*, 1996). Embryos mutant for *zfh-1* and *pointed* exhibit subtle defects in the somatic musculature (Klambt, 1993; Lai *et al.*, 1993). By comparison, precursors for several mesodermally-derived tissues are present in embryos mutant for *mef2*, but the tissues exhibit defects at later stages of differentiation (Bour *et al.*, 1995; Lilly *et al.*, 1995; Ranganayakulu *et al.*, 1995), *tinman* is widely expressed in the early mesoderm but, as mentioned previously, the primary defects observed in *tinman* mutant embryos are in formation of the heart and visceral musculature (Azpiazu and Frasch, 1993; Bodmer, 1993).

C. Muscle Patterning

The term "patterning," in the context of the larval body wall musculature, refers to formation of a segmentally-repeated array of 30 unique fibers per abdominal hemisegment that develops during embryogenesis and will be used by the crawling larva (Crossley, 1978; Bate, 1993). Studies have suggested that this pattern is established very early in the myogenic program, prior to or coincident with the earliest evidence of muscle differentiation. The first morphological sign of somatic myogenesis in the embryo is the formation of binucleate and trinucleate cells that appear in characteristic positions at the onset of germ band retraction late in stage 11, at about 7.5 hr AEL (Bate, 1990). The pattern of these muscle precursors is complete by the end of germ band shortening, such that an individual fiber observed later in development is represented by a single precursor at this stage (Bate, 1990). Therefore, these precursors possess and respond to information that specifies the unique identity and subsequent differentiation program of each larval muscle. Although the exact mechanisms by which these cells are set aside and specified are not entirely clear, and the genes implementing their unique myogenic differentiation programs are not fully understood, research in the last few years has provided critical insights into these issues.

1. The Founder Cell Model

A useful framework for considering muscle development and patterning has been provided by studies in the grasshopper showing the presence of large, morphologically distinct, mononucleate mesodermal cells, termed muscle "pioneers" (Ho *et al.*, 1983). Laser ablation of individual pioneers eliminated formation of specific muscle fibers, suggesting that the mononucleate pioneers contained information for directing the subsequent differentiation program of the muscle fibers as myoblast fusion proceeded (Ball *et al.*, 1985). Although morphologically distinct mononucleate cells have not been observed in *Drosophila* embryos, studies have shown the existence of binucleate and trinucleate cells in characteristic and reproducible positions, reminiscent of the muscle pioneers (Bate, 1990). More recent studies have suggested that these multinucleate cells, termed muscle "precursors," arise from founder cells that "seed" the fusion process (Rushton *et al.*, 1995) and have supported the model that muscle development in *Drosophila* requires the presence of two distinct cell types. According to this model (Fig. 1), a small number of founder cells, one for each muscle fiber, mediate fusion with a larger group of myoblasts that are not committed to a particular muscle fiber type. In its simplest interpretation, muscle fiber diversification is specified by information contained within each founder cell and transferred to the unspecified fusion-competent cells as fusion progresses. Support for this hypothesis has been provided by the morphological examination of myogenesis in embryos defective for myoblast fusion (Rushton *et al.*, 1995). In

The Founder Cell Model for Muscle Patterning

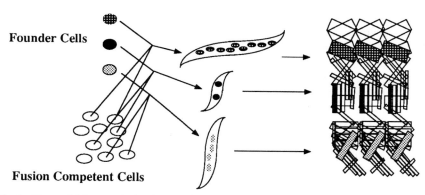

Founder Cells

Fusion Competent Cells

Fig. 1 The model for muscle development in *Drosophila* requires the presence of two distinct cell types (Bate, 1990; Rushton *et al.*, 1995). According to this model, a unique founder cell is present for each muscle fiber and mediates fusion with a larger group of myoblasts that are not committed to a particular muscle fiber type. Positional information contained within each founder cell is thought to be transferred to the unspecified fusion-competent cells as fusion progresses.

these studies, small binucleate and trinucleate cells were shown to take up positions reminiscent of specific differentiated muscle fibers, span their future territory, and attach to the epidermis in characteristic locations. Therefore, these binucleate muscle precursors possess information that directs their unique morphological behavior and are probably derived from specialized founder cells.

2. Specification and Segregation of Founder Cells

As described previously, signaling molecules such as WG and DPP clearly play a role in the specification and segregation of the progenitors of the somatic muscles (Staehling-Hampton *et al.*, 1994; Volk and VijayRaghavan, 1994; Baylies *et al.*, 1995; Maggert *et al.*, 1995; Azpiazu *et al.*, 1996; Ranganayakulu *et al.*, 1996). These presumptive myoblasts are derived primarily from ventrally and posteriorly located domains of mesodermal cells that are marked by high levels of TWI expression (Dunin-Borkowski *et al.*, 1995; Baylies and Bate, 1996). In addition, however, results have implicated the bHLH-containing protein encoded by the *l'sc* gene in development of the musculature. Examination of the expression pattern of *l'sc* has revealed clusters of *l'sc*-expressing cells in particular domains within the somatic mesoderm. These clusters are found in locations similar to those of the cells that express high levels of TWI. Moreover, each cluster contains a unique number of cells and is located in a particular region within the somatic mesoderm (Carmena *et al.*, 1995). These studies further show that a single cell, characterized by markedly higher levels of *l'sc*, is selected from each cluster and gives rise to one or more founder cells for particular muscle fibers (Carmena et al., 1995). This model is consistent with earlier studies invoking the neurogenic genes in the specification and segregation of founder cells. In these studies, domains of several different subsets of specific muscle progenitors were shown to be enlarged in embryos mutant for various neurogenic genes (Corbin et al., 1991; Bate et al., 1993; Baker and Schubiger, 1996), suggesting that the neurogenic genes may be involved in the selection of founder cells. The function of the neurogenic genes has been studied most extensively in the ectoderm, where they mediate a cell fate decision between epidermis and neurogenic ectoderm (reviewed in Lehmann et al., 1983; Artavanis-Tsakonas, 1988; Campos-Ortega, 1988; Greenspan, 1990). Results of these studies are consistent with the possibility that the neurogenic genes play an analogous role in myogenesis, mediating a cell fate decision between founder cells and fusion-competent myoblasts (Corbin et al., 1991; Bate et al., 1993; Baker and Schubiger, 1996). As suggested for l'sc (Carmena et al., 1995), this effect appears to be, at least in part, through the autonomous action of the neurogenic genes in the mesoderm rather than through their presence in overlying ectodermal cells (Corbin et al., 1991; Baker and Schubiger, 1996). These data also seem to argue that the specification of each founder cell is controlled within a unique cluster of l'sc-expressing cells. Therefore, current evidence does not seem to support the view that a population of homogenous founder cells, which will later be given unique characteristics, segregates from the somatic mesoderm.

In discussing the molecular basis of muscle patterning in the Drosophila embryo, one must also consider the segment-specific differences that are observed in this pattern (described in Crossley, 1978; Bate, 1993). These differences suggest additional levels of control by homeotic genes such as those encoded by the bithorax complex, for which a role in the ectoderm has been established (reviewed in Martinez Arias, 1993; Lawrence and Morata, 1994). Mutations in the genes of the bithorax complex, for example, cause partial transformation of the abdominal pattern of larval muscles to that of more anterior segments (Hooper, 1986, and references therein). Moreover, transformations in the pattern of muscle precursors by homeotic genes argue that segmental identity is established early in muscle development, either prior to or coincident with the specification of founder cells (Greig and Akam, 1993; Michelson, 1994). These studies have also shown that the pattern of muscle precursors can be transformed by misexpression of bithorax complex–encoded genes within the mesoderm itself. Therefore, the segmental diversity of the muscle pattern appears to arise, in part, through the autonomous action of homeotic genes within the mesoderm.

3. Determination and Differentiation of Individual Founder Cells

The lines of evidence that have been described argue that the muscle pattern is established by the time of formation of the muscle precursors, in their predecessor founder cells, and apparently in the unique clusters of cells from which each founder cell is selected. The last few years have resulted in significant progress in understanding the segregation and specification of founder cells. However, the mechanisms through which the unique identity of each *Drosophila* founder cell is manifested are not yet understood. The ultimate consequences of subdivision of the somatic mesoderm and specification of muscle precursors appears to be the activation of factors that mark subsets of the *Drosophila* founder cells and can be followed into subsets of differentiated muscle fibers (reviewed in Bate, 1993; Abmayr *et al.*, 1995). Factors such as these may be necessary for implementation of the unique differentiation programs of each muscle founder cell. Several genes exhibit such patterns of expression, including *S59* and *vestigial* (*vg*) (Dohrmann *et al.*, 1990; Bate and Rushton, 1993). *S59* is a homeobox-containing gene that is initially detected during stage 11 in a small number of mononucleate cells in each segment (Dohrmann *et al.*, 1990; Carmena *et al.*, 1995; Rushton *et al.*, 1995) and can be followed into at least four distinct abdominal muscle fibers. Once fusion has occurred, S59 protein is observed in all nuclei within the syncitium, such that any patterning information provided by *S59* has been communicated to all nuclei. Persistent expression of the bHLH-containing TWI protein (discussed earlier) may occur in myoblasts that go on to express *S59*, suggesting that *twist* may regulate *S59* (Bate, 1993). By comparison, expression of the novel nuclear VG protein is evident late in stage 11 in cells internal to the central nervous system that appear to be precursors of the ventral longitudinal muscles (Rushton *et al.*, 1995). VG is later observed in a subset of dorsal, lateral, and ventral muscles (Bate *et al.*, 1993). The *muscle segment homeobox* (*msh*) gene is expressed in a subset of the somatic muscles, and misexpression of MSH appears to alter the pattern of muscle precursors (Lord *et al.*, 1995). Intriguingly, the spatial and temporal pattern of *msh* expression has similarities to that of *l'sc*, consistent with a role in the specification of founder cells (Carmena *et al.*, 1995; D'Alessio and Frasch, 1996). Expression of *apterous* (*ap*), another homeobox-containing gene, is observed in several segmentally-repeated mesodermal cells during stage 11 (Bourgouin *et al.*, 1992). Although endogenous *apterous* expression decreases before muscle fibers are evident, *apterous*-directed β-galactosidase can be followed into a subset of muscles. At least some of these muscles are missing in *apterous*-mutant embryos and appear to be duplicated in response to misexpression of the *apterous*-encoded protein, consistent with a role in development of a subset of muscle fibers (Bourgouin *et al.*, 1992). Lastly *nautilus* (*nau*), described in detail later, is expressed in a subset of muscle precursors and differentiated muscle fibers (Michelson *et al.*, 1990; Paterson *et al.*, 1991; Abmayr *et al.*, 1992) and appears to be capable of driving cells into a myogenic program of differentiation (Keller *et al.*, 1997).

While *S59*, *vestigial*, *apterous*, and *nautilus* have been the most extensively studied markers for muscle precursors and the founder cells from which they are derived, the expression patterns of these four genes do not include the precursors of all muscle fibers. Therefore, additional genes must function in the development of other muscle fibers. These probably include the homeobox-encoding *even-skipped* gene (Frasch *et al.*, 1987), which is expressed both in the pericardial cells and in cells that give rise to one of the dorsal muscles (Bodmer, 1993; Bour *et al.*, 1995). The homeodomain-containing Kruppel protein is also expressed in a subset of muscle fibers and their precursors (Gaul *et al.*, 1987; Gisselbrecht *et al.*, 1996). *Gooseberry* and *gooseberry-neuro*, two closely linked genes, both appear to be expressed transiently in the mesoderm (Gutjahr *et al.*, 1993). Lastly, *pox-meso* transcripts are observed in the somatic mesoderm in a segmentally repeated pattern during stages 10 and 11 (Bopp *et al.*, 1989).

The majority of these genes encode protein domains that support their role as transcriptional regulators, and overlapping patterns of expression suggest the potential for combinatorial modes of action and/or functional redundancy between many of these gene products. For example, both *S59* and *apterous* are expressed in muscle 27 (Dohrmann *et al.*, 1990; Bourgouin *et al.*, 1992). Interestingly, mutations in *apterous* do not appear to have a dramatic effect on the development of this muscle, consistent with the possibility that *S59* may be capable of replacing the mutant *apterous* gene. Similarly, the expression pattern of *pox-meso* appears to overlap with that of *gooseberry* (Bopp *et al.*, 1989). The potential for functional redundancy in the somatic mesoderm is also suggested by the expression patterns of genes such as *tinman* (Azpiazu and Frasch, 1993; Bodmer, 1993), *zfh-1* (Lai *et al.*, 1991, 1993) and *pointed* (Klambt, 1993), which are detected throughout the mesoderm at earlier stages but only affect subsets of muscle fibers later. While its sequence and pattern of expression remain unknown, mutations of the *not enough muscles* gene also exhibit variable defects in the somatic musculature (Burchard *et al.*, 1995). It is enticing to consider that the differentiation program of each founder cell becomes specialized by the expression of a unique combination of regulatory factors within the cell. Molecules that encode transcriptional regulatory proteins and are expressed in, or affect, subsets of muscle precursors seem to be excellent candidates for proteins that implement the distinct myogenic differentiation programs of each muscle fiber.

D. Muscle Differentiation

In the previous section we discussed the issue of muscle patterning and described molecules that are expressed in subsets of founder cells and may function to initiate their unique differentiation programs. We now turn to a discussion of molecules involved in muscle differentiation itself. Such molecules include proteins that control the morphological changes common to all somatic muscle fibers as well as proteins that influence the unique characteristics of each particular muscle fiber.

As in vertebrate skeletal muscle, the somatic muscles of a *Drosophila* larva are striated, multinucleate syncitia that contain myofilaments arranged in repeating units called sarcomeres. The protein composition of the resulting myofibrils includes filamentous actin, tropomyosin, α-actinin, myosin heavy chain, myosin light chain, paramyosin, and troponins (reviewed in Bernstein *et al.*, 1993). Distinct isoforms of these proteins contribute to the generation of different muscle fiber types and have been analyzed in detail in the adult fly. These isoforms are generated by a combination of mechanisms that include activation of distinct genes and alternative splicing of the primary transcript from a single gene. The organization of these genes, as well as general issues of muscle ultrastructure and function, have been described in great detail and therefore are not discussed here (reviewed in Crossley, 1978; Bernstein *et al.*, 1993).

The genes encoding the major muscle-specific proteins of the myofibril are likely targets of proteins that activate the myogenic differentiation program. One such regulator is MEF2, the only apparent *Drosophila* homologue of the *my*ocyte-specific *e*nhancer *f*actor-2 family. This protein is expressed in, and required for, the differentiation of all somatic, visceral, and cardiac mesoderm (Lilly *et al.*, 1994, 1995; Nguyen *et al.*, 1995). *mef2* encodes a transcription factor with an evolutionarily-conserved MADS box–containing DNA-binding domain in addition to a MEF2-specific domain. It is expressed throughout the mesoderm of stage 11 embryos and later in the somatic and visceral muscles as well as in the cardiac cells of the heart. Mutant analysis has revealed that the early stages of myogenesis occur normally, as evidenced by the presence of β3-tubulin-positive myoblasts and *nautilus*-expressing muscle founder cells (Bour *et al.*, 1995; Lilly *et al.*, 1995). By contrast, the later stages of muscle differentiation are seriously impaired in *mef2* mutants (Bour *et al.*, 1995; Lilly *et al.*, 1995). The most severe defect is a lack of myosin-heavy chain (MHC)–expressing myoblasts and differentiated somatic muscle fibers. Evidence suggests that *mef2* in both *Drosophila* and vertebrates directly activates structural genes associated with myogenic differentiation (Olson *et al.*, 1995; Lin *et al.*, 1996, 1997). For example, the MEF2 binding site is both necessary and sufficient for MEF2-dependent muscle-specific transcription of the tropomyosin promoter (Lin *et al.*, 1996, 1997). While direct MEF2 regulation of other *Drosophila* muscle-specific promoters has not been shown, the absence of transcripts for these proteins in *mef2*-mutant embryos, in combination with the extensive analysis of *mef2* in vertebrate systems, provides strong support for the further regulation of muscle-specific promoters by MEF2.

Other molecules that are likely to play a role in general muscle differentiation include the two muscle-specific LIM proteins, Mlp60A and Mlp84B (Arber *et al.*, 1994; Stronach *et al.*, 1996). Expression of both genes is limited to the somatic, visceral, and pharyngeal muscles, but the exact localization patterns of their respective proteins differ. Mlp60A, or DMLP1 (Arber *et al.*, 1994) is expressed throughout muscle fibers, while Mlp84B appears to be concentrated at their terminal ends. Although no mutations have been reported in the *Drosophila* homologues, overexpression of vertebrate MLPl in C2C12 myoblasts was found to stimulate myogenic differentiation (Arber *et al.*, 1994).

In *Drosophila* (Campos-Ortega and Hartenstein, 1985; Doberstein *et al.*, 1997), as in many other species (Fischman, 1972), syncitial muscle fibers are generated by the fusion of mononucleate myoblasts. The muscle fibers of a *Drosophila* larva are generated by the fusion of as few as 3–5 cells or as many as 20–25 cells (Bate, 1990), depending on the identity of the muscle fiber. For the most part, the earliest fusion events temporally precede the activation of genes such as muscle-specific myosin heavy chain (Bate, 1993; Drysdale *et al.*, 1993). However, studies in both *Drosophila* embryos and vertebrate tissue culture cells have shown that myoblast fusion is not a prerequisite for expression of muscle-specific structural genes (Emerson and Beckner, 1975; Holtzer, Rubinstein, *et al.*, 1975; Holtzer, Strahs, *et al.*, 1975; Moss and Strohman, 1976; Trotter and Nameroff, 1976; Vertel and Fischman, 1976; Nguyen *et al.*, 1983; Endo and Nadal-Ginard, 1987; Corbin *et al.*, 1991; Drysdale *et al.*, 1993; Luo *et al.*, 1994; Paululat *et al.*, 1995; Rushton *et al.*, 1995; Doberstein *et al.*, 1997). Indeed, several genetic mutations have now been identified in which the process of myoblast fusion is defective, with no apparent effect on the ability of the unfused myoblasts to express myosin (Luo *et al.*, 1994; Paululat *et al.*, 1995; Rushton *et al.*, 1995; Doberstein *et al.*, 1997). Thus, mechanisms that regulate the formation of multinucleate syncitia appear to function independently from mechanisms involved in the transcriptional initiation of the muscle-specific structural genes.

The biochemical mechanisms controlling myoblast fusion are poorly understood in *Drosophila*, and molecules involved in this process are only beginning to be identified. One may anticipate that membrane-associated proteins and cell-surface molecules will be directly involved. Studies in vertebrate systems have demonstrated the participation of cell adhesion molecules, cytoskeletal components, calcium and calcium-regulated molecules, metalloproteases, and lipids. (Menko and Boettiger, 1987; Knudsen, McElwee, *et al.*, 1990; Knudsen, Myers, *et al.*, 1990; Peck and Walsh, 1993; reviewed in Bischoff, 1978; Wakelam, 1985; White and Blobel, 1989; Knudsen, 1992). In *Drosophila*, fusion appears to occur in cells that are in close proximity to the ectoderm, with concomitant depletion of more internal cells (Bate, 1990), and analysis of mutations in the neurogenic genes has suggested that the epidermis and/or central nervous system is critical for this process (Bate *et al.*, 1993; Baker and Schubiger, 1995). Disruption of fusion can be caused by altered forms of DRAC1, a small guanosine tri-phosphatase that is expressed throughout the mesoderm, suggesting a potential link with the cytoskeleton (Luo *et al.*, 1994). More recent studies have confirmed this suggestion since (1) a dominant negative form of DRAC1 disrupts a variety of processes that require an intact cytoskeleton (Harden *et al.*, 1995) and (2) DRAC1 is essential for assembly of actin at adherens junctions (Eaton *et al.*, 1995). Although its coding sequence has not revealed a clear function, the recently cloned *blown fuse* gene is essential for myoblast fusion (Doberstein *et al.*, 1997). Additional loci that have been identified on the basis of a lack-of-fusion phenotype include *rolling stone* (Paululat *et al.*, 1995), *myoblast city* (Rushton *et al.*, 1995), *sticks and stones* (Abmayr *et al.*, 1994), and *singles bar* (Maeland *et al.*, 1996). Finally, several regions of the X chromosome appear to have defects in myogenesis that including myoblast fusion, but these have not been fully characterized (Drysdale *et al.*, 1993).

Following the earliest fusion events, the developing myotubes send out filopodia and begin to migrate over the epidermis to their attachment sites (Bate, 1990). An important aspect of migration is the maintenance of myotube cell shape during extension. Studies have identified roles for both MSP-300 and laminin in this process (Yarnitzky and Volk, 1995; Rosenberg-Hasson *et al.*, 1996). MSP-300, a dystrophin-like protein, is a member of the spectrin superfamily and is expressed at the leading edge of migrating muscles (Volk, 1992). Laminin, by contrast, is an extracellular matrix glycoprotein that is deposited along the basement membranes of a variety of tissues (Fessler and Fessler, 1989; Montell and Goodman, 1989; Kusche-Gullberg *et al.*, 1992). Analysis of embryos homozygous for both mutations suggests that MSP-300 and laminin interact functionally to maintain myotube cell shape (Rosenberg-Hasson *et al.*, 1996). Additional studies have suggested a role for the ectodermal segment border cells in migration (Volk and VijayRaghavan, 1994). Mutations in several segment-polarity genes, including *wingless*, *naked*, and *patched*, cause an absence or ectopic appearance of the segment border cells, along with disorganized muscle fibers. Importantly, these results suggest that the segment border cells normally guide migrating muscles to their attachment sites. Interpretation of the muscle defects in these mutant backgrounds is complicated, however, by an earlier requirement for genes such as *wingless* in specification of muscle precursors, discussed previously.

The extended myotubes subsequently select and attach to specific sites on the ectoderm. While the precise mechanisms through which muscles recognize their attachment sites are unclear, one study has suggested that genes such as *derailed* (*drl*), which encodes a receptor tyrosine kinase, are involved in this process (Callahan *et al.*, 1996). *derailed* was earlier shown to play a role in axon guidance (Callahan *et al.*, 1995). It is also expressed in a subset of muscle fibers and their attachment sites, and it appears to be required for attachment site selection (Callahan *et al.*, 1996). In the absence of *derailed*, these muscles bypass their normal attachment site and frequently attach elsewhere. This defect can be rescued in *derailed* mutants by selective expression of DRL in the muscles, suggesting that the epidermal cells may not play a major role in selection of the attachment site.

The actual process of muscle attachment appears to involve cell surface receptors specific to both the myotube and the ectoderm. The position-specific integrins, αPS1βPS and αPS2βPS, are expressed in complementary patterns at the sites of muscle attachment, in the epidermal cell membrane and muscle cell membrane, respectively (Bogaert *et al.*, 1987; Leptin *et al.*, 1989; Wehrli *et al.*, 1993; reviewed in Hynes, 1992; Brown, 1993). Embryos mutant for the βPS subunit, *lethal (1) myospheroid* [*l(1)mys*], contain multinucleate syncitia that appear to form normally but detach from the epidermis as contraction occurs (Wright, 1960; Newman and Wright, 1981). By comparison, the muscles of embryos mutant for the αPS2 subunit, *inflated*, are relatively normal in stage 16 embryos but detach from the epidermis shortly thereafter (Brown, 1994). Studies have indicated that the αPS1 subunit, encoded by the *multiple edematous wings* (*mew*) gene, does not play a role equivalent to that of the αPS2 subunit (Brower *et al.*, 1995; Roote and Zusman, 1995). Mutations in *mew* result in only minor muscle attachment defects compared with those observed in *inflated* mutants. The variation in mutant phenotype among the three position-specific integrins suggests that additional molecules, perhaps those of the extracellular matrix, are involved in this interaction.

Results have also shown that the novel secreted serine protease–like molecule encoded by *masquerade* (*mas*) is required for muscle attachment (Murugasu-Oei *et al.*, 1995). *masquerade* encodes a preproprotein that is cleaved to generate two polypeptides, one of which accumulates at all apodemes, or muscle attachment sites. Loss of *masquerade* function results in defective muscle attachment, suggesting that MAS is involved in the stabilization of cell–matrix interactions (Murugasu-Oei *et al.*, 1995). The structural protein β1-tubulin is expressed at all apodemes, possibly to provide a rigid structure that can withstand the tensile forces of muscle contraction (Buttgereit *et al.*, 1991). The recently described *alien* gene is also expressed at the apodemes, although no functional data have been reported (Goubeaud *et al.*, 1996).

Not surprisingly, additional molecules that are involved in myotube migration and attachment are specifically expressed in the epidermal muscle attachment cells. The *stripe* (*sr*) gene is expressed exclusively in these epidermal cells and their precursors and encodes alternatively spliced forms of a zinc finger–containing transcription factor (Lee *et al.*, 1995; Frommer *et al.*, 1996). In embryos mutant for *stripe*, expression of several markers for muscle attachment is reduced and, while the early stages of myogenesis appear to occur normally, the differentiated fibers are erratic and many myoblasts remain unfused (Frommer *et al.*, 1996). Although a mutant phenotype has not yet been described, the bHLH-containing gene *delilah* is also expressed in epidermal cells at the sites of muscle attachment and could play a role in the differentiation of these cells (Armand *et al.*, 1994). In addition, epidermal cells at the segmental groove also express the protein Groovin on their lateral and basal surfaces (Volk and VijayRaghavan, 1994).

Finally, formation of the neuromuscular junction is critical to muscle function. It is important to note, however, that most aspects of muscle patterning and differentiation appear to be independent of motor neuron innervation. Studies have demonstrated that motor neuron outgrowth begins only after the completion of germ band retraction (Johansen *et al.*, 1989; Broadie and Bate, 1993a), well after the muscle pattern is established (Bate, 1990). Perhaps more importantly, myoblast fusion and muscle attachment occur normally in the absence of motor neuron innervation, as shown in embryos mutant for *prospero* (Broadie and Bate, 1993b). Conversely, multinucleate muscle fibers are not required for proper innervation, and functional neuromuscular synapses can form in the absence of myoblast fusion on mononucleate myoblasts (Prokop *et al.*, 1996). Although the neuromuscular synapses fail to form, appropriate target recognition occurs in *mef2*-mutant embryos. These results suggest that target recognition is separable from the formation of the neuromuscular junction and that the latter is dependent on aspects of muscle differentiation that are under *mef2* control (Prokop *et al.*, 1996).

The ability of motor neurons to locate their appropriate targets is thought to occur through recognition of specific molecular markers (reviewed in Keshishian, 1993). For example, the membrane glycoproteins Fasciclin III (FasIII), Connectin, and Toll are expressed on the surfaces of various subsets of muscle fibers and the motor neurons that innervate them (Halpern *et al.*, 1991; Nose *et al.*, 1992; Broadie and Bate, 1993a). Moreover, each of these molecules is capable of mediating cell adhesion when expressed in cultured cells (Snow *et al.*, 1989; Keith and Gay, 1990; Nose *et al.*, 1992). However, target recognition in *fasIII*-mutant embryos is essentially normal, suggesting the existence of functionally redundant molecules (Chiba *et al.*, 1995). In addition to cell surface markers that are expressed in unique subsets of muscle fibers, proteins that are expressed in all muscles may also play a role in neuromuscular specificity. One such example is the *abrupt* (*ab*) gene which encodes a BTB-zinc finger–containing transcription factor (Hu *et al.*, 1995). Although AB appears to be present in all muscles, a subset of muscles that express relatively high levels are more dramatically affected in *abrupt*-mutant embryos. This observation implies that variations in the level of AB protein in individual muscles may be functionally significant (Hu *et al.*, 1995).

III. *nautilus*

NAU represents the only apparent *Drosophila* homologue of the vertebrate family of myogenic bHLH factors that includes MyoD (Davis *et al.*, 1987), MRF4 (Rhodes and Konieczny, 1989; Braun *et al.*, 1990; Miner and Wold, 1990), myf5 (Braun, Buschhausen-Denker, *et al.*, 1989), and myogenin (Edmondson and Olson, 1989; Wright *et al.*, 1989) (see later discussion). This situation is not unique among invertebrates, as only a single member of the myogenic regulatory family has been identified in *C. elegans* (Krause *et al.*, 1990), the sea urchin (Venuti *et al.*, 1991), or the ascidian, *Halocynthia roretzi*

```
CeMyoD (e)    TLVAGANAPRR--TKLDRRKAATMRERRRLRKVNEAFEVVKQRTCPNPNQRLPKVEILRSAIDYINTLERML
SUM1 (s)      CLMWACKACKRKNVAVDKRKAATLRERRRLRKVNEAFEALKRHTCANPNQRLPKVEILRNAIEYIEKLERLL
AMD1 (a)      CLVWACKACKRKTGPHDRRRAATLRERRRLKRVNEAYENLKRCACSNPNQRLPKVEILRNAITYIENLQRIL

NAU (d)       CLTWACKACKKKSVTVDRRKAATMRERRRLRKVNEAFEILKRRTSSNPNQRLPKVEILRNAIEYIESLEDLL

herculin (m)  CLIWACKTCKRKSAPTDRRKAATLRERRRLKKINEAFEALKRRTVANPNQRLPKVEILRSAISYIERLQDLL
MRF4 (c)      CLIWACKTCKRKSAPTDRRKAATLRERRRLKKINEAFEALKRRTVANPNQRLPKVEILRSAISYIERLQDLL
MRF4 (x)      CLIWACKTCKRKSAPTDRRKAATLRERRRLKKINEAFEALKRRTVANPNQRLPKVEILRSAINYIERLQDLL
myf5 (m)      CLMWACKACKRKSTTMDRRKAATMRERRRLKKVNQAFETLKRCTTTNPNQRLPKVEILRNAIRYIESLQELL
myf5 (c)      CLMWACKACKRKSTTMDRRKAATMRERRRLKKVNQAFETLKRCTTANPNQRLPKVEILRNAIRYIESLQELL
myf5 (n)      CLLWACKACKRKSSTMDRRKAATMRERRRLKKVNSAFETLKRCTTANPNQRLPKVEILRNAISYIESLQELL
MyoD (m)      CLLWACKACKRKTTNADRRKAATMRERRRLSKVNEAFETLKRCTSSNPNQRLPKVEILRNAIRYIEGLQALL
CMD1 (c)      CLLWACKACKRKTTNADRRKAATMRERRRLSKVNEAFETLKRCTSTNPNQRLPKVEILRNAIRYIESLQALL
MyoD (z)      CLLWACKACKRKTTNADRRKAATMRERRRLSKVNDAFETLKRCTSTNPNQRLPKVEILRNAISYIESLQALL
myogenin (m)  CLPWACKVCKRKSVSVDRRRAATLREKRRLKKVNEAFEALKRSTLLNPNQRLPKVEILRSAIQYIERLQALL
myogenin (c)  CLPWACKICKRKTVSIDRRRAATLREKRRLKKVNEAFEALKRSTLLNPNQRLPKVEILRSAIQYIERLQSLL
myogenin (p)  CLPWACKVCKRKSVSVDRRRAATLREKRRLKKVNEAFEALKRSTLLNPNQRLPKVEILRSAIQYIERLQALL
```

Fig. 2 Comparisons of the bHLH domain of NAU with those of various members of the myogenic regulatory factor family revealed significant homology to all family members. ClustalW was used to align the bHLH domain of NAU with that of select members of the myogenic regulatory family. Included are NAU, CeMyoD, SUM1, AMD1, and various homologues of MRF4 (herculin), myf5, MyoD, and myogenin. e, *C. elegans*; s, sea urchin; a, ascidian; d, *Drosophila*; m, mouse; c, chicken; x, *Xenopus*; n, newt; z, zebrafish; p, pig.

(Araki *et al.*, 1994). While at first glance this organization may seem to suggest that the single factor present in invertebrate species is likely to serve a purpose analogous to the sum of all four vertebrate factors, analysis of mutant phenotypes (discussed later) has not supported this expectation. Nevertheless, *nautilus* has provided a valuable tool for the further characterization of muscle development in *Drosophila* and may influence our thinking about the mechanism of action of its vertebrate counterparts. In this section, we describe studies that have focused on the behavior of *nautilus* and *nautilus*-expressing cells.

A. *nautilus*: Isolation and Pattern of Expression

As mentioned, *nautilus* encodes a muscle-specific bHLH-containing protein that was isolated on the basis of homology to the MRF family (Michelson *et al.*, 1990; Paterson *et al.*, 1991). It appears to represent the only homologue of this gene family, since neither low-stringency nor polymerase chain reaction–based efforts have yet revealed additional family members. Fig. 2 shows an alignment of the bHLH domain that includes select family members. As initially reported (Michelson *et al.*, 1990), NAU exhibits approximately equal homology to all four of the vertebrate genes throughout this region. Comparisons of the bHLH domain of NAU to each of the vertebrate homologues isolated to date has yielded a narrow range of similarity values, from 91.7% for myf5 and NAU to 84.7% for MyoD and NAU. No significant regions of homology were found outside of the bHLH region, with the exception of two short stretches amino- and carboxy-terminal to the bHLH domain. These regions have been noted previously in the vertebrate factors (Braun, Buschhausen-Denker, *et al.*, 1989; Braun *et al.*, 1990), although their functional significance is not known. Each of these regions shares similarity between NAU and myf5, whereas only one of these regions appears to be common between NAU and either MyoD or MRF4. By contrast, NAU and myogenin do not seem to share any sequence similarity outside of the bHLH domain. Since functions have not been assigned to these short stretches of homology, there is no present indication that NAU is more related to one member of the vertebrate family than another. Comparisons of the bHLH domains of NAU and the invertebrate factors revealed a broader range of similarity values than those obtained with the vertebrate members. These ranged from a high of 93.1% for the sea urchin homologue SUM1 and NAU to a low of 79.1% for the *C. elegans* homologue CeMyoD and NAU. Interestingly, CeMyoD appears to be less related to the vertebrate factors than NAU, with similarity values ranging from 76.4% to 70.8% (see Fig. 2). These lower values appear to be the result of greater divergence in the amino terminal region of the bHLH. The functional significance of this observation is not clear.

The earliest observed expression of *nautilus* is found in mesodermal cells that are in positions reminiscent of the earliest embryonic muscle precursors described by Bate (1990) (Michelson *et al.*, 1990; Paterson *et al.*, 1991; Abmayr *et al.*, 1992). It has not been detected in nonmesodermal derivatives or, as anticipated, in *twist*-mutant embryos that lack all mesodermally-derived tissues (Michelson *et al.*, 1990). Neither transcript nor protein has been detected in the visceral or cardiac musculature, but both are clearly expressed at high levels in mature pharyngeal muscles (Michelson *et al.*, 1990; Paterson *et al.*, 1991; Abmayr *et al.*, 1992; Keller *et al.*, 1997). The pattern of expression of NAU protein is identical to that of the mRNA (Abmayr *et al.*, 1992) (Fig. 3), indicating that *nautilus* expression is not regulated at the level of translation. Thus, *nautilus* expression is restricted to cells of the somatic mesoderm. NAU is present in the nucleus of single cells and can be followed into several nuclei within a mature syncitia (Abmayr *et al.*, 1992). The NAU-expressing cells therefore behave similarly to the S59- and VG-expressing founder cells described by Rushton *et al.* 1995). These cells are found in characteristic and reproducible positions in the embryo, where they appear to seed the fusion process. As fusion progresses, additional nuclei within the developing syncitia take on the characteristics of the NAU-expressing founder cell, including the expression of NAU.

Consistent with the suggestion that NAU marks founder cells, examination of NAU-expressing cells in embryos that are defective for myoblast fusion revealed behaviors similar to those of the S59- and VG-expressing cells reported earlier (Rushton *et al.*, 1995). In this mutant background, for example, a single NAU-expressing cell is present within each segment in the same position as the original founder cell (Fig. 4). In the absence of fusion, this cell becomes morphologically distinct and elongated (data not shown). This behavior seems to suggest that *nautilus* is expressed in muscle founder cells prior to overt myogenesis and continues to be expressed as these cells differentiate into muscle fibers.

As mentioned earlier, each abdominal hemisegment includes 30 specific muscle fibers that are generated by fusion of as few as 3 to as many as 25

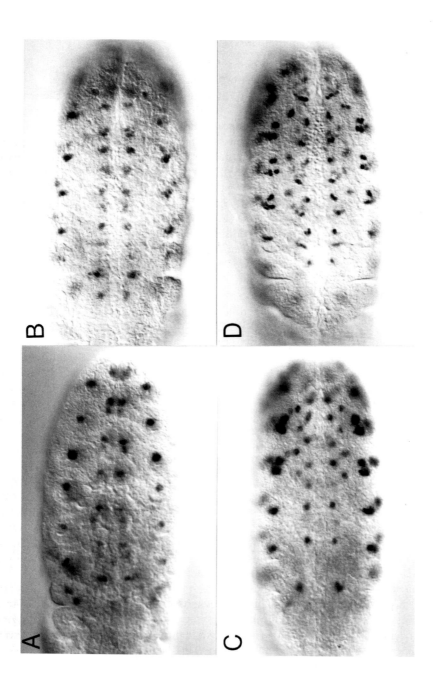

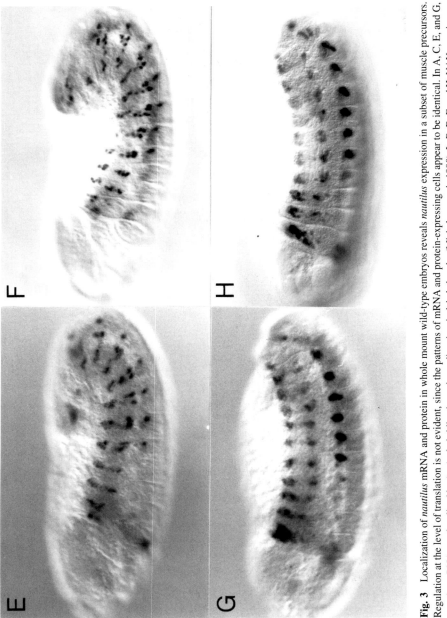

Fig. 3 Localization of *nautilus* mRNA and protein in whole mount wild-type embryos reveals *nautilus* expression in a subset of muscle precursors. Regulation at the level of translation is not evident, since the patterns of mRNA and protein-expressing cells appear to be identical. In A, C, E, and G, *nautilus* mRNA was localized by *in situ* hybridization using a digoxigenin-labeled probe (Michelson *et al.*, 1990). In B, D, F, and H, NAU protein was detected using a polyclonal rat antiserum. Stages were evaluated based on Campos-Ortega and Hartenstein (1985). Anterior is to the left; dorsal is at the top in E–H. (A,B) Ventral view at late stage 11. (C,D) Ventral view at early stage 12. (E,F) Lateral view at late stage 12. (G,H) Lateral view at stage 13. (From Abmayr, *et al.*, 1992, copyright © Lippincott-Raven Press, Ltd.)

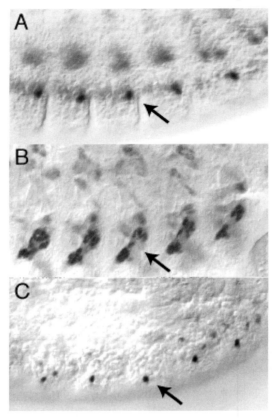

Fig. 4 Comparison of NAU-expressing cells in wild-type and fusion defective *mbc* mutant embryos revealed features similar to those of the S59- and VG-expressing founder cells observed by Rushton *et al.* (1995). This behavior suggests that NAU marks a subset of differentiated muscle fibers and the founder cells from which these muscles are derived. NAU protein was detected using a polyclonal rat antiserum. Embryos are positioned with the anterior to the left and dorsal to the top. (A) Wild type, late stage 12. (B) Wild type, late stage 15. (C) *mbc*-mutant embryo, late stage 15. Arrows denote the position of muscle 26 and its NAU-expressing precursor.

myoblasts (Bate, 1990). Therefore, the total number of myoblasts contributing to the musculature in each hemisegment may include several hundred cells. From Fig. 3, it is apparent that *nautilus* is not expressed in all progenitors of the somatic musculature. Since the number of founder cells must total 30 in each abdominal hemisegment, it also seems apparent that *nautilus* is expressed in only a subset of the founder cells. Consistent with this early expression pattern, only a subset of mature muscles are observed to express NAU later in development (Keller *et al.*, 1997). Therefore, in contrast to the broad role for *nautilus* in the myogenesis of all larval body wall muscles that may have been anticipated, its pattern of expression seems more consistent with a role in only a subset of muscles and their progenitors.

B. Loss of Function

In situ hybridization to the polytene chromosomes of third-instar *Drosophila* larvae with the use of a labeled fragment has localized the cytological position of *nautilus* to region 95A/B on the right arm of the third chromosome (Paterson *et al.*, 1991; Abmayr *et al.*, 1992). Two marked transposable element insertions in this region provided an entry point for the design of genetic screens to isolate deletions in *nautilus*. This genetic analysis was hampered, however, by the identification of a nearby gene that also plays a role in myogenesis. This locus, termed *mbc*, is essential for myoblast fusion and has been described elsewhere (Rushton *et al.*, 1995). Initially, all genetic deficiencies in this region removed both *nautilus* and *mbc*, so their respective mutant phenotypes could not be separated. More recently, we have isolated genetic deletions that remove *nautilus* entirely but leave the *mbc* locus intact. This makes it possible to examine the mutant phenotype of embryos homozygous for these deficiencies to get a preliminary idea of the role that *nautilus* plays in myogenesis.

The embryos shown in Fig. 5B are homozygous for a genetic deficiency that removes the entire *nautilus* coding sequence as well as several surrounding genes. No NAU expression is detected in these deficiency embryos (data not shown). By contrast, the nearby *mbc* gene is completely intact (M. Erickson and S. Abmayr, unpublished data). Consistent with the NAU pattern of expression, it is apparent from these embryos that *nautilus* is not essential for the formation of all muscle fibers. A subset of muscle fibers appear to be defective or missing in many of the mutant embryos. Among these are the dorsal oblique and acute muscles (Fig. 5B). Similar defects are observed in embryos mutant for *daughterless* (*da*), a potential heterodimer partner of NAU (discussed later). It is noteworthy that the affected muscles appear to be ones that normally express *nautilus* (S. Abmayr, unpublished observation), suggesting that *nautilus* may be responsible for their formation. However, it is still unclear whether loss of *nautilus* is actually responsible for the muscle defects seen in these mutant embryos, and it remains a possibility that these defects are a consequence of the loss of a nearby gene. These data clearly establish that many muscles are formed in the absence of *nautilus* and are consistent with the earlier interpretation that *nautilus* expression marks only a subset of muscle fibers and their precursors.

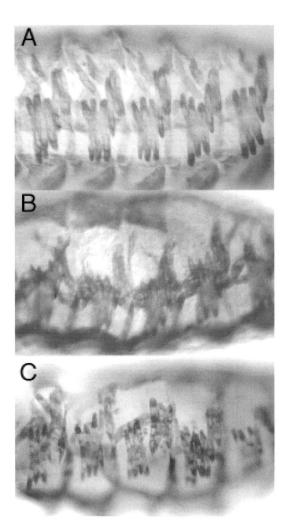

Fig. 5 Many muscle fibers are evident in embryos homozygous for a deficiency that removes the *nautilus gene*. Therefore, it is not essential for the formation of all muscles. A subset of muscle fibers do appear to be defective or missing in these embryos, although the deletion of other genes in the region may contribute to this mutant phenotype. However, similar defects are observed in embryos mutant for *daughterless*, a potential heterodimer partner of NAU. The muscle pattern of late stage 15 or stage 16 embryos was visualized with a monoclonal antibody to muscle myosin heavy chain (gift of D. Keihart). Embryos are positioned with the anterior to the left and dorsal to the top. (A) Wild type. (B) Embryo homozygous for a deficiency that removes *nautilus*. (C) Embryo null for the *daughterless* gene.

C. Consequences of Inappropriate Expression of *nautilus*

As detailed later, the invertebrate myogenic bHLH factors SUM1 and CeMyoD are actually capable of converting mouse $10T\frac{1}{2}$ cells to a myogenic phenotype (Venuti *et al.*, 1991; Krause *et al.*, 1992). Although no reports have described the ability of the *Drosophila* NAU protein to transform mouse $10T\frac{1}{2}$ cells, these studies provide encouraging evidence that invertebrate homologues can function in a manner analogous to that of the vertebrate MRFs. Since the vertebrate myogenic bHLH proteins appear to be in heterodimeric complexes with the ubiquitously expressed bHLH protein E12, the inability of the NAU protein to function in this assay may simply reflect its failure to form heterodimers with E12.

An alternative method to address the ability of *nautilus* to direct a myogenic program of differentiation is to express it in *Drosophila* cells in which it is not normally found. The effects of inappropriate *nautilus* expression have recently been examined with the use of the directed approach provided by the GAL4 system of Brand and Perrimon (1993) (Keller *et al.*, 1997). In this study, NAU protein was expressed throughout the mesoderm, including the cardiac cells of the heart, as well as all cells of the somatic musculature.

In wild-type *Drosophila* embryos, the heart consists of several cell types, including two rows of myosin-expressing cardioblasts. These are surrounded by pericardial cells and attached to the overlying epidermis by the alary muscles (Campos-Ortega and Hartenstein, 1985; Bate, 1993; Rugendorff *et al.*, 1994). The mesodermally derived cardioblasts do not normally express NAU. However, inappropriate expression of NAU within these cells induced a morphological change in which the cells extended beyond their normal structure and appeared to undergo limited fusion (Keller *et al.*, 1997). They also expressed the high levels of muscle myosin protein and actin mRNA that are characteristic of somatic muscles. Finally, these transformed cardiac cells initiated expression of Mlp60A, a LIM domain–containing protein that is expressed in the visceral, somatic, and pharyngeal muscles of wild-type embryos (Arber *et al.*, 1994; Stronach *et al.*, 1996) but is not normally expressed in the heart. Misexpression of *nautilus*, then, appears to drive the cardiac cells into a developmental program in which they exhibit properties and patterns of gene expression reminiscent of a somatic muscle cell. Of note, cardioblasts that did not exhibit a dramatic morphological change frequently did express increased levels of myosin and actin 57B, as well as detectable Mlp60A, suggesting that lower levels of NAU protein are sufficient to alter transcriptional patterns but that higher levels are required for morphological transformation.

Therefore *nautilus* is capable, *in vivo*, of diverting cells originally destined to contribute to the heart to a somatic muscle differentiation program. As with its vertebrate counterparts, this observation would seem to suggest a general function in myoblast determination and/or differentiation. However, *nautilus* appears to be expressed in only a subset of muscle founder cells, and many muscle fibers are formed in its absence. Therefore *nautilus* is unlikely to be inducing myogenic differentiation in all muscle precursors. Of note, inappropriate expression of *nautilus* in muscle precursors in which it is not normally expressed alters their differentiation program (Keller *et al.*, 1997). Although muscles that normally express *nautilus* are not significantly altered by increased levels, differentiation of muscles that do not normally express *nautilus* is perturbed. These muscles are often in incorrect locations, do not attach, or appear to be duplicates of muscles that normally do express *nautilus* (Keller *et al.*, 1997). The appearance of duplicated muscles suggests that *nautilus* has either induced the formation of a second founder cell or has appropriated an already existing founder cell to a different developmental plan.

D. Potential Interacting Molecules

Numerous studies in the last decade have identified molecules that interact with the vertebrate MRF proteins (reviewed in Lassar and Munsterberg, 1994; Rudnicki and Jaenisch, 1995; Molkentin and Olson, 1996). These include the ubiquitously expressed bHLH proteins E12 and E47, the products of the *E2A* gene (Aronheim *et al.*, 1993); the HLH-containing Id protein (Benezra *et al.*, 1990); and the MADS box DNA-binding protein MEF2 (Lilly *et al.*, 1994; Nguyen *et al.*, 1994). Heterodimer formation between the MRFs and E12/E47 has been shown to occur both *in vitro* and *in vivo* (Lassar, Thayer, *et al.*, 1989; Lassar *et al.*, 1991; Brennan and Olson, 1990). Id, which contains an HLH domain in the absence of a basic region, interferes with the action of the MRF proteins by sequestering E12/E47 in nonproductive complexes (Benezra *et al.*, 1990). Moreover, overexpression of Id has been shown to inhibit the myogenic differentiation program *in vivo* through interaction with *E2A*-encoded proteins (Jen *et al.*, 1992). Finally, the myocyte enhancer factor MEF2 has been shown to interact with myogenic bHLH proteins (Molkentin *et al.*, 1995).

In comparison with vertebrate systems, identification and biochemical analysis of molecules that interact with NAU has been limited. Nevertheless, homologues of E12/E47, Id, and MEF2 have been identified in *Drosophila* and are likely candidates for such interactions. The widely expressed *daughterless* gene encodes a nuclearly localized bHLH-containing protein that has approximately 89% homology to E12 within the bHLH domain and therefore is a candidate heterodimer partner for NAU (Caudy, Grell, *et al.*, 1988; Caudy, Vassin, *et al.*, 1988; Murre, McCraw, *et al.*, 1989: Murre, Schonleber, *et al.*, 1989; Cronmiller and Cummings, 1993). This level of homology does not, however, extend outside the bHLH, because Da exhibits only 39.0% overall similarity to human E12. Previous studies have shown that *daughterless* interacts in the nervous system with the bHLH-encoding members of the *achaete-scute* complex (Cabrera and Alonso, 1991; Van Doren *et al.*, 1991) and is critical for the differentiation of neuronal precursors, presumably through these interactions (Vaessin *et al.*, 1994).

Consistent with the possibility that Da functions as the heterodimer partner of NAU, *daughterless* mutants have significant muscle defects (Caudy, Grell, *et al.*, 1988). Among the visible muscle defects are the absence of several of the dorsal oblique and acute muscles and defects in the lateral longitudinal muscles, both notable similarities to deficiencies that remove *nautilus* (Fig. 5B and 5C). It should be noted, however, that *daughterless* is expressed throughout the embryo, and embryos mutant for *daughterless* also exhibit significant defects in the nervous system (Caudy, Grell, *et al.*, 1988; Caudy, Vassin, *et al.*, 1988; Cronmiller and Cummings, 1993). It therefore remains a formal possibility that the muscle defects are an indirect consequence of the absence of Da in overlying ectodermal tissue rather than a direct requirement in the mesoderm as a partner for NAU.

The HLH-containing extramacrochaete protein EMC, like Id, is lacking the basic domain, cannot bind to DNA, and antagonizes sequence-specific DNA binding by other bHLH complexes such as that of DA and proteins of the *achaete-scute* complex in the nervous system (Ellis *et al.*, 1990; Garrell and Modolell, 1990; Cabrera and Alonso, 1991; Van Doren *et al.*, 1991; Martinez *et al.*, 1993; Cabrera *et al.*, 1994; Cubas *et al.*, 1994; Ellis, 1994). EMC may have a similar antagonistic affect on Da-containing heterodimers in the mesoderm. Finally, the importance of MEF2 in *Drosophila* myogenesis has already been discussed. Its mutant phenotype clearly indicates a more broad role in myogenesis than that of *nautilus*. Nevertheless, the possibility that it interacts with NAU in a subset of founder cells to induce their further differentiation, in a complex reminiscent of that shown for mammalian MEF2 and MyoD, has not yet been explored.

IV. The Myogenic Regulatory Family in Other Organisms

The family of MRFs that encode bHLH-containing proteins now includes homologues from organisms as diverse as sheep, rainbow trout, and ascidians. In most vertebrates, the MRF family includes four members with a high degree of homology throughout the bHLH domain. By contrast, the invertebrates *D. melanogaster*, sea urchin, *C. elegans*, and *H. roretzi* appear to contain only one factor with a high degree of homology throughout the conserved bHLH domain. Several comprehensive reviews discussing aspects of the isolation, induction, patterns of expression, and modes of action of these molecules have been written, and interested readers are encouraged to examine these references for more detailed information (Buckingham, 1992, 1994; Weintraub, 1993; Lassar and Munsterberg, 1994; Olson and Klein, 1994; Krause, 1995; Rudnicki and Jaenisch, 1995; Cossu *et al.*, 1996; Molkentin and Olson, 1996; Venuti and Cserjesi, 1996). Herein, we focus on specific aspects of the other vertebrate and nonvertebrate factors that provide insight into potential parallels and differences with *nautilus*, the *Drosophila* member of this family.

A. Isolation, Biochemical Characterization, and Analysis of Expression Patterns

1. Mammals

The mammalian members of the MRF gene family, including MyoD (Davis *et al.*, 1987), MRF4 (Rhodes and Konieczny, 1989; Braun *et al.*, 1990; Miner and Wold, 1990), myf5 (Braun, Buschhausen-Denker, *et al.*, 1989), and myogenin (Edmondson and Olson, 1989; Wright *et al.*, 1989), are the most extensively studied of the bHLH-containing myogenic factors. All encode DNA-binding proteins with a high degree of homology throughout the bHLH structural motif. Molecular and biochemical characterization of the mammalian genes has shown that they encode nuclear factors in which the basic domain mediates DNA binding while the HLH domain is responsible for protein–protein interactions (Tapscott *et al.*, 1988; Braun, Bober, *et al.*, 1989; Lassar, Buskin, *et al.*, 1989; Lassar *et al.*, 1991; Brennan and Olson, 1990; Davis *et al.*, 1990). The MRFs have been shown to form heterodimers with ubiquitously expressed bHLH proteins such as E12 and are able to activate transcription from muscle-specific promoters through consensus E-boxes. In cultured cells, all of them also are able to induce a myogenic program of differentiation that culminates in the formation of multinucleate muscle fibers (reviewed in Emerson, 1990; Olson, 1990; Buckingham, 1992, 1994; Weintraub, 1993; Lassar and Munsterberg, 1994; Olson and Klein, 1994; Krause, 1995; Rudnicki and Jaenisch, 1995; Cossu *et al.*, 1996; Molkentin and Olson, 1996; Venuti and Cserjesi, 1996).

Although the four MRF genes appear to function somewhat interchangeably in cultured cells, they exhibit distinct patterns of expression in mouse embryos (reviewed in Sassoon, 1993; Buckingham, 1994; Ontell *et al.*, 1995; Venuti and Cserjesi, 1996). The earliest MRF expression in the mouse is in the somites, which are derived from paraxial mesoderm. The somites later subdivide into several compartments, the sclerotome, dermatome, and myotome, which give rise to several mesodermal derivatives, including bone, cartilage, dermal tissue, and skeletal muscle. Two different populations of myoblasts originate in the somite: the myotomal myoblasts, from which the trunk muscles are derived, and migratory myoblasts, which leave the somite and later give rise to limb and body wall muscles. Myotomal myoblasts begin to express the myogenic bHLH proteins almost immediately. By contrast, migratory myoblasts do not express these factors until they have reached their destination.

Myf5 is the earliest factor expressed in the somite (Ott *et al.*, 1991), and the only one expressed before compartmentalization and myogenic differentiation (Sassoon *et al.*, 1989; Tajbakhsh and Buckingham, 1994). It is also the earliest detected in the limb bud (Ott *et al.*, 1991). However, studies have shown that myogenic precursors are present in the limb bud prior to myf5 expression (Tajbakhsh and Buckingham, 1994; Tajbakhsh, Bober, *et al.*, 1996), seemingly favoring a model in which the bHLH proteins are necessary for myogenic differentiation but not for determination. Myogenin transcripts appear slightly later than myf5 in the somite, at the same time as formation of the myotome and coincident with the appearance of myosin heavy chain (Sassoon *et al.*, 1989). Myogenin expression continues in differentiating cells and follows that of myf5 in the limb bud. MyoD protein is detected more than a day after myf5 in the developing somites, and it increases significantly in the next few days. MyoD is expressed at the same time as myogenin in the limb bud less than a day after detection of myf5 (Sassoon *et al.*, 1989; Ott *et al.*, 1991). Finally, MRF4, which is only transiently expressed during embryonic development, is expressed at high levels from fetal development through birth (Rhodes and Konieczny, 1989; Braun *et al.*, 1990; Miner and Wold, 1990; Bober *et al.*, 1991; Hinterberger *et al.*, 1992).

In addition to temporal differences in their expression patterns, the four mammalian genes appear to have distinct and complex spatial expression patterns within the somite (Smith *et al.*, 1994). For example, the initial expressions of myf5 and MyoD appear to mark different subdomains within the mouse somites (Smith *et al.*, 1994), perhaps as a consequence of the combination of inducers provided by nearby tissues (Munsterberg *et al.*, 1995; Munsterberg and Lassar, 1995; reviewed in Cossu *et al.*, 1996; Molkentin and Olson, 1996). This observation suggests the possibility that these factors distinguish two populations of myoblasts, consistent with the results of Braun and Arnold (1996), in which ablation of the myf5-expressing cells did not eliminate MyoD-expressing cells from the population. Interestingly, fiber type diversity in the chick (described later) originates, at least in part, from the commitment of distinct myoblasts to formation of specific fiber types (DiMario *et al.*, 1993). Moreover, studies by Hughes *et al.* (1993) are consistent with the possibility that the bHLH factors play different roles in fiber type specification, such that MyoD and myogenin are associated with development of fast twitch and slow twitch muscles, respectively. However, other studies using a β-galactosidase reporter construct targeted to the endogenous myf5 gene clearly showed that all muscle cells express myf5 (Tajbaskhsh, Bober, *et al.*, 1996). While this apparent conflict may reflect subtle technical differences in the experimental conditions, the issue of myf5 expression in all or only a subset of myogenic cells appears to be unresolved. Finally, the transient expression of MRF4 described earlier is restricted to a small domain within the somite, while myogenin is found throughout the myotome (Smith *et al.*, 1994, and references therein).

These differences in the temporal and spatial patterns of expression of the mammalian bHLH factors provide indirect data in support of the hypothesis that they play different roles in myogenesis, perhaps affecting different populations of myoblasts. However, genetic studies that address their roles *in vivo* have not, at first glance, revealed differences in their function. Muscle defects are not apparent in either MyoD- or myf5-mutant mice (Braun *et al.*, 1992; Rudnicki *et al.*, 1992; reviewed in Rudnicki and Jaenisch, 1995). In fact, the lack of a muscle phenotype appears to reflect the ability of these two genes to replace each other's function, since MyoD$^-$/myf5$^-$ mice are missing muscle fibers as well as their myogenic precursors (Rudnicki *et al.*, 1993). While this observation seems to be in some conflict with the expectation that they mark subsets of myoblasts, studies have suggested that the apparent redundancy occurs at the level of cell proliferation rather than gene regulation. In these studies, ablation of the myf5 population does not appear to eliminate the presence of MyoD-expressing cells (Braun and Arnold, 1996). Moreover, a myf5-independent myogenic pathway can be activated in myf5-mutant embryonic stem cells (Braun and Arnold, 1994). In one study, in the absence of myf5 expression, the cells that normally express myf5 in the limb bud adopted nonmuscle cell fates, suggesting that other myogenic factors are not present within the cells to compensate for the loss of myf5 (Tajbakhsh, Rocancourt, *et al.*, 1996). The authors of that study suggested that the aberrant development of these cells contributes to the skeletal and rib defects observed in myf5-mutant mice. Consistent with this interpretation, compensation for this rib defect is accomplished by early expression of myogenin in the same cells, apparently inducing them to continue a myogenic program (Wang *et al.*, 1996). By comparison, mice lacking the myogenin gene appear to be missing many muscle fibers even though the initial events of somite differentiation appear to occur normally (Hasty *et al.*, 1993; Nabeshima *et al.*, 1993; Rawls *et al.*, 1995; Venuti *et al.*, 1995). MyoD expression appears to be unaffected, suggesting both the presence of myoblasts and the inability of MyoD to compensate for the loss of myogenin. Moreover, normal myoblasts appear to be present in the myogenin-mutant embryos, since cells isolated from the muscle-forming regions are capable of differentiating in culture (Nabeshima *et al.*, 1993). Therefore, it appears that myogenin acts at a later stage in myogenesis than MyoD or myf5. Results have established that its functions do not overlap with those of either MyoD or myf5, since the mutant phenotypes of both myogenin$^-$/MyoD$^-$ and myogenin$^-$/myf5$^-$ mice are more severe than those of the individual genes (Rawls *et al.*, 1995). However, these doubly mutant embryos still contain some differentiated muscle fibers. Finally, while mutations targeted to MRF4 have yielded somewhat variable mutant phenotypes in the hands of different investigators, it seems clear that even the most severe mutant does not have serious muscle abnormalities (Braun and Arnold, 1995; Patapoutian *et al.*, 1995; Zhang *et al.*, 1995; Olson *et al.*, 1996).

One final issue that merits discussion is whether any of the bHLH-containing MRF proteins function in later aspects of myogenesis, including the formation and differentiation of satellite cells, the stem cells of adult skeletal muscle. In response to muscle stress and injury, the descendants of these cells proliferate and either fuse with preexisting muscle fibers or form new fibers. To address the role of MyoD in these cells, mice were generated in which both the MyoD and dystrophin genes were mutated, so that regeneration of the defective muscles could be examined (Megeney *et al.*, 1996). The severity of the myopathy caused by loss of dystrophin was greatly increased in the absence of a functional MyoD gene, suggesting that MyoD is essential for later stages of myogenesis that require functional myogenic stem cells for muscle regeneration and repair.

2. Avian Organisms

Homologues to all four MRFs have been isolated from avian species (Lin *et al.*, 1989; de la Brousse and Emerson, 1990; Fujisawa-Sehara *et al.*, 1990, 1992; Pownall and Emerson, 1992; Saitoh *et al.*, 1993). Like the mouse MRFs, quail MRFs are detected immediately following somite formation. In contrast to the mouse, however, the quail MyoD homologue, qmf1, is the first to be expressed, supporting the hypothesis that qmf1 plays the same role in avian myogenesis as myf5 in mouse myogenesis (Pownall and Emerson, 1992). The qmf1 transcript is expressed during the rostral caudal sequence of somite morphogenesis and is abundantly expressed in medially located cells within the dermomyotome (de la Brousse and Emerson, 1990). This is followed by expression of qmf3 and qmf2, the homologues of myf5 and myogenin, respectively (Pownall and Emerson, 1992).

The chick homologue of MyoD, CMD1, has been shown to transform $10T\frac{1}{2}$ mouse fibroblasts to a myogenic program of differentiation (Lin *et al.*, 1989). Northern analysis of mRNAs from isolated tissues has shown that both CMD1 and the chick MRF4 homologue are expressed in breast muscle by embryonic day 10. The levels of these transcripts increase gradually with development, becoming maximal in adult muscle tissue (Fujisawa-Sehara *et al.*, 1992). By comparison, expression of myf5 is detectable by embryonic day 8 and persists until day 7 after hatching (Saitoh *et al.*, 1993). Finally, chick myogenin is present at high levels in breast muscle on embryonic days 10 through 12, then gradually decreases (Fujisawa-Sehara *et al.*, 1990; Saitoh *et al.,* 1993).

Interestingly, transplantation of marked myoblasts from different types of chicken muscle tissue has shown that skeletal muscle fiber type diversity originates in part from the commitment of distinct myoblasts to the formation of specific fiber types (DiMario *et al.*, 1993). This observation seems consistent with the findings of Hughes *et al.* (1993) in the mouse, in which different MRFs are associated with slow versus fast fibers, and the possibility that different MRFs specify distinct subsets of myoblasts.

3. Xenopus

Structural homologues of all four of the mammalian members of the MRF gene family have been identified in *Xenopus*, including two recently-duplicated homologues of MyoD (Hopwood *et al.*, 1989, 1991; Harvey, 1990; Scales *et al.*, 1990; Jennings, 1992). The expression pattern of these genes in the frog embryo differs somewhat from that in mouse and avian embryos. In particular, MRF expression is observed in presomitic mesoderm, prior to compartmentalization of the somite and overt myogenesis (Hopwood *et al.*, 1989, 1991, 1992; Harvey, 1990).

MyoD is the first factor to be expressed in the *Xenopus* embryo. Indeed, its initial expression is reflected in a small maternal component found throughout the embryo (Hopwood *et al.*, 1989; Harvey, 1990), the consequence of which is unclear (see discussion in Hopwood *et al.*, 1992). Both of the MyoD homologues are zygotically activated at the midblastula transition and continue to be detected at low levels throughout the embryo (Rupp and Weintraub, 1991).The zygotic transcript accumulates specifically in the mesoderm of midgastrula-stage embryos (Hopwood *et al.*, 1989; Scales *et al.*, 1990), an apparent consequence of its stabilization in cells of the marginal zone in response to mesoderm induction (Rupp and Weintraub, 1991). At this stage, therefore, the transcript is restricted to cells in the location of the muscle precursors (Hopwood *et al.*, 1989; Rupp and Weintraub, 1991), and it is found within the somites at late gastrula. This transcript is observed in the mesoderm of midgastrula-stage embryos (Hopwood *et al.*, 1989), about 2 hr before expression of markers associated with muscle differentiation, and one-half day before the appearance of myofibrils. The pattern of expression of MyoD protein has been examined in detail, and it essentially parallels that of the zygotic transcript (Harvey, 1992; Hopwood *et al.*, 1992). The earliest expression of protein occurs at midgastrula stage, in the positions occupied by cells that will contribute to the somitic mesoderm. Localization of the protein to the nucleus appears to be a regulated process that occurs in response to mesoderm induction (Rupp *et al.*, 1994). It appears to be present in the nuclei of virtually all cells of the myotome, accumulating first in the anterior somitic mesoderm. In contrast to the mouse and chick embryo, in which MyoD is expressed after somite formation, this expression in the presomitogenic phase of development is reminiscent of the pattern of MyoD expression in the zebrafish embryo. In some contrast, however, MyoD is expressed throughout the entire paraxial mesoderm in *Xenopus*.

Myf5 transcripts are observed in a pattern similar to that of MyoD (Hopwood *et al.*, 1991), accumulating in the prospective somite region of early gastrulas. MRF4 is expressed much later in development, in cells that are already differentiating into muscles (Jennings, 1992). Therefore, as with its mammalian counterpart, *Xenopus* MRF4 does not appear to be involved in early stages of muscle commitment or differentiation. Finally, while a single homologue of myogenin has been identified, there is no evidence that it is a functional gene (Jennings, 1992). This somewhat surprising result suggests that the various homologues do not serve identical functions in different vertebrate organisms, since myogenin appears to be the only MRF family member in the mouse that is absolutely essential for myogenesis.

4. Zebrafish

To date, both MyoD and myogenin homologues have been described in zebrafish (Weinberg *et al.*, 1996). The MyoD transcript is expressed in the presomitogenic phase in a small subset of the paraxial (the adaxial) mesodermal cells, far in advance of overt myogenic differentiation. This supports a role in specification of myogenic precursors rather than their differentiation. This is in marked contrast to both mouse and chick, in which expression of the MyoD homologue is not observed until after somite formation. While it is quite similar to the expression in *Xenopus*, it is noteworthy that zebrafish MyoD is expressed in a more restricted region within the somite than is its *Xenopus* counterpart.

By comparison, the myogenin homologue is detected in developing somites at approximately 10.5 hr in the zebrafish embryo, roughly 1–2 hr after MyoD. Its later expression, at 12–12.5 h, is restricted to somites that have already formed. In addition, myogenin remains high after MyoD has decline in older somites. Of note, myogenin is detected in only a subset of the MyoD-expressing cells (Weinberg *et al.*, 1996). These include the adaxial cells early in development and cells that extend laterally from the adaxial cells later. Thus it appears that MyoD is not sufficient for activation of myogenin, since all cells that express MyoD do not express myogenin. However, myogenin is expressed only in cells that have expressed MyoD, rather than in a mutually exclusive cell population.

In summary, then, the expression patterns of zebrafish MyoD and myogenin suggest that they play somewhat different roles than their mammalian counterparts. The MyoD pattern of expression supports a broad role in the specification of myogenic precursor cells rather than in their differentiation. By contrast, the more localized pattern of myogenin transcripts later in development may be related to the differentiation of a subset of the MyoD-expressing myogenic cells.

5. *Caenorhabditis elegans*

As mentioned earlier, a single member of the MRF family, termed CeMyoD, has also been identified in *C. elegans* (Krause *et al.*, 1990). Its pattern of expression has been examined at the level of protein (Krause *et al.*, 1990; Chen *et al.*, 1992) and with the use of a construct in which the *hlh-1* promoter drives expression of a β-galactosidase reporter (Krause *et al.*, 1990). Early in development, both are expressed in a rather broad domain that includes nonmuscle lineages. Later, they are expressed primarily in cells destined to form the body wall musculature and in six nonmuscle glial cells (Krause *et al.*, 1990; Krause, 1995). While CeMyoD does not appear to be expressed in the pharyngeal or reproductive muscles, it is expressed in some blastomeres that give rise to pharyngeal muscles (Chen *et al.*, 1992; Krause *et al.*, 1992).

One may anticipate that the *C. elegans* gene plays a similar functional role to its vertebrate counterparts, since it is capable of converting mammalian cells in culture to a myogenic phenotype (Krause *et al.*, 1992). This issue has been addressed by mutational studies in which the phenotype of deficiencies and a point mutation in *hlh-1* have been examined. These studies have revealed that *hlh-1* is not required for myosin expression or for the formation of body wall muscles (Chen *et al.*, 1992, 1994). Nevertheless, its loss does result in lethality of mutant individuals. These results are consistent with the suggestion that CeMyoD is necessary for complete morphogenesis of a subset of muscles in the worm, the body wall muscles. The pharyngeal muscles do not express CeMyoD and, not surprisingly, appear to be unaffected in the mutants.

These data are therefore relatively consistent with the apparent role of *nautilus*, in that only a subset of muscles appears to require CeMyoD and this requirement is likely to be at the level of morphogenesis and differentiation of the muscles rather than specification of muscle precursors. Of note, the muscles of *C. elegans* are not multinucleate and do not seem to be patterned in a fashion reminiscent of a *Drosophila* embryo. It may be that all body wall muscles in *C. elegans* are interchangeable with respect to morphology and structure. If this suggestion is correct, multiple factors would not be needed for specification of unique muscle pioneers or morphogenesis of different body wall muscles, as in *Drosophila*.

6. Sea Urchin

The single MRF homologue in sea urchin, SUM1, is expressed approximately 15 hr prior to overt myogenic differentiation, as determined by expression of the myosin heavy chain gene (Venuti *et al.*, 1991). While this early expression of protein appears to be restricted to presumptive myoblasts, expression of the SUM1 transcript is observed in nonmuscle cell types (see Venuti and Cserjesi, 1996). Like many of the MRF family members, including CeMyoD, this single gene is capable of activating muscle promoters and inducing the formation of myotubes when expressed in $10T\frac{1}{2}$ cells (Venuti *et al.*, 1991). Interestingly, premature expression of SUM1 in the mesenchymal cells of blastula-stage sea urchin embryos was sufficient to activate expression from muscle-specific enhancer elements (Venuti *et al.*, 1993). This inappropriate expression of SUM1 did not, however, activate the endogenous MHC promoter or convert cells to a myogenic phenotype morphologically.

7. Ascidians

A single MRF homologue, AMD1, has been identified in the ascidian *H. roretzi* (Araki *et al.*, 1994). While it has not been characterized in detail, it encodes two distinct transcripts that differ in their 3' untranslated and polyadenylation sites. The AMD1 transcript is clearly expressed prior to overt myogenic differentiation, as measured by expression of myosin heavy chain, as well as in the adult body wall muscles. It is not detected in the heart or in other nonmuscle tissues.

B. Analysis of Inappropriate Expression and Overexpression *in Vivo*

Numerous studies have established that the vertebrate myogenic bHLH factors are capable of inducing a myogenic program of differentiation in cultured mouse

$10T\frac{1}{2}$ cells (reviewed in Olson, 1990; Weintraub *et al.*, 1991). These factors can also induce a myogenic phenotype when expressed in cultured cells of non-mesodermal origin (Lin *et al.*, 1989; Weintraub *et al.*, 1989; Choi *et al.*, 1990). In addition to altered patterns of gene expression characterized by the transcriptional induction of muscle-specific structural genes, such myogenic transformations include myoblast fusion and muscle contraction. Of note, the invertebrate homologues from *C. elegans* and sea urchin are also able to change patterns of gene expression when expressed in mouse cells, although they are unable to induce a full myogenic program (Venuti *et al.*, 1991; Krause *et al.*, 1992).

A more limited number of studies have addressed the ability of the myogenic bHLH proteins to divert cells to a myogenic developmental program when expressed in intact tissues (Hopwood and Gurdon, 1990; Hopwood *et al.*, 1991; Miner *et al.*, 1992; Faerman *et al.*, 1993; Santerre *et al.*, 1993; Ludolph *et al.*, 1994). In some instances, misexpression of bHLH proteins was shown to change patterns of gene expression in a manner consistent with the previous studies in cultured cells but did not transform cells to skeletal muscle morphologically (Hopwood and Gurdon, 1990; Hopwood *et al.*, 1991; Miner *et al.*, 1992; Rupp *et al.*, 1994). In a few tissues, however, misexpression of bHLH proteins appeared to induce the full repertoire of myogenesis (Santerre *et al.*, 1993; Ludolph *et al.*, 1994).

V. Summary and Conclusions

Several aspects of muscle development appear to be conserved between *Drosophila* and vertebrate organisms. Among these is the conservation of genes that are critical to the myogenic process, including transcription factors such as *nautilus*. From a simplistic point of view, *Drosophila* therefore seems to be a useful organism for the identification of molecules that are essential for myogenesis in both *Drosophila* and in other species.

nautilus, the focal point of this review, appears to be involved in the specification and/or differentiation of a specific subset of muscle founder cells. As with several of its vertebrate and invertebrate counterparts, it is capable of inducing a myogenic program of differentiation reminiscent of that of somatic muscle precursors when expressed in other cell types. We therefore favor the model that *nautilus* implements the specific differentiation program of these founder cells, rather than their specification. Further analyses are necessary to establish the validity of this working hypothesis. Studies have revealed a critical role for Pax-3 in specifying a particular subset of myogenic cells, the progenitors of the limb muscles. These myogenic cells migrate from the somite into the periphery of the organism, where they differentiate. These myoblasts do not express MyoD or myf5 until they have arrived at their destination and begin the morphologic process of myogenesis (Bober *et al.*, 1994; Goulding *et al.*, 1994; Williams and

Ordahl, 1994). They then begin to express these genes, possibly to put the myogenic plan into action. Thus, as with *nautilus*, MyoD and myf5 may be necessary for the manifestation of a muscle-specific commitment that has already occurred.

By comparison with vertebrates, it was anticipated that the single *Drosophila* gene would serve the purpose of all four vertebrate genes. However, its restricted pattern of expression and apparent loss-of-function phenotype are inconsistent with this expectation. It remains to be determined whether *nautilus* functions in a manner similar to just one of the vertebrate genes. Since the myf5- and MyoD-expressing myoblasts are proliferative, the loss of one cell type appears to be compensated by proliferation of the remaining cell type. This apparent plasticity may obscure differences in mutant phenotype resulting from the loss of particular cells that express each of these genes. In *Drosophila*, by comparison, *nautilus*-expressing cells committed to the myogenic program undergo few, if any, additional cell divisions, and thus no other cells are available to compensate for the loss of *nautilus*. Therefore, the apparent differences between the *Drosophila* *nautilus* gene and its vertebrate counterparts may reflect, at least in part, differences in the developmental systems rather than differences in the function of the genes themselves.

References

Abmayr, S. M., Erickson, M. R., Bour, B. A., and Kulp, M. (1994). Genetic analysis of muscle formation in *Drosophila*. *J. Cell. Biochem.* (Suppl.) **18D,** 474.

Abmayr, S. M., Erickson, M. S., and Bour, B. A. (1995). Embryonic development of the larval body wall musculature of *Drosophila melanogaster*. *Trends Genet.* **11,** 153–159.

Abmayr, S. M., Michelson, A. M., Corbin, V., Young, M. W., and Maniatis, T. (1992). *nautilus*, a *Drosophila* member of the myogenic regulatory gene family. *In* "Neuromuscular Development and Disease" (H. M. Blau and A. M. Kelly, Eds.), pp. 1–16. Raven Press, New York.

Araki, I., Saiga, H., Makabe, K. W., and Satoh, N. (1994). Expression of *AMD1*, a gene for a MyoD1-related factor in the ascidian *Halocynthia roretzi*. *Roux's Arch. Dev. Biol.* **203,** 320–327.

Arber, S., Halder, G., and Caroni, P. (1994). Muscle LIM protein, a novel essential regulator of myogenesis, promotes myogenic differentiation. *Cell* **79,** 221–231.

Armand, P., Knapp, A. C., Hirsh, A. J., Wieschaus, E. F., and Cole, M. D. (1994). A novel basic helix-loop-helix protein is expressed in muscle attachment sites of the *Drosophila* epidermis. *Mol. Cell. Biol.* **14,** 4145–4154.

Aronheim, A., Shiran, R., Rosen, A., and Walker, M. D. (1993). The E2A gene product contains two separable and functionally distinct transcription activation domains. *Proc. Natl. Acad. Sci. U.S.A.* **90,** 8063–8067.

Artavanis-Tsakonas, S. (1988). The molecular biology of the *Notch* locus and the fine tuning of differentiation in *Drosophila*. *Trends Genet.* **4,** 95–100.

Azpiazu, N., and Frasch, M. (1993). *tinman* and *bagpipe*: Two homeo box genes that determine cell fates in the dorsal mesoderm of *Drosophila*. *Genes Dev.* **7,** 1325–1340.

Azpiazu, N., Lawrence, P. A., Vincent, J. P., and Frasch, M. (1996). Segmentation and specification of the mesoderm. *Genes Dev.* **10,** 3183–3194.

Baker, R., and Schubinger, G. (1995). Ectoderm induces muscle-specific gene expression in *Drosophila* embryos. *Development* **121,** 1387–1398.

Baker, R., and Schubiger, G. (1996). Autonomous and nonautonomous Notch functions for embryonic muscle and epidermis in *Drosophila. Development* **122,** 617–626.

Ball, E. E., Ho, R. K., and Goodman, C. S. (1985). Muscle development in the grasshopper embryo. I. Muscles, nerves, and apodemes in the metathoracic leg. *Dev. Biol.* **111,** 383–398.

Bate, M. (1990). The embryonic development of larval muscles in *Drosophila. Development* **110,** 791–804.

Bate, M. (1993). The mesoderm and its derivates. *In* "The Development of *Drosophila melanogaster*" (M. Bate and A. Martinez Arias, Eds.), vol. 2, pp. 1013–1090. Cold Spring Harbor Laboratory Press, Cold Spring Harbor, New York.

Bate, M., and Rushton, E. (1993). Myogenesis and muscle patterning in *Drosophila. C. R. Acad. Sci. Paris* **316,** 1055–1061.

Bate, M., Rushton, E., and Currie, D. A. (1991). Cells with persistent *twist* expression are the embryonic percursors of adult muscles in *Drosophila. Development* **113,** 79–89.

Bate, M., Rushton, E., and Frasch, M. (1993). A dual requirement for neurogenic genes in *Drosophila* myogenesis. *Development* (Suppl.), 149–161.

Baylies, M. K., and Bate, M. (1996). *twist*: A myogenic switch in *Drosophila. Science* **272,** 1481–1484.

Baylies, M. K., Martinez Arias, A., and Bate, M. (1995). *wingless* is required for the formation of a subset of muscle founder cells during *Drosophila* embryogenesis. *Development* **121,** 3829–3837.

Beer, J., Technau, G. M., and Campos-Ortega, J. A. (1987). Lineage analysis of transplanted individual cells in embryos of *Drosophila melanogaster*. IV. Commitment and proliferative capabilities of mesodermal cells. *Roux's Arch. Dev. Biol.* **196,** 222–230.

Benezra, R., Davis, R. L., Lockshon, D., Turner, D. L., and Weintraub, H. (1990). The protein Id: A negative regulator of helix-loop-helix DNA binding proteins. *Cell* **61,** 49–59.

Bernstein, S. I., O'Donnell, P. T., and Cripps, R. M. (1993). Molecular genetic analysis of muscle development, structure and function in *Drosophila. Int. Rev. Cytol.* **143,** 63–152.

Bieman, M., Shilo, B. Z., and Volk, T. (1996). Heartless, a *Drosophila* FGF receptor homolog, is essential for cell migration and establishment of several mesodermal lineages. *Genes Dev.* **10,** 2993–3002.

Bischoff, R. (1978). Myoblast fusion. *In* "Membrane Fusion" (G. Poste and G. L. Nicolson, Eds.), pp. 127–179. North-Holland Publishing Company, New York.

Bischoff, R., and Holtzer, H. (1969). Mitosis and the processes of differentiation of myogeneic cells *in vitro. J. Cell Biol.* **41,** 188–200.

Bober, E., Franz, T., Arnold, H. H., Gruss, P., and Tremblay, P. (1994). Pax-3 is required for the development of limb muscles: A possible role for the migration of dermomyotomal muscle progenitor cells. *Development* **120,** 603–612.

Bober, E., Lyons, G. E., Braun, T., Cossu, G., Buckingham, M., and Arnold, H. H. (1991). The muscle regulatory gene Myf-6 has a biphasic pattern of expression during early mouse development. *J. Cell Biol.* **113,** 1255–1265.

Bodmer, R. (1993). The gene *tinman* is required for specification of the heart and visceral muscles in *Drosophila. Development* **118,** 719–729.

Bodmer, R., Jan, L. Y., and Jan, Y., N. (1990). A new homeobox-containing gene, *msh-2*, is transiently expressed early during mesoderm formation of *Drosophila. Development* **110,** 661–669.

Bogaert, T., Brown, N., and Wilcox, M. (1987). The *Drosophila* P2 antigen is an invertebrate integrin that, like the fibronectin receptor, becomes localized to muscle attachments. *Cell* **51,** 929–940.

Bopp, D., Jamet, E., Baumgartner, S., Burri, M., and Noll, M. (1989). Isolation of two tissue-specific *Drosophila* paired box genes, *Pox meso* and *Pox neuro. EMBO J.* **8,** 3447–3457.

Bour, B. A., O'Brien, M. A., Lockwood, W. L., Goldstein, E. S., Bodmer, R., Taghert, P. H., Abmayr, S. M., and Nguyen, H. T. (1995). *Drosophila* MEF2, a transcription factor that is essential for myogenesis. *Genes Dev.* **9**, 730–741.

Bourgouin, C., Lundgren, S. E., and Thomas, J. B. (1992). *apterous* is a *Drosophila* LIM domain gene required for the development of a subset of embryonic muscles. *Neuron* **9**, 549–561.

Brand, A. H., and Perrimon, N. (1993). Targeted gene expression as a means of altering cell fates and generating dominant phenotypes. *Development* **118**, 401–415.

Braun, T., and Arnold, H. H. (1994). ES-cells carrying two inactivated myf-5 alleles form skeletal muscle cells: Activation of an alternative myf-5-independent differentiation pathway. *Dev. Biol.* **164**, 24–36.

Braun, T., and Arnold, H. H. (1995). Inactivation of Myf-6 and Myf-5 gene in mice leads to alterations in skeletal muscle development. *EMBO J.* **14**, 1176–1186.

Braun, T., and Arnold, H. H. (1996). *myf-5* and *myoD* genes are activated in distinct mesenchymal stem cells and determine different skeletal muscle cell lineages. *EMBO J.* **15**, 310–318.

Braun, T., Bober, E., Buschhausen-Denker, G., Kotz, S., Grzeschik, K.-H., and Arnold, H. H. (1989). Differential expression of myogenic determination genes in muscle cells: Possible auto-activation by the *Myf* gene products. *EMBO J.* **8**, 3617–3625.

Braun, T., Bober, E., Winter, B., Rosenthal, N., and Arnold, H. H. (1990). Myf-6, a new member of the human gene family of myogenic determination factors: Evidence for a gene cluster on chromosome 12. *EMBO J.* **9**, 821–831.

Braun, T., Buschhausen-Denker, G., Bober, E., Tannich, E., and Arnold, H. H. (1989). A novel human muscle factor related to but distinct from MyoD1 induces myogenic conversion in 10T$\frac{1}{2}$ fibroblasts. *EMBO J.* **8**, 701–709.

Braun, T., Rudnicki, M. A., Arnold, H.-H., and Jaenisch, R. (1992). Targeted inactivation of the muscle regulatory gene *Myf-5* results in abnormal rib development and perinatal death. *Cell* **71** 369–382.

Brennan, T. J., and Olson, E. N. (1990). Myogenin resides in the nucleus and acquires high affinity for a conserved enhancer element on heterodimerization. *Genes Dev.* **4**, 582–595.

Broadie, K., and Bate, M. (1993a). Innervation directs receptor synthesis and localization in *Drosophila* embryo synaptogenesis. *Nature* **361**, 350–353.

Broadie, K., and Bate, M. (1993b). Muscle development is independent of innervation during *Drosophila* embryogenesis. *Development* **119**, 533–543.

Brower, D. L., Bunch, T. A., Mukai, L., Adamson, T. E., Wehrli, M., Lam, S., Friedlander, E., Roote, C. E., and Zusman, S. (1995). Nonequivalent requirements for PS1 and PS2 integrin at cell attachments in *Drosophila*: Genetic analysis of the αPS1 integrin subunit. *Development* **121**, 1311–1320.

Brown, N. (1993). Integrins hold *Drosophila* together. *Bioessays* **15**, 383–390.

Brown, N. H. (1994). Null mutations in the αPS2 and βPS integrin subunit genes have distinct phenotypes. *Development* **120**, 1221–1231.

Buckingham, M. (1992). Making muscle in mammals. *Trends Genet.* **8**, 144–149.

Buckingham, M. E. (1994). Muscle: The regulation of myogenesis. *Curr. Opin. Genet. Dev.* **4**, 745–751.

Burchard, S., Paululat, A., Hunz, U., and Renkawitz-Pohl, R. (1995). The mutant *not enough muscles* (*nem*) reveals reduction of the *Drosophila* embryonic muscle pattern. *J. Cell Sci.* **108**, 1443–1454.

Buttgereit, D., Leiss, D., Michiels, F., and Renkawitz-Pohl, R. (1991). During *Drosophila* embryogenesis the β1 tubulin gene is specifically expressed in the nervous system and the apodemes. *Mech. Dev.* **33**, 107–118.

Cabrera, C. V., and Alonso, M. C. (1991). Transcriptional activation by heterodimers of the *acheate-scute* and *daughterless* gene products of *Drosophila*. *EMBO J.* **10**, 2965–2973.

Cabrera, C. V., Alonso, M. C., and Huikeshoven, H. (1994). Regulation of scute function by extramacrochaete in vitro and in vivo. *Development* **120**, 3595–3603.

Callahan, C. A., Bonkovsky, J. L., Scully, A. L., and Thomas J. B. (1996). *derailed* is required for muscle attachment site selection in *Drosophila*. *Development* **122**, 2761–2767.

Callahan, C. A., Muralidhar, M. G., Lundgren, S. E., Scully, A. L., and Thomas, J. B. (1995). Control of neuronal pathway selection by a *Drosophila* receptor protein-tyrosine kinase family member. *Nature* **376**, 171–174.

Campos-Ortega, J. A. (1988). Cellular interactions during early neurogenesis of *Drosophila melanogaster. Trends Neurosci.* **9**, 400–405.

Campos-Ortega, J. A., and Hartenstein, V. (1985). "The Embryonic Development of *Drosophila melanogaster.*" Springer-Verlag, Berlin, Germany.

Carmena, A., Bate, M., and Jimenez, F. (1995). *lethal of scute,* a proneural gene, participates in the specification of muscle progenitors during *Drosophila* embryogenesis. *Genes Dev.* **9**, 2373–2383.

Caudy, M., Grell, E. H., Dambly-Chaudiere, C., Ghysen, A., Yeh Jan, L., and Nung Jan, Y. (1988). The maternal sex determination gene *daughterless* has zygotic activity necessary for the formation of peripheral neurons in *Drosophila. Genes Dev.* **2**, 843–852.

Caudy, M., Vassin, H., Brand, M., Tuma, R., Jan, L. Y. (1988). *daughterless,* a *Drosophila* gene essential for both neurogenesis and sex determination, has sequence similarities to *myc* and the *achaete-scute* complex. *Cell* **55**, 1061–1067.

Chen, L., Krause, M., Draper, B., Weintraub, H., and Fire, A. (1992). Body-wall muscle formation in *Caenorhabditis elegans* embryos that lack the MyoD homolog *hlh-1. Science* **256**, 240–243.

Chen, L., Krause, M., Sepanski, M., and Fire, A. (1994). The *Caenorhabditis elegans* MYOD homologue HLH-1 is essential for proper muscle function and complete morphogenesis. *Development* **120**, 1631–1641.

Chiba, A., Snow, P., Keshishian, H., and Hotta, Y. (1995). Fasciclin III as a synaptic target recognition molecule in *Drosophila. Nature* **374**, 166–168.

Choi, J., Costa, M. L., Mermelstein, C. S., Chagas, C., Holtzer, S., and Holtzer, H. (1990). MyoD converts primary dermal fibrolasts, chrondoblasts, smooth muscle, and retinal pigmented epithelial cells into striated mononucleated myoblasts and multinucleated myotubes. *Proc. Natl. Acad. Sci. U.S.A.* **87**, 7988–7992.

Corbin, V., Michelson, A. M., Abmayr, S. M., Neel, V., Alcamo, E., Maniatis, T., and Young, M. W. (1991). A role for the *Drosophila* neurogenic genes in mesoderm differentiation. *Cell* **67**, 311–323.

Cossu, G., Tajbakhsh, S., and Buckingham, M. (1996). How is myogenesis initiated in the embryo? *Trends Genet.* **12**, 218–223.

Cronmiller, C., and Cummings, C. A. (1993). The *daughterless* gene product is a nuclear protein that is broadly expressed throughout development. *Mech. Dev.* **42**, 159–169.

Crossley, A. C. (1978). The morphology and development of the *Drosophila* muscular system. *In* "The Genetics and Biology of *Drosophila 2b*" (M. Ashburner and T. R. F. Wright, Eds.), vol. 2b, pp. 499–560. Academic Press, London.

Cubas, P., Modolell, J., and Ruiz-Gomez, M. (1994). The helix-loop-helix extramacrochaete protein is required for proper specification of many cell types in the *Drosophila* embryo. *Development* **120**, 2555–2565.

D'Alessio, M., and Frasch, M. (1996). *msh* may play a conserved role in dorsoventral patterning of the neuroectoderm and mesoderm. *Mech. Dev.* **58**, 217–231.

Davis, R. L., Cheng, P.-F., Lassar, A. B., and Weintraub, H. (1990). The MyoD DNA binding domain contains a recognition code for muscle-specific gene activation. *Cell* **60**, 733–746.

Davis, R. L., Weintraub, H., and Lassar, A. B. (1987). Expression of a single transfected cDNA converts fibroblasts to myoblasts. *Cell* **51**, 987–1000.

de la Brousse, F. C., and Emerson, C. P. (1990). Localized expression of a myogenic regulatory gene, qmfl, in the somite dermatome of avian embryos. *Genes Dev.* **4**, 567–581.

DiMario, J. X., Fernyak, S. E., and Stockdale, F. E. (1993). Myoblasts transferred to the limbs of embryos are committed to specific fibre fates. *Nature* **362**, 165–167.

Doberstein, S. K., Fetter, R. D., Mehta, A. Y., and Goodman, C. S. (1997). Genetic analysis of myoblast fusion: *blown fuse* is required for progression beyond the prefusion complex. *J. Cell. Biol.* **136**, 1249–1261.

Dohrmann, C., Azpiazu, N., and Frasch, M. (1990). A new *Drosophila* homeo box gene is expressed in mesodermal precursor cells of distinct muscles during embryogenesis. *Genes Dev.* **4**, 2098–2111.

Drysdale, R., Rushton, E., and Bate, M. (1993). Genes required for embryonic muscle development in *Drosophila melanogaster. Roux's Arch. Dev. Biol.* **202**, 276–295.

Dunin-Borkowski, O. M., Brown, N. H., and Bate, M. (1995). Anterior-posterior subdivision and the diversification of the mesoderm in *Drosophila. Development* 121, 4183–4193.

Eaton, S., Auvinen, P., Luo, L., Jan, Y. N., and Simons, K. (1995). CDC42 and Rac1 control different actin-dependent processes in the *Drosophila* wing disc epithelium. *J. Cell Biol.* **131**, 151–164.

Edmondson, D. G., and Olson, E. N. (1989). A gene with homology to the myc similarity region of MyoD1 is expressed during myogenesis and is sufficient to activate the muscle differentiation program. *Genes Dev.* **3**, 628–640.

Ellis, H. M. (1994). Embryonic expression and function of the *Drosophila* helix-loop-helix gene, *extramacrochaete. Mech. Dev.* **47**, 65–72.

Ellis, H. M., Spann, D. R., and Posakony, J. W. (1990). *extramacrochaete,* a negative regulator of sensory organ development in *Drosophila,* defines a new class of helix-loop-helix proteins. *Cell* **61**, 27–38.

Emerson, C. P. (1990). Myogenesis and developmental control genes. *Curr. Opin. Cell Biol.* **2**, 1065–1075.

Emerson, C. P., and Beckner, S. K. (1975). Activation of myosin synthesis in fusing and mononucleated myoblasts. *J. Mol. Biol.* **93**, 431–447.

Endo, T., and Nadal-Ginard, B. (1987). Three types of muscle-specific gene expression in fusion-blocked rat skeletal muscle cells: Translational control in EGTA-treated cells. *Cell* **49**, 515–526.

Faerman, A., Pearson-White, S., Emerson, C., and Shani, M. (1993). Ectopic expression of MyoD1 in mice causes prenatal lethalities. *Mech. Dev.* **196**, 165–173.

Fessler, J. H., and Fessler, L. I. (1989). *Drosophila* extracellular matrix. *Annu. Rev. Cell Biol.* **5**, 309–339.

Fischman, D. A. (1972). Development of striated muscle. *In* "The Structure and Function of Muscle," vol. 1 (G. H. Bourne, Ed.), pp. 75–148. Academic Press, New York.

Frasch, M. (1995). Induction of visceral and cardiac mesoderm by ectodermal Dpp in the early *Drosophila* embryo. *Nature* **374**, 464–467.

Frasch, M., Hoey, T., Rushlow, C., Doyle, H., and Levine, M. (1987). Characterization and localization of the *even-skipped* protein of *Drosophila. EMBO J.* **6**, 749–759.

Frommer, G., Vorbruggen, G., Pasca, G., Jackle, H., and Volk, T. (1996). Epidermal egr-like zinc finger protein of *Drosophila* participates in myotube guidance. *EMBO J.* **15**, 1642–1649.

Fujisawa-Sehara, A., Nabeshima, Y., Hosoda, Y., Obinata, T., and Nabeshima, Y. I. (1990). Myogenin contains two domains conserved among myogenic factors. *J. Biol. Chem.* **265**, 15219–15223.

Fujisawa-Sehara, A., Nabeshima, Y., Komiya, T., Uetsuki, T., Asakura, A., and Nabeshima, Y. I. (1992). Differential trans-activation of muscle-specific regulatory elements including the myosin light chain box by chicken MyoD, myogenin, and MRF4. *J. Biol. Chem.* **267**, 10031–1038.

Garrell, J., and Modolell, J. (1990). The *Drosophila extramacrochaete* locus, an antigonist of proneural genes that, like these genes, encodes a helix-loop-helix protein. *Cell* **61**, 39–48.

Gaul, U., Seifert, E., Schuh, R., and Jackle, H. (1987). Analysis of *Kruppel* protein distribution during early *Drosophila* development reveals posttranscriptional regulation. *Cell* **50**, 639–647.

Gisselbrecht, S., Skeath, J. B., Doe, C. Q., and Michelson, A. M. (1996). *heartless* encodes a fibroblast growth factor receptor (DFR1/DFGF-R2) involved in the directional migration of early mesodermal cells in the *Drosophila* embryo. *Genes Dev.* **10**, 3003–3017.

Goubeaud, A., Knirr, S., Renkawitz-Pohl, R., and Paululat, A. (1996). The *Drosophila* gene *alien* is expressed in the muscle attachment sites during embryogenesis and encodes a protein highly conserved between plants, *Drosophila* and vertebrates. *Mech. Dev.* **57**, 59–68.

Goulding, M., Lumsden, A., and Paquette, A. J. (1994). Regulation of Pax-3 expression in the dermomyotome and its role in muscle development. *Development* **120**, 957–971.

Govind, S., and Steward, R. (1991). Dorsoventral pattern formation in *Drosophila*: Signal transduction and nuclear targeting. *Trends Genet.* **7**, 119–125.

Greenspan, R. J. (1990). The *Notch* gene, adhesion, and developmental fate in the *Drosophila* embryo. *The New Biologist* **2**, 595–600.

Greig, S., and Akam, M. (1993). Homeotic genes autonomously specify one aspect of pattern in the *Drosophila* mesoderm. *Nature* **362**, 630–362.

Gutjahr, T., Patel, N. H., Li, X., Goodman, C. S., and Noll, M. (1993). Analysis of the *gooseberry* locus in *Drosophila* embryos: *gooseberry* determines the circular pattern and activates *gooseberry neuro*. *Development* **118**, 21–31.

Halpern, M. E., Chiba, A., Johansen, J., and Keshishian, H. (1991). Growth cone behavior underlying the development of stereotypic synaptic connections in *Drosophila* embryos. *J. Neurosci.* **11**, 3227–3238.

Harden, N., Loh, H. Y., Chia, W., and Lim, L. (1995). A dominant inhibitory version of the small GTP-binding Rac disrupts cytoskeletal structurs and inhibits developmental cell shape changes in *Drosophila*. *Development* **121**, 903–914.

Harvey, R. (1990). The *Xenopus* MyoD gene: an unlocalized maternal mRNA predates lineage-restricted expression in the early embryo. *Development* **108**, 669–680.

Harvey, R. P. (1992). MyoD protein expression in *Xenopus* embryos closely follows a mesoderm induction-dependent amplification of *MyoD* transcription and is synchronous across the future somite axis. *Mech. Dev.* **37**, 141–149.

Hasty, P., Bradley, A., Morris, J. H., Edmondson, D. G., Venuti, J. M., Olson, E. N., and Klein, W. H. (1993). Muscle deficiency and neonatal death in mice with a targeted mutation in the myogenin gene. *Nature* **364**, 501–506.

Hinterberger, T. J., Mays, J. L., and Konieczny, S. F. (1992). Structure and myofiber-specific expression of the rat muscle regulatory gene MRF4. *Gene* **117**, 201–207.

Ho, R. K., Ball, E. B., and Goodman, C. S. (1983). Muscle pioneers: Large mesodermal cells that erect a scaffold for developing muscles and motoneurones in grasshopper embryos. *Nature* **301**, 66–69.

Holtzer, H., Rubinstein, N., Fellini, S., Yeoh, G., Chi, J., Birnbaum, J., and Okayama, M. (1975). Lineages, quantal cell cycles, and the generation of cell diversity. *Q. Rev. Biophys.* **8**, 523–557.

Holtzer, H., Strahs, K., Biehl, J., Somlyo, A. P., and Ishikawa, H. (1975). Thick and thin filaments in postmitotic, mononucleated myoblasts. *Science* **188**, 943–945.

Hooper, J. (1986). Homeotic gene function in the muscles of *Drosophila* larvae. *EMBO J.* **5**, 2321–2329.

Hopwood, N. D., and Gurdon, J. B. (1990). Activation of muscle genes without myogenesis by ectopic expression of MyoD in frog embryo cells. *Nature* **347**, 197–200.

Hopwood, N. D., Pluck, A., and Gurdon, J. B. (1989). MyoD expression in the forming somites is an early response to mesoderm induction in *Xenopus* embryos. *EMBO J.* **8**, 3409–3417.

Hopwood, N. D., Pluck, A., and Gurdon, J. B. (1991). *Xenopus* Myf-5 marks early muscle cells and can activate muscle genes ectopically in early embryos. *Development* **111**, 551–560.

Hopwood, N. D., Pluck, A., Gurdon, J. B., and Dilworth, S. M. (1992). Expression of XMyoD protein in *Xenopus laevis* embryos. *Development* **114**, 31–38.

Hu, S., Fambrough, D., Atashi, J. R., Goodman, C. S., and Crews, S. T. (1995). The *Drosophila* *abrupt* gene encodes a BTB-zinc finger regulatory protein that controls the specificity of nueromuscular connections. *Genes Dev.* **9**, 2936–2948.

Hughes, S. M., Taylor, J. M., Tapscott, S. J., Gurley, C. M., Carter, W. J., and Peterson, C. A. (1993). Selective accumulation of MyoD and Myogenin mRNAs in fast and slow adult skeletal muscle is controlled by innervation and hormones. *Development* **118**, 1137–1147.

Hynes, R. O. (1992). Integrins: Versatility, modulation, and signaling in cell adhesion. *Cell* **69**, 11–25.

Jen, Y., Weintraub, H., and Benezra, R. (1992). Overexpression of Id protein inhibits the muscle differentiation program: in vivo association of Id with E2A proteins. *Genes Dev.* **6**, 1466–1479.

Jennings, C. G. B. (1992). Expression of the myogenic gene MRF4 during *Xenopus* development. *Dev. Dyn.* **150**, 121–132.

Johansen, J., Halpern, M. E., and Keshishian, H. (1989). Axonal guidance and the development of muscle fiber-specific innervation in *Drosophila* embryos. *J. Neurosci.* **9**, 4318–4332.

Keith, F. J., and Gay, N. J. (1990). The *Drosophila* membrane receptor Toll can function to promote cellular adhesion. *EMBO J.* **9**, 4299–4306.

Keller, C. A., Erickson, M. S., and Abmayr, S. M. (1997). Misexpression of *nautilus* induces myogenesis in cardioblasts and alters the pattern of somatic muscle fibers. *Dev. Biol.* **181**, 197–212.

Keshishian, H. (1993). Making the right connections. *Nature* **361**, 299–300.

Klambt, C. (1993). The *Drosophila* gene pointed encodes two ETS-like proteins which are involved in the development of the midline glial cells. *Development* **117**, 163–176.

Knudsen, K. A. (1992). Fusion of myoblasts. *In* "Membrane Fusion" (J. Wilschut and D. Hoekstra, Eds.), pp. 601–626. Marcel Dekker, New York.

Knudsen, K. A., McElwee, S. A., and Myers, L. (1990). A role for the neural cell adhesion molecule, NCAM, in myoblast interaction during myogenesis. *Dev. Biol.* **138**, 159–168.

Knudsen, K. A., Myers, L., and McElwee, S. A. (1990). A role for the Ca^{2+}-dependent adhesion molecule, *N*-cadherin, in myoblast interaction during myogenesis. *Dev. Biol.* **188**, 175–184.

Kosman, D., Ip, Y. T., Levine, M., and Arora, K. (1991). Establishment of the mesoderm-neuroectoderm boundary in the *Drosophila* embryo. *Science* **254**, 118–122.

Krause, M. (1995). MyoD and myogenesis in *C. elegans*. *Bioessays* **17**, 219–228.

Krause, M., Fire, A., Harrison, S. H., Priess, J., and Weintraub, H. (1990). CeMyoD accumulation defines the body wall muscle cell fate during *C. elegans* embryogenesis. *Cell* **63**, 907–919.

Krause, M., Fire, A., White-Harrison, S., Weintraub, H., and Tapscott, S. (1992). Functional conservation of nematode and vertebrate myogenic regulatory factors. *J. Cell Sci.* (Suppl.) **16**, 111–115.

Kusche-Gullberg, M., Garrison, K., Mackrell, A. J., Fessler, L. I., and Fessler, J. H. (1992). Laminin A chain: Expression during *Drosophila* development and genomic sequences. *EMBO J.* **11**, 4519–4527.

Lai, Z., Fortini, M. E., and Rubin, G. M. (1991). The embryonic expression patterns of zfh-1 and zfh-2, two *Drosophila* genes encoding novel zinc-finger homeodomain proteins. *Mech. Dev.* **34**, 123–134.

Lai, Z.-C., Rushton, E., Bate, M., and Rubin, G. M. (1993). Loss of function of the *Drosophila* zfh-1 gene results in abnormal development of mesodermally derived tissues. *Proc. Natl. Acad. Sci. U.S.A.* **90**, 4122–4126.

Lassar, A., and Munsterberg, A. (1994). Wiring diagrams: Regulatory circuits and the control of skeletal myogenesis. *Curr. Opin. Cell Biol.* **6**, 432–442.

Lassar, A. B., Buskin, J. N., Lockshon, D., Davis, R. L., Apone, S., Haushka, S. D., and Weintraub, H. (1989). MyoD is a sequence-specific DNA binding protein requiring a region of myc homology to bind to the muscle creatine kinase enhancer. *Cell* **58**, 823–831.

Lassar, A. B., Davis, R. L., Wright, W. E., Kadesch, T., Murre, C., Voronova, A., Baltimore, D., and Weintraub, H. (1991). Functional activity of myogenic HLH proteins requires hetero-oligomerization with E12/47-like proteins *in vivo. Cell* **49**, 741–752.

Lassar, A. B., Thayer, M. J., Overell, R. W., and Weintraub, H. (1989). Transformation by activated *ras* or *fos* prevents myogenesis by inhibiting expression of MyoD1. *Cell* **58**, 659–667.

Lawrence, P. A. (1992). "The Making of a Fly: The Genetics of Animal Design." Blackwell Scientific Publications, Oxford.

Lawrence, P. A., Bodmer, R., and Vincent, J. P. (1995). Segmental patterning of heart precursors in *Drosophila. Development* **121**, 4303–4308.

Lawrence, P. A., Johnston, P., and Vincent, J. P. (1994). Wingless can bring about a mesoderm-to-ectoderm induction in *Drosophila* embryos. *Development* **120**, 3355–3359.

Lawrence, P. A., and Morata, G. (1994). Homeobox genes: Their function in *Drosophila* segmentation and pattern formation. *Cell* **78**, 181–189.

Lee, J. C., VijayRaghavan, K., Celniker, S. E., and Tanouye, M. A. (1995). Identification of a *Drosophila* muscle development gene with structural homology to mammalian early growth response transcription factors. *Proc. Natl. Acad. Sci. U.S.A.* **92**, 10344–10348.

Lehmann, R., Jimenez, F., Dietrich, U., and Campos-Ortega, J. A. (1983). On the phenotype and development of mutants of early neurogenesis in *Drosophila melanogaster. Roux's Arch. Dev. Biol.* **192**, 62–74.

Leptin, M. (1991). *twist* and *snail* as positive and negative regulators during *Drosophila* mesoderm development. *Genes Dev.* **5**, 1568–1576.

Leptin, M., Bogaert, T., Lehmann, R., and Wilcox, M. (1989). The function of PS integrins during *Drosophila* embryogenesis. *Cell* **56**, 401–408.

Lilly, B., Galewsky, S., Firulli, A. B., Schulz, R. A., and Olson, E. N. (1994). D-MEF2: A MADS box transcription factor expressed in differentiating mesoderm and muscle cell lineages during *Drosophila* embryogenesis. *Proc. Natl. Acad. Sci. U.S.A.* **91**, 5662–5666.

Lilly, B., Zhao, B., Ranganayakulu, G., Paterson, B. M., Schulz, R. A., and Olson, E. N. (1995). Requirement of MADS domain transcription factor D-MEF2 for muscle formation in *Drosophila. Science* **267**, 688–693.

Lin, M. H., Bour, B. A., Abmayr, S. M., and Storti, R. V. (1997). Ectopic expression of MEF2 in the epidermis induces epidermal expression of muscle genes and abnormal muscle development in *Drosophila. Dev. Biol.* **182**, 240–255.

Lin, M.-H., Nguyen, H. T., Dybala, C., and Storti, R. V. (1996). Myocyte-specific enhancer factor 2 acts cooperatively with a muscle activator region to regulate *Drosophila* tropomyosin gene muscle expression. *Proc. Natl. Acad. Sci. U.S.A.* **93**, 4623–4628.

Lin, Z.-Y., Dechesne, C. A., Eldridge, J., and Paterson, B. M. (1989). An avian muscle factor related to MyoD1 activates muscle-specific promoters in nonmuscle cells of different germ layer origin and in BrdU-treated myoblasts. *Genes Dev.* **3**, 986–996.

Lord, P. C. W., Lin, M., Hales, K. H., and Storti, R. V. (1995). Normal expression and the effects of ectopic expression of the *Drosophila muscle segment homeobox* (*msh*) gene suggest a role in differentiation and patterning of embryonic muscles. *Dev. Biol.* **171**, 627–640.

Ludolph, D. C., Neff, A. W., Mescher, A. L., Malacinski, G. M., Parker, M. A., and Smith, R. C. (1994). Overexpression of XMyoD or XMyf5 in *Xenopus* embryos induces the formation of enlarged myotomes through recruitment of cells of nonsomitic lineage. *Dev. Biol.* **166**, 18–33.

Luo, L., Liao, Y. J., Jan, L. Y., and Jan, Y. N. (1994). Distinct morphogenetic functions of similar small GTPases: *Drosophila* Drac1 is involved in axonal outgrowth and myoblast fusion. *Genes Dev.* **8**, 1787–1802.

Maeland, A. D., Bloor, J. W., and Brown, N. H. (1996). Characterization of *singles bar*, a new mutant in *Drosophila melanogaster* required for muscle development. *Mol. Biol. Cell* **7**, 39a.

Maggert, K., Levine, M., and Frasch, M. (1995). The somatic-visceral subdivision of the embryonic mesoderm is initiated by dorsal gradient thresholds in *Drosophila*. *Development* **121**, 2107–2116.

Martinez Arias, A. (1993). Development and patterning of the larval epidermis of *Drosophila*. *In* "The Development of *Drosophila melanogaster*" (M. Bate and A. Martinez Arias, Eds.), vol. I, pp. 517–608. Cold Spring Harbor Laboratory Press, Cold Spring Harbor, New York.

Martinez, C., Modolell, J., and Garrell, J. (1993). Regulation of the proneural gene *achaete* by helix-loop-helix proteins. *Mol. Cell. Biol.* **13**, 3514–3521.

Megeney, L. A., Kablar, B., Garrett, K., Anderson, J. E., and Rudnicki, M. A. (1996). MyoD is required for myogenic stem cell function in adult skeletal muscle. *Genes Dev.* **10**, 1173–1183.

Menko, A. S., and Boettiger, D. (1987). Occupation of the extracellular matrix receptor, integrin, is a control point for myogenic differentiation. *Cell* **51**, 51–57.

Michelson, A. M. (1994). Muscle pattern diversification in *Drosophila* is determined by the autonomous function of homeotic genes in the embryonic mesoderm. *Development* **120**, 755–768.

Michelson, A. M., Abmayr, S. M., Bate, M., Martinez Arias, A., and Maniatis, T. (1990). Expression of a MyoD family member prefigures muscle pattern in *Drosophila* embryos. *Genes Dev.* **4**, 2086–2097.

Miner, J. H., Miller, J. B., and Wold, B. J. (1992). Skeletal muscle phenotypes initiated by ectopic MyoD in transgenic mouse heart. *Development* **114**, 853–860.

Miner, J. H., and Wold, B. (1990). Herculin, a fourth member of the MyoD family of myogenic regulatory genes. *Proc. Natl. Acad. Sci. U.S.A.* **87**, 1089–1093.

Molkentin, J. D., Black, B. L., Martin, J. F., and Olson, E. N. (1995). Cooperative activation of muscle gene expression by MEF2 and myogenic bHLH proteins. *Cell* **83**, 1125–1136.

Molkentin, J. D., and Olson, E. N. (1996). Defining the regulatory networks for muscle development. *Curr. Opin. Genet. Dev.* **6**, 445–453.

Montell, D. J., and Goodman, C. S. (1989). *Drosophila* laminin: Sequence of B2 subunit and expression of all three subunits during embryogenesis. *J. Cell Biol.* **109**, 2241–2453.

Moss, P. S., and Strohman, R. C. (1976). Myosin synthesis by fusion-arrested chick embryo myoblasts in cell culture. *Dev. Biol.* **48**, 431–437.

Munsterberg, A. E., Kitajewski, J., Bumcrot, D. A., McMahon, A. P., and Lassar, A. B. (1995). Combinatorial signaling by Sonic hedgehog and Wnt family members induces myogenic bHLH gene expression in the somite. *Genes Dev.* **9**, 2911–2922.

Munsterberg, A. E., and Lassar, A. B. (1995). Combinatorial signals from the neural tube, floor plate and notochord induce myogenic bHLH gene expression in the somite. *Development* **121**, 651–660.

Murre, C., McCraw, P. S., Vaessin, H., Caudy, M., Yan, L. Y., Yan, Y. N., Cabrera, C. V., Buskin, J. N., Hauschka, S. D., Lassar, A. B., Weintraub, H., and Baltimore, D. (1989). Interactions between heterologous helix-loop-helix proteins generate complexes that bind specifically to a common DNA sequence. *Cell* **58**, 537–544.

Murre, C., Schonleber McCaw, P., and Baltimore, D. (1989). A new DNA binding and dimerization motif in immunoglobin enhancer binding, *daughterless, MyoD*, and *myc* proteins. *Cell* **56**, 777–783.

Murugasu-Oei, B., Rodrigues, V., Yang, X., and Chia, W. (1995). Masquerade: A novel secreted serine protease–like molecule is required for somatic muscle attachment in the *Drosophila* embryo. *Genes Dev.* **9**, 139–154.

Nabeshima, Y., Hanaoka, K., Hayasaka, M., Esumi, E., Li, S., Nonaka, I., and Nabeshima, Y. (1993). Myogenin gene disruption results in perinatal lethality because of severe muscle defect. *Nature* **364,** 532–535.

Newman, J. S. M., and Wright, T. R. F. (1981). A histological and ultrastructural analysis of developmental defects produced by the mutation, *lethal(1)myospheroid*, in *Drosophila melanogaster. Dev. Biol.* **86,** 393–402.

Nguyen, H. T., Bodmer, R., Abmayr, S. M., McDermott, J. C., Spoerel, N. A., and Nadal-Ginard, B. (1994). D-mef2: A new *Drosophila* mesoderm-specific MADS box-containing gene with a bi-modal expression profile during embryogenesis. *Proc. Natl. Acad. Sci. U.S.A.* **91,** 7520–7524.

Nguyen, H. T., Medford, R. M., and Nadal-Ginard, B. (1983). Reversibility of muscle differentiation in the absence of commitment: Analysis of a myogenic cell line temperature-sensitive for commitment. *Cell* **34,** 281–293.

Nose, A., Mahajan, V. B., and Goodman, C. S. (1992). Connectin: A homophilic cell adhesion molecule on a subset of muscles and the motoneurons that innervate them in *Drosophila. Cell* **70,** 553–567.

Olson, E. N. (1990). MyoD family: A paradigm for development? *Genes Dev.* **4,** 1454–1461.

Olson, E. N., Arnold, H. H., Rigby, P. W. J., and Wold, B. J. (1996). Know your neighbors: Three phenotypes in null mutants of the myogenic bHLH gene MRF4. *Cell* **85,** 1–4.

Olson, E. N., and Klein, W. H. (1994). bHLH factors in muscle development: Dead lines and committments, what to leave in and what to leave out. *Genes Dev.* **8,** 1–8.

Olson, E. N., Perry, M., and Schulz, R. A. (1995). Regulation of muscle differentiation by the MEF2 family of MADS box transcription factors. *Dev. Biol.* **172,** 2–14.

Ontell, M., Ontell, M. P., and Buckingham, M. (1995). Muscle-specific gene expression during myogenesis in the mouse. *Microsc. Res. Tech.* **30,** 354–365.

Ott, M. O., Bober, E., Lyons, G., Arnold, H., and Buckingham, M. (1991). Early expression of the myogenic regulatory gene, myf-5, in precursor cells of skeletal muscle in the mouse embryo. *Development* **111,** 1097–1107.

Padgett, R. W., St. Johnson, R. D., and Gelbart, W. M. (1987). A transcript from a *Drosophila* pattern gene predicts a protein homologous to the transforming growth factor-beta family. *Nature* **325,** 81–84.

Park, M., Wu, X., Golden, K., Axelrod, J. D., and Bodmer, R. (1996). The wingless signaling pathway is directly involved in *Drosophila* heart development. *Dev. Biol.* **177,** 104–116.

Patapoutian, A., Yoon, J. K., Miner, J. H., Wang, S., Stark, K., and Wold, B. (1995). Disruption of the mouse MRF4 gene identifies multiple waves of myogenesis in the myotome. *Development* **121,** 3347–3358.

Paterson, B. M., Walldorf, U., Eldridge, J., Dubendorfer, A., Frasch, M., and Gehring, W. J. (1991). The *Drosophila* homologue of vertebrate myogenic-determination genes encodes a transiently expressed nuclear protein marking primary myogenic cells. *Proc. Natl. Acad. Sci. U.S.A.* **88,** 3782–3786.

Paululat, A., Burchard, S., and Renkawitz-Pohl, R. (1995). Fusion from myoblasts to myotubes is dependent on the *rolling stone* gene (*rost*) of *Drosophila. Development* **121,** 2611–2620.

Peck, D. and Walsh, F. S. (1993). Differential effects of over-expressed neural cell adhesion molecule isoforms on myoblast fusion. *J. Cell Biol.* **123,** 1587–1595.

Pownall, M. E., and Emerson, C. P. (1992). Sequential activation of three myogenic regulatory genes during somite morphogenesis in quail embryos. *Dev. Biol.* **151,** 67–79.

Prokop, A., Landgraf, M., Rushton, E., Broadie, K., and Bate, M. (1996). Presynaptic development at the *Drosophila* neuromuscular junction: Assembly and localization of presynaptic active zones. *Neuron* **17,** 617–626.

Ranganayakulu, G., Schulz, R. A., and Olson, E. N. (1996). Wingless signaling induces *nautilus* expression in the ventral mesoderm of the *Drosophila* embryo. *Dev. Biol.* **176,** 143–148.

Ranganayakulu, G., Zhao, B., Dokidis, A., Molkentin, J. D., Olson, E. N., and Schulz, R. A. (1995). A series of mutations in the D-MEF2 transcription factor reveal multiple functions in larval and adult myogenesis in *Drosophila*. *Dev. Biol.* **171,** 169–181.

Rawls, A., Morris, J. H., Rudnicki, M., Braun, T., Arnold, H. H., Klein, W. H., and Olson, E. N. (1995). Myogenin's functions do not overlap with those of MyoD or Myf-5 during mouse embryogenesis. *Dev. Biol.* **172,** 37–50.

Rhodes, S. J., and Konieczny, S. F. (1989). Identification of MRF4: A new member of the muscle regulatory gene family. *Genes Dev.* **3,** 2050–2061.

Roote, C. E., and Zusman, S. (1995). Functions for PS integrins in tissue adhesion, migration, and shape changes during early embryonic development ion *Drosophila*. *Dev. Biol.* **169,** 322–336.

Rosenberg-Hasson, Y., Renert-Pasca, M., and Volk, T. (1996). A *Drosophila* dystrophin-related protein, MSP-300, is required for embryonic muscle morphogenesis. *Mech. Dev.* **65,** 83–94.

Rudnicki, M., and Jaenisch, R. (1995). The MyoD family of transcription factors and skeletal myogenesis. *Bioessays* **17,** 203–209.

Rudnicki, M. A., Braun, T., Hinuma, S., and Jaenisch, R. (1992). Inactivation of MyoD in mice leads to up-regulation of the myogenic HLH gene Myf-5 and results in apparently normal muscle development. *Cell* **71,** 383–390.

Rudnicki, M. A., Schnegelsberg, P. N. J., Stead, R. H., Braun, T., Arnold, H. H., and Jaenisch, R. (1993). MyoD or Myf-5 is required for the formation of skeletal muscle. *Cell* **75,** 1351–1359.

Rugendorff, A., Younossi-Hartenstein, A., and Hartenstein, V. (1994). Embryonic origin and differentiation of the *Drosophila* heart. *Roux's Arch. Dev. Biol.* **203,** 266–280.

Rupp, R. A. W., Snider, L., and Weintraub, H. (1994). *Xenopus* embryos regulate the nuclear localization of XMyoD. *Genes Dev.* **8,** 1311–1323.

Rupp, R. A. W., and Weintraub, H. (1991). Ubiquitous MyoD transcription at the midblastula transition precedes inductive-dependent MyoD expression in presumptive mesoderm of X. *laevis*. *Cell* **65,** 927–937.

Rushton, E., Drysdale, R., Abmayr, S. M., Michelson, A. M., and Bate, M. (1995). Mutations in a novel gene, *myoblast city*, provide evidence in support of the founder cell hypothesis for *Drosophila* muscle development. *Development* **121,** 1979–1988.

Saitoh, O., Fujisawa-Sehara, A., Nabeshima, Y. I., and Periasamy, M. (1993). Expression of myogenic factors in denervated chicken breast muscle: isolation of the chicken Myf5 gene. *Nucleic Acids Res.* **21,** 2503–2509.

Santerre, R. F., Bales, K. R., Janney, M. J., Hannon, K., Fisher, L. F., Bailey, C. S., Morris, J., Ivarie, R., and Smith II, C. K. (1993). Expression of bovine *myf5* induces ectopic skeletal muscle formation in transgenic mice. *Mol. Cell. Biol.* **13,** 6044–6051.

Sassoon, D. A. (1993). Myogenic regulatory factors: Dissecting their role and regulation during vertebrate embryogenesis. *Dev. Biol.* **156,** 11–23.

Sassoon, D., Lyons, G., Wright, W. E., Lin, V., Lassar, A., Weintraub, H., and Buckingham, M. (1989). Expression of two myogenic regulatory factors myogenin and MyoD1 during mouse embryogenesis. *Nature* **341,** 303–307.

Scales, J. B., Olson, E. N., and Perry, M. (1990). Two distinct *Xenopus* genes with homology to MyoD1 are expressed before somite formation in early embryogenesis. *Mol. Cell. Biol.* **10,** 1516–1524.

Shishido, E., Higashijima, S.-I., Emori, Y., and Saigo, K. (1993). Two FGF-receptor homologues of *Drosophila*: One is expressed in mesodermal primordium in early embryos. *Development* **117,** 751–761.

Smith, T. H., Kachinsky, A. M., and Miller, J. B. (1994). Somite subdomains, muscle cell origins, and the four muscle regulatory factor proteins. *J. Cell Biol.* **127,** 95–105.

Snow, P. M., Bieber, A. J., and Goodman, C. S. (1989). Fasciclin III: A novel homophilic adhesion molecule in *Drosophila*. *Cell* **59,** 313–323.

Staehling-Hampton, K., Hoffman, F. M., Baylies, M. K., Rushton, E., and Bate, M. (1994). dpp induces mesodermal gene expression in *Drosophila*. *Nature* **372,** 783–786.

Stronach, B. E., Siegrist, S. E., and Beckerle, M. C. (1996). Two muscle-specific LIM proteins in *Drosophila*. *J. Cell Biol.* **134,** 1179–1195.

Tajbakhsh, S., Bober, E., Babinet, C., Pournin, S., Arnold, H., and Buckingham, M. (1996). Gene targeting the *myf-5* locus with *nLacZ* reveals expression of this myogenic factor in mature skeletal muscle fibers as well as early embryonic muscle. *Dev. Dyn.* **206,** 291–300.

Tajbakhsh, S., and Buckingham, M. E. (1994). Mouse limb muscle is determined in the absence of the earliest myogenic factor myf-5. *Proc. Natl. Acad. Sci. U.S.A.* **91,** 747–751.

Tajbakhsh, S., Rocancourt, D., and Buckingham, M. (1996). Muscle progenitor cells failing to respond to positional cues adopt non-myogenic fates in *myf-5* null mice. *Nature* **384,** 266–384.

Tapscott, S. J., Davis, R. L., Thayer, M. J., Cheng, P.-F., Weintraub, H., and Lassar, A. B. (1988). MyoD1: A nuclear phosphoprotein requiring a myc homology region to convert fibroblasts to myoblasts. *Science* **242,** 405–411.

Thisse, B., El Messal, M., and Perin-Schmitt, F. (1987). The *twist* gene: Isolation of a *Drosophila* zygotic gene necessary for the establishment of dorsoventral pattern. *Nucleic Acids Res.* **15,** 3439–3452.

Thisse, B., Stoetzel, C., Gorostiza-Thisse, C., and Perrin-Schmitt, F. (1988). Sequence of the *twist* gene and nuclear localization of its protein in endomesodermal cells of early *Drosophila* embryos. *EMBO J.* **7,** 2175–2183.

Trotter, J. A., and Nameroff, M. (1976). Myoblast differentiation *in vitro*: Morphological differentiation of mononucleated myoblasts. *Dev. Biol.* **49,** 548–555.

Vaessin, H., Brand, M., Jan, L. Y., and Jan, Y. N. (1994). *daughterless* is essential for neuronal precursor differentiation but not for initiation of neuronal precursor formation in *Drosophila* embryo. *Development* **120,** 935–945.

Van Doren, M., Ellis, H. M., and Posakony, J. W. (1991). The *Drosophila extramachrochaete* protein antagonizes sequence-specific DNA binding by *daughterless/achaete-scute* protein complexes. *Development* **113,** 245–255.

Venuti, J. M., and Cserjesi, P. (1996). Molecular embryology of skeletal myogenesis. *In* "Current Topics in Developmental Biology," vol. 34 (R. A. Pedersen and G. P. Schatten, Eds.), pp. 169–206. Academic Press, New York.

Venuti, J. M., Goldberg, L., Chakraborty, T., Olson, E. N., and Klein, W. (1991). A myogenic factor from sea urchin embryos capable of programming muscle differentiation in mammalian cells. *Proc. Natl. Acad. Sci. U.S.A.* **88,** 6219–6223.

Venuti, J. M., Kozlowski, M. T., Gan, L., and Klein, W. H. (1993). Developmental potential of muscle cell progenitors and the myogenic factor SUM-1 in the sea urchin embryo. *Mech. Dev.* **41,** 3–14.

Venuti, J. M., Morris, J. H., Vivian, J. L., Olson, E. N., and Klein, W. H. (1995). Myogenin is required for late but not early aspects of myogenesis during mouse development. *J. Cell Biol.* **128,** 563–576.

Vertel, B. M., and Fischman, D. A. (1976). Myosin accumulation in mononucleated cells of chick muscle cultures. *Dev. Biol.* **48,** 438–446.

Volk, T. (1992). A new member of the spectrin superfamily may participate in the formation of embryonic muscle attachments in *Drosophila*. *Development* **116,** 721–730.

Volk, T., and VijayRaghavan, K. (1994). A central role for epidermal segment border cells in the induction of muscle patterning in the *Drosophila* embryo. *Development* **120,** 59–70.

Wakelam, M. J. O. (1985). The fusion of myoblasts. *Biochem. J.* **228,** 1–12.

Wang, Y., Schnegelsberg, P. N. J., Dausman, J., and Jaenisch, R. (1996). Functional specificity of the muscle-specific transcription factors Myf-5 and myogenin. *Nature* **379,** 823–825.

Wehrli, M., DiAntonio, A., Fearnley, I. M., Smith, R. J., and Wilcox, M. (1993). Cloning and characterization of αPS1, a novel *Drosophila melanogaster* integrin. *Mech. Dev.* **43,** 21–36.

Weinberg, E. S., Allende, M. L., Kelly, C. S., Abdelhamid, A., Murakami, T., Andermann, P., Doerre, O. G., Grunwald, D. J., and Riggleman, B. (1996). Developmental regulation of zebrafish *MyoD* in wildtype, *no tail* and *spadetail* embryos. *Development* **122**, 271–280.

Weintraub, H. (1993). The MyoD family and myogenesis: Redundancy, networks, and thresholds. *Cell* **75**, 1241–1244.

Weintraub, H., Davis, R., Tapscott, S., Thayer, M., Krause, M., Benezra, R., Blackwell, T. K., Turner, D., Rupp, R., Hollenberg, S., Zhuang, Y., and Lassar, A. (1991). The *myoD* gene family: Nodal point during specification of the muscle cell lineage. *Science* **251**, 761–766.

Weintraub, H., Tapscott, S. J., Davis, R. L., Thayer, M. J., Adam, M. A., Lassar, A. B., and Miller, A. D. (1989). Activation of muscle-specific genes in pigment, nerve, fat, liver, and fibroblast cell lines by forced expression of MyoD. *Proc. Natl. Acad. Sci. U.S.A.* **86**, 5434–5438.

White, J. M., and Blobel, C. P. (1989). Cell-to-cell fusion. *Curr. Opin. Cell Biol.* **1**, 934–939.

Williams, B. A., and Ordahl, C. P. (1994). Pax-3 expression in segmental mesoderm marks early stages in myogenic cell specification. *Development* **120**, 785–796.

Wright, T. R. F. (1960). The phenogenetics of the embryonic mutant *lethal myospheriod,* in *Drosophila melaogaster. J. Exp. Zool.* **143**, 77–99.

Wright, W. E., Sassoon, D. A., and Lin, V. K. (1989). Myogenin, a factor regulating myogenesis has a domain homologous to MyoD. *Cell* **56**, 607–617.

Wu, X., Golden, K., and Bodmer, R. (1995). Heart development in *Drosophila* requires the segment polarity gene *wingless. Dev. Biol.* **169**, 619–628.

Yarnitzky, T., and Volk, T. (1995). Laminin is required for heart, somatic muscles, and gut development in the *Drosophila* embryo. *Dev. Biol.* **169**, 609–618.

Zhang, W., Behringer, R. R., and Olson, E. N. (1995). Inactivation of the myogenic bHLH gene *MRF4* results in up-regulation of myogenin and rib abnormalities. *Genes Dev.* **9**, 1388–1399.

3

Hydrozoa Metamorphosis and Pattern Formation

Stefan Berking
Zoological Institute
University of Cologne
D-50923 Cologne, Germany

I. Introduction

The most popular hydrozoon, *Hydra*, has been studied for more than 250 years. The initial interest in *Hydra* was its astonishingly high ability to regenerate. Small parts of the body column were observed to regenerate all missing structures, leading to a smaller but complete animal. This was known for plants but not for animals. *Hydra* has a simple architecture (Fig. 1). The body is a tube with two differently developed ends, a mouth opening surrounded by 5–7 tentacles and a sticky basal plate or foot. The body column is made out of two tissue layers separated by an extracellular matrix, the mesogloea. The tissue contains only a few cell types, whose distribution and dynamics have been well investigated. *Hydra* has a long history as a model for the understanding of basic features of pattern formation. Reasons for this are the radial symmetry of the body, which reduces much of the patterning to changes in one dimension, and that the animal appears to be potentially immortal. Cell proliferation and patterning goes on continuously. Browne (1909) detected the phenomenon of induction in *Hydra*, which became well known after the experiments with amphibians by Spemann and Mangold. Turing (1952), Wolpert (1969), and Gierer and Meinhardt (1972) developed their influential models on pattern formation on the basis of experiments with *Hydra*.

Current Topics in Developmental Biology, Vol. 38

Stefan Berking

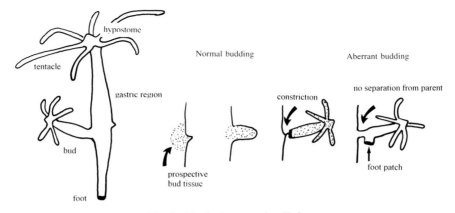

Fig. 1   The freshwater polyp *Hydra*.

Pattern formation in marine hydrozoa has been studied for more than 100 years (see Tardent, 1963, 1978). More recently, some marine hydrozoa have received attention because they allow an easy access to the whole life cycle. *Hydractinia*, a rather typical hydrozoan with a simple structure, is the best studied. There exist female and male animals, which release sperm and oocytes, respectively, into the surrounding water. Following fertilization, cleavage takes place, leading to different cell types and an elongated body shape (Fig. 2). Two days later a spindle-shaped larva has developed that is about 1mm long and is covered by cilia that allow a slow movement. This larva is competent to develop into a polyp. The process of that transformation, called metamorphosis, takes less than 1 day. The resulting polyp is able to feed, whereas the larva depends on stored yolk. At the base of the polyp, which is fixed to the substrate, tubes grow out. These stolons can branch and reunite, and secondary polyps (hydranths) can form on their tops. Therewith a colony develops. Typical of hydrozoa is that either polyps or stolons produce medusae, and the medusae produce gametes by which the life cycle is closed. *Hydractinia* does not produce the typical medusa. It produces a reduced form which is unable to separate from the polyps and to swim. While sitting on a special type of polyp, the so-called gonozooid, this reduced form produces the gametes.

This paper concentrates on two topics, metamorphosis from the larval to the polyp state and pattern formation in larvae, polyps, and colonies. Metamorphosis and pattern formation in larvae and colonies have been studied primarily by experiments with *Hydractinia*, and pattern formation in polyps, by experiments with *Hydra*.

In *Hydractinia*, the larval state is stable until external cues trigger the onset of metamorphosis. Our interest is in the net of biochemical pathways involved in the processes from triggering to the final change of the morphology. We would like to learn how a genome is able to produce two quite different architectures

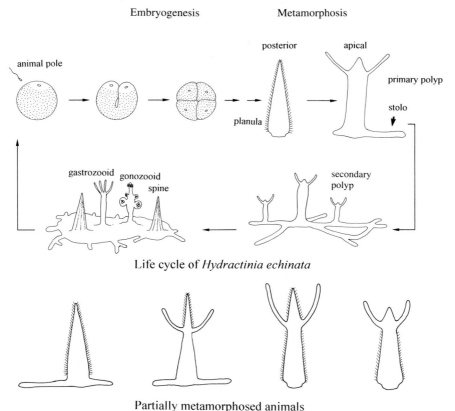

Life cycle of *Hydractinia echinata*

Partially metamorphosed animals

Fig. 2   The marine hydrozoan *Hydractinia*.

(three, in some species in which medusae are formed). What are the differences between the pattern-forming systems of larva and polyp, and what takes place when one system looses influence while the other takes over? *Hydractinia* appears to be suited for study of these questions of general importance.

Pattern formation has been investigated with great success in many organisms, including *Caenorhabditis*, *Drosophila*, *Xenopus*, zebrafish, and mice. Why should one, in addition, study such a simple and exotic animal as *Hydra*? Is it interesting to find genes in *Hydra* that are closely related to developmentally relevant genes in other species? *Hydra* and its relatives are exotic not only because they are close to the basis of the animal kingdom but also because they have some strange features: never ending pattern formation and morphogenesis. Tissue of the upper part of the body transforms into head structures, while tissue of the head is constantly lost at the apical end. Nerve cells are replaced continuously. The same is true for the foot. Further, cell aggregates of dissociated tissue of adult animals transform into complete and ultimately normally structured animals. From almost homogenous starting conditions, in a process of self-organization, order is created. Is this a type of morphogenesis completely different from that of the other so-called model organisms in developmental biology, or is this a general patterning process? Morphogenesis of a higher organism is a complex process that we can hope to understand only as a combination of elementary mechanisms. Hydrozoa provide an access to some elementary processes, including regulation of cell proliferation and differentiation as well as interaction of and communication between cells. At present we are confronted with a huge and increasing mass of experimental results, and particular efforts must be made to discover rules of pattern formation in order to handle these results. These models also are expected to help in the understanding of pattern-forming processes in other species, and they are expected to increase the understanding of what substances and genes are involved in that process in hydrozoa and in similar processes in other species.

II. Control of Metamorphosis

A. Settlement

1. Substrate Selection

Colonies of *Hydractinia* (like those of many benthic marine invertebrates) are found only at specific sites. Almost all colonies of *Hydractinia* cover gastropod shells inhabited by a hermit crab. Larvae do not metamorphose when kept in Petri dishes under sterile conditions. Therefore it was concluded that a larva needs a special external cue for metamorphosis to commence. That cue should be present almost exclusively on shells inhabited by a crab.

In almost all cases the specific cues for settlement and metamorphosis of benthic marine invertebrates are thought to be delivered by microorganisms. For organisms that settle on a solid surface, the inducing agent is thought to be present in biofilms that cover the respective surfaces. There is also strong evidence that the substrate displays specific inducing signals, possibly in interaction with the biofilm. In *Hydractinia*, bacteria of the genus *Alteromonas* are proposed to deliver that specific cue (Müller, 1973; Leitz and Wagner, 1993). An open question was how these bacteria become restricted to shells inhabited by a crab. In a reinvestigation, larvae of *Hydractinia* were exposed to various substrates collected in the intertidal zone close to a site where adult *Hydractinia* were found, including stones, empty mollusk shells, and the algae *Laminaria*, *Fucus*, and *Ulva*. Metamorphosis was found to start immediately on all substrates but not on a clean polystyrene dish. Metamorphosis was induced following the application of various bacteria including *Serratia marinorubra*, *Pseudomonas* spp., *Alteromonas* spp., *Staphylococcus aureus*, and *Echerichia coli* (M. Kroiher and S. Berking, unpublished data). Close to a site where *Hydractinia* were found, we collected shells which were inhabited by snails and found them to be colonized by the hydroids *Podocoryne* and *Gonothyrea* but not by *Hydractinia*, although larvae of *Hydractinia*, when placed on these shells, started metamorphosis immediately. Therefore the specific place where adult *Hydractinia* are found does not reflect the differential distribution of specific cues. Substrate selection by the larva appears to play a minor role, but selection by the crab appears to play a major role. A crab feeds on almost everything it can get, including the hydroids on its own shell, obtained by scratching over the shell's surface with the sharp ends of its legs. In contrast to *Podocoryne* and *Gonothyrea*, which are on the shells when they are inhabited by the snail, *Hydractinia echinata* produces a rigid chitinous mat on which polyps develop after an initial phase of stolon formation. This mat produces spines (therefore the name *echinata*) behind which the polyps can hide (see Fig. 2). *Hydractinia* colonies are very dense and produce a huge mass of gametes per day. The food necessary for that production may be easier to get close to the crab than at other sites. Therefore the reason for that special place where *Hydractinia* colonies are found appears to be that they can survive close to the crab.

In many marine benthic invertebrates, settlement and metamorphosis are coupled: a larva becomes fixed to a suited substrate and then it begins metamorphosis. This occurs in many cnidarians, in particular in the athecate hydrozoa, including *Hydractinia*. However, larvae of some species behave completely differently. Larvae of members of the tubulariidae (hydrozoa) develop inside of the gonophore. The planula develops tentacles which are the first tentacles of the polyp. Only then the larva, now called actinula, is released from the gonophore and starts to settle on a substrate. Here, obviously, settlement follows the initial steps of metamorphosis. In anthozoa this sequence is common. In tachomedusae, the planula develops directly into a medusa, and the polyp stage is missing. The well known medusae of the scyphozoan *Aurelia* produces planula larvae which, when exposed to artificial substrates in the laboratory, in most cases fastened to the underside of the objects and metamorphosed (Brewer, 1978). In the natural environment polyps of related species, including cubozoa, are generally found fixed to substrates in that orientation. Therefore, the substrate may be argued to deliver a specific cue for metamorphosis to commence. However, metamorphosis is also initiated when a larva reaches the surface of the water: the basal part adheres to the water–air interphase, tentacles and mouth opening facing the bottom (M. Kroiher and S. Berking, unpublished data). Specific bacteria are unlikely to exist there as a cue for metamorphosis induction; the water–air interface itself appears to be the cue. Under laboratory conditions, when air is bubbled vigorously through the aquarium, the polyps are forced to leave the surface and to attach to the glass or to stones at the bottom. We argue that in the natural environment metamorphosis usually takes place in the open sea at the water–air interface. Polyps that reach the tumbling water at the shore may be forced to leave the surface and settle on the rocks, preferentially from the under-side. It appears that in *Aurelia* final settlement follows metamorphosis. In medu-sae of scyphozoans that live in the open sea, such a strategy appears to be favorable.

Planulae, like many larvae of other species, are unable to feed; they are lecitrophic. Their task is to find a place where the adult can feed and survive. Biofilms may indicate such a place. Biofilms include various microorganisms in an extracellular matrix. The microorganisms survive by degrading organic debris contained in the water following attachment to this matrix. A larva is expected to stick like debris and to be attacked by the microorganisms. The larva generally survives that attack by excretion of mucus, precursors of chitin synthesis, and possibly enzymes that degrade the bacteria. There are several possibilities for the metamorphosis-inducing signal to be produced during that process. It may result from interaction of the microorganisms with the substrate. Bacteria may release the signal substance, possibly as a result of attack by the larva's enzymes. Larvae may even contain their own (highly specific) metamorphosis inductor in an inactive form, stored in specific cells or as a component of the larva's extracellu-lar matrix. Following contact between the larva and the biofilm the inducing agent becomes free, possibly as a result of the attack by the microorganisms. This may explain why larvae are often found to help settlement and metamorphosis of larvae of their own species. If adults contain that agent as well, they can favor the settlement of larvae of their own species in close vicinity.

2. Attachment and Signal Transmission

In many species including *Hydractinia*, a firm contact to the substrate, even in the absence of a biofilm, is a prerequisite for metamorphosis induction. Larvae of *Hydractinia* can be induced to metamorphose by a treatment with seawater containing the excellently soluble cesium chloride (CsCl). The larvae attach to the Petri dish or to the water–air interface and start metamorphosis. If attachment is hindered, the larvae metamorphose only when much higher concentrations of the Cs^+ ions are applied (Berking and Walther, 1994). The physical parameter "contact" is somehow transformed into a chemical cue that supports meta-morphosis. An obvious assumption is that where there is firm contract there is a reduced exchange of substances between larval tissue and the surrounding sea-water. It was argued that the signal in question is a changed membrane potential, produced at the site of firm contact to the substrate (Yool *et al.*, 1986; Pechenik and Heyman, 1987; Freeman, 1993). However, in *Hydractinia* the treatment and the contact has to last for hours in order to be efficient, whereas a breakdown of membrane potential occurs in a matter of seconds. Thus, a slower process must be envisaged.

Ammonia was proposed to play a key role. Externally applied ammonia is able to cause metamorphosis in many species, including *Hydractinia* (Berking and Schüle, 1987; Berking, 1988a; Coon *et al.*, 1990; Berking and Herrmann, 1990;

Gilmour, 1991). Bacteria in the biofilm may contribute to the rise of the ammonia concentration in the larva, in particular when enclosed between the larva and the substrate. In *Hydractinia* larvae, ammonia is produced constantly due to degradation of the yolk. The ways of release include potassium channels and antiports. When the export is hindered, as by a firm contact to the substrate, the internal concentration should rise. By doubling of the internal concentration metamorphosis commences (Berking, 1988a; Walther *et al.*, 1996). Ammonia is a very small molecule that spreads easily in the tissue, reaching every site in a comparatively short time. Thus, it appears to be suited to play a role as an internal signal assisting metamorphosis induction.

B. Biochemical Pathways Involved in Metamorphosis Induction and Maintenance of the Larval State

Several agents are known to induce metamorphosis, but only a few of them are expected to be present in the larva or its natural habitat (Table I). Nevertheless, these agents are excellent tools to get access to the biochemical pathways linked to metamorphosis control. About 500 substances have been up to now applied to *Hydractinia* larvae under controlled conditions; only a few of them induce metamorphosis. These few substances provide an insight into the net of pathways involved in metamorphosis induction.

1. Protein Kinase C–Like Enzymes

Phorbolesters, including TPA, PDB, RPA, EPA, and PDD (Müller, 1985) and some diacylglycerols (DAGs) (Leitz and Müller, 1987) were found to induce

Table I. Substances and Conditions That Induce Metamorphosis in *Hydractinia*

Metamorphosis is inducible by	Reference
Alteromonas-lipid	Spindler and Müller, 1972
1,2-Diacyl-sn-glycerol (DAG)	Leitz and Müller, 1987
Phorbolesters TPA, PDB, RPA, EPA, PDD	Müller, 1985
Vanadate	Leitz and Wirth, 1991
Li^+ m K^+	Spindler and Müller, 1972
Cs^+ m Rb^+	Müller and Buchal, 1973
NH_4^+, methylamine, tetraethylammonium, (TEA$^+$) BA^{2+}, Sr^{2+}, amiloride	Berking, 1988b
Mg^{2+} deficiency	Müller, 1985
Temperature shift	Kroiher *et al.*, 1992
Methylglyoxale-*bis*-(guanylhydrazone) (MGBG)	Walther *et al.*, 1996
Metamorphosin (an LWamide peptide)	Leitz *et al.*, 1994
Putrescine + (*bis*-(cyclohexylammonium) sulfate	Walther and Berking, in press

metamorphosis. Dioctanoylglycerol (diC8) was found to be the most effective DAG. Consistently, inhibitors of protein kinase C enzymes (PKCs), including sphingosine and K252a, antagonize induction by diC8. One of the most effective artificial DAGs in mammalian experiments, oleyl acetylglycerol, is ineffective. Inositolphospates and Ca^{2+} ions do not induce metamorphosis. Very puzzling is the finding that the concentration of the inducing agent has to be rather high, 10 μM/L, and that the treatment has to last for some hours in order to induce metamorphosis. Such a treatment is expected to have a strong influence on many biochemical pathways linked to PKCs. Further, high concentrations of PKC-stimulating agents such as TPA are able to cause the enzyme's degradation when applied for a long time. There is evidence that metamorphosis was not irreversibly started during a 3-hr treatment with the inducing agents but became fixed immediately after treatment when the animals were transferred back into pure seawater (Berking and Walther, 1994). Therefore it is possible that a decreased PKC activity following overstimulation is the decisive signal.

Whatever the mechanism, it is clear that a strong interference with the PKC signal transduction pathway induces metamorphosis. However, in the natural pathway such an overstimulation may not take place. Induction of metamorphosis by exposure of larvae to a bacterial film of *Alteromonas espejiana* only caused an immediate and transient increase of PKC activity for some minutes (Schneider and Leitz, 1994). But it should be kept in mind that larvae consist of different cell types, all of which are expected to make use of DAGs as second messengers, so an increase of enzyme activity in those cells that are responsible for induction may be overlooked or the observed increase may be overinterpreted. One should also keep in mind that it is not trivial that overstimulation of such a central second messenger pathway results in an ordered development after the end of treatment.

Ouabain, which blocks the sodium–potassium pump, was found to antagonize induction by Cs^+, Rb^+, NH_4^+, Li^+, vanadate, and seawater in which the Mg^{2+} concentration was reduced (Müller and Buchal, 1973; Berking, 1988a; Leitz and Wirth, 1991; Müller, 1985); most important, ouabain antagonized the inducing influence of the bacterial film (Müller, 1973). In contrast, ouabain stimulated induction by diC8 (Berking and Walther, 1994). It therefore appears unlikely that bacteria or the respective bacterial products directly stimulate PKC activity.

2. Inorganic Ions and Ammonia

Several alkali and earth alkali ions induce metamorphosis in larvae of different species including *Hydractinia*: Li^+, K^+, Cs^+, Rb^+, Ba^{2+}, Sr^{2+}, (but not Ca^{2+}), TEA^+, NH_4^+, and methylamine (see Table I). This has led to the hypothesis that membrane depolarization is the initial event in metamorphosis induction (Yool *et al.*, 1986; Pechenik and Heyman, 1987; Freeman, 1993). A problem with this assumption is the long time of treatment necessary for induction and the optimum in all dose-response curves. For instance, K^+ ions have to be applied in a concentration of 180 mM for about 3 hr to induce metamorphosis efficiently. At slightly greater concentrations, the larvae remain larvae. But membrane depolarization is a matter of seconds and is expected to be caused by other concentrations as well. Seawater enriched with 50 mM of CsCl applied for 3 hr displays almost the same efficiency as a concentration of 5 mM applied for 15 hr. It appears difficult to explain that on the basis of a changed membrane potential.

Based on these problems, it was proposed that NH_4^+ plays a central role in metamorphosis induction. As outlined previously, ammonia is produced constantly and has to be released constantly into the surrounding seawater. The export in the form of NH_3 cannot be hindered, but the export in the form of NH_4^+ can be hindered by blocking potassium channels. Cs^+, Rb^+, Ba^{2+}, Sr^{2+}, and TEA^+ are found to induce metamorphosis (see Table I) and to block potassium channels (Stanfield, 1983; Latorre et al., 1984). In some systems, activators of PKC were shown to do the same. Amiloride and Li^+ ions, which also induce metamorphosis, are able to reduce the export of ammonia through sodium–hydrogen antiports. Applied Cs^+ ions were found to cause a doubling of the internal concentration of ammonia (Walther et al., 1996). Both an increase of the internal concentration of NH_4^+ and metamorphosis induction were obtained by raising the temperature to 28°C for 3 hr (Kroiher et al., 1992); in addition, heat shock proteins were produced. However, the heat shock proteins do not appear to be responsible for metamorphosis induction, because they were produced only in heat-treated larvae and in larvae treated with Cs^+ or ammonia but not in larvae that were induced by diC8 or seawater in which the Mg^{2+} concentration was reduced. In these larvae the internal concentration of ammonia was not increased significantly, either.

The metamorphosis-inducing influence of ammonia is widespread among marine invertebrate larvae. It has been found in echinoids (Gilmour, 1991) and in tunicata (Ciona) (Berking and Herrmann, 1990). In the scyphozoan Cassiopea it induces partial metamorphosis (Berking and Schüle, 1987). In larvae of mollusca (oyster) it induces settlement behavior; for metamorphosis to commence, additional cues appear to be necessary (Coon et al., 1990). Concentrations of 10 mM have been measured in interstitial water from marine sediments (Bruland, 1983). Concentrations of 300–400 μM have been found in samples taken from the surfaces of oyster shells (Fitt and Coon, 1992). High ammonia concentrations resulting from reduced mixing may be the cause for the preference of many marine invertebrate larvae to settle in grooves and pits in the substrate.

How ammonia acts is yet unclear. Treatment with ammonia changes the pH of cells. Because ouabain antagonizes the inducing influence of ammonia, it is argued that ammonia is taken up in form of NH_4^+. This should cause a transient acidification. Consistently, organic acids generally stimulate metamorphosis (Berking, 1991a). However, during an inducing treatment, which has to last for about 3 hr, the cells are expected to come back to their normal pH. Therefore we do not expect a change of the internal pH to be the main switch for metamorphosis to commence.

3. Neuroactive Substances

The larval anterior of many cnidarian species was hypothesized to contain neurosecretory cells that sense the environmental cues triggering attachment and metamorphosis (Thomas *et al.*, 1987). Edwards *et al.* (1987) found exogenous catecholamines to cause metamorphosis in *Halocordyle*. They proposed the bacterial inducer to be sensed by neurosecretory cells of the RFamide type. Such cells were found in the anterior region of *Hydractinia* (Plickert, 1989). However, RFamides applied exogenously have not been found to induce metamorphosis in *Hydractinia* or other species. In *Hydractinia* epinephrine and dopamine did not cause induction, and they did not even stimulate induction by Cs^+ ions (Walther *et al.*, 1996). Serotonin induced metamorphosis in *Phialidium* (McCauley, 1996). In *Hydractinia* immunostainings indicate serotonin-containing nerve cells in the anterior ends of larvae. Inhibition of serotonin synthesis strongly antagonizes induction, and externally applied serotonin stimulates induction but does not induce metamorphosis when applied to *Hydractinia* larvae (Walther *et al.*, 1996).

Leitz *et al.* (1994) isolated a peptide from anthozoa which ends as LWamide. This peptide, called metamorphosine, induces metamorphosis in *Hydractinia* when applied for some hours. By molecular cloning techniques Gajewski *et al.* (1996) found a cDNA encoding a preprotein for two peptides with such a C-terminus. Similar peptides have now been found in several species including anthozoa (Gajewski *et al.*, 1996; Leviev and Grimmelikhuijzen, 1995) and *Hydra* (Takahashi *et al.*, 1997). The LWamides are localized in nerve cells (Leitz and Lay, 1995; Gajewski *et al.*, 1996). In larvae of *Hydractinia* they are predominant in the anterior region, where most nerve cells are found. The neurites project to the posterior.

4. Stabilization of the Larval State

Several agents were found to stabilize the larval state in the presence of inducers. A systematic study using Cs^+ ions as inducer and some simple organic compounds including aliphatic straight chain hydrocarbons and cyclohydrocarbons, monocyclic and polycyclic aromates, and lower alcohols, showed that the more lipophilic a substance is the lower is the concentration that reduces induction by one half. The larvae were not killed by the treatment (Chicu and Berking, 1997). Thus a low concentration of a metamorphosis-inhibiting substance does not guarantee that it acts specifically.

In a search for endogenous substances that antagonize induction, four activities were detected and analyzed to be *N*-methylpicolinic acid (homarine), *N*-methylnicotinic acid (trigonelline), *N*-trimethylglycine (betaine), and β-aminoethansulfonic acid (taurine). All these substances are present in larval tissue in rather high concentrations (about 90 mM for taurine, 3–25 mM for the others), and they are able to antagonize induction in micromolar concentrations in the medium (Berking, 1986a, 1987, 1988b). The noted substances are very hydrophilic. An unspecific inhibition due to their lipophilic character was excluded. Therefore it was proposed that they are involved in stabilization of the larval state.

Taurine was found to have several functions in various animals: it serves as an osmoregulating substance, it is able to affect phosphorylation in membrane proteins, and it acts as a putative inhibitory neurotransmitter by opening potassium channels. If in *Hydractinia* it also acts in the latter way, one may understand its metamorphosis-inhibiting activity: it may hinder accumulation of ammonia.

Homarine, trigonelline, and betaine have one feature in common. They have a labile methyl group which can be transferred to suitable acceptors. Related substances lacking this methyl group (e.g., picolinic acid, nicotinic acid) are unable to antagonize induction. The methyl group of betaine is transferred to homocysteine, resulting in methionine. Methionine and adenosine triphosphate give rise to S-adenosylmethionine (SAM), the most important methyldonor in living tissue. In shrimps, the methyl group of homarine was found to be transferred via SAM to several targets, including ammonia (Netherton and Gurin, 1982). It was hypothesized that these three molecules stabilize the larval state by their ability to methylate a specific target via SAM. Consistently, methionine antagonized induction as well, and homocysteine did not. Experiments with inhibitors of SAM synthesis and transmethylation supported the proposition. Cycloleucine (1-amino-cyclopentane-1-carboxylic acid) antagonized the inhibitory influence of methionine and of the other methyldonors, probably because of its similarity with methionine. Sinefungin had the same influence, probably because of its similarity with SAM. However, both agents were not found to induce metamorphosis. It appears that a general inhibition of transmethylation is not sufficient to induce metamorphosis (Berking, 1986b).

The noted methyldonors antagonize induction in 24-hr treatments with low concentrations of Cs$^+$ but most do not with short treatments (3 hr). This has focused our view on the polyamine pathway. SAM is used not only to deliver methyl groups but also to deliver aminopropyl groups. Aminopropylation of the diamine putrescine leads to spermidine and then to spermine. From bacteria to humans, these polyamines are indispensable for cell proliferation, growth, differentiation, and development (for review, see Tabor and Tabor, 1984). Both spermine and spermidine antagonize induction. Inhibitors of polyamine synthesis such as *bis*-(cyclohexylammonium) sulfate, methylthioadenosine, and methylglyoxal-*bis*-(guanylhydrazone) increase the frequency of inductions when applied simultaneously with an inducer. The latter was found to induce metamorphosis, however with low frequency, when applied for 1 day (Walther *et al.*, 1996). Putrescine stimulates metamorphosis induction. When simultaneously applied with *bis*-(cyclohexylammonium) sulfate, which prevents the step from putrescine to the polyamines, metamorphosis is induced (with low frequency) (Walther and Berking, in press). Therefore one may argue that the diamine putrescine and the polyamines act on the same target in an antagonistic manner.

5. Signal Uptake and Signal Transmission

To explain the occurrence of completely metamorphosed animals, one has to assume that either the respective inducing signal is taken up at every site of the larva or that an internal signal is generated, triggered by the external one, and spreads into all body regions. A reduced efflux of ammonia at the larval anterior end following tight attachment and a subsequent increase of ammonia in the whole body is one possible explanation (see previous discussion). Schwoerer-Böhning (1990) postulated an internal signal generated at the anterior end that spread along the body column following induction. The model is mainly based on sectioning experiments. They showed that anterior parts of sectioned larvae can be induced by various agents to metamorphose, while posterior agents remain in the larval state. Only seawater in which the Mg^{2+} concentration was reduced caused posterior parts to metamorphose as well. Anterior parts gave rise to stolons and a gastric region, and, with low frequency, also to a small head, while posterior parts transformed into a polyp's head and gastric region. Grafting of stimulated anterior fragments to nonstimulated posterior fragments caused metamorphosis of the whole construct. Leitz and Lay (1995) proposed the peptide metamorphosine to be the signal generated at the anterior end. Metamorphosine would reach all larval parts not by short range diffusion from nerve fibers in their vicinity but by diffusion or transport from the anterior end.

Müller *et al.* (1977) found by experiments at a marine station (Helgoland, North Sea) that posterior parts can be induced by Cs^+ ions as well, but with low frequency. We obtained the same results with larvae from the same station but not with larvae obtained from animals cultured in the laboratory for months (M. Kroiher and S. Berking, unpublished data). A possible explanation may be that sectioning and onset of regeneration renders a larva refractory to respond to inducing signals. The strength of this signal and the time course of its propagation may show some variability. Posterior parts became rapidly insensitive after cutting, even to a treatment with Mg^{2+}-reduced sea water (Berking and Walther, 1994). It appears that the arguments for a signal generated exclusively at the anterior end of a larva on induction are not strong.

There is a different way to question whether or not a larval anterior is necessary for metamorphosis induction. Some larvae undergo a partial metamorphosis in response to certain treatments, including Li^+, NH_4^+, TEA^+, vanadate, and inhibitors of protein synthesis (Spindler and Müller, 1972; Berking, 1988a; Leitz and Wirth, 1991; Kroiher *et al.*, 1991) (see Fig. 2). These animals consist of larval and polyp tissue, including the respective structures. Some animals possess a larval anterior and a polyp's head, or just the opposite (i.e., stolons and a larval posterior). Some others develop a ring of tentacles flanked by larval tissue anteriorly and posteriorly. Scanning electron micrographs show a sharp border between larval and polyp ectodermal cells (Berking and Walther, 1994). We used partially metamorphosed animals that bore stolons instead of a larval anterior and a larval posterior. When these animals were treated with inducers such as NH_4^+ or DAG they completed metamorphosis, indicating that the larval anterior is not the only site for signal uptake. Obviously larval posterior tissue can sense the inducing signal by itself.

Leitz and Lay (1995) found that such partially metamorphosed animals and also polyps contain metamorphosine. Therefore it may be possible that the various treatments cause the release of metamorphosine. Consistently, metamorphosing larvae were found to release a metamorphosis-stimulating agent (Berking, 1988a).

Externally applied ammonia induces metamorphosis within a narrow concentration range. Above and below that concentration, larvae remain in the larval stage. The larvae do not appear to suffer at superoptimal concentrations. They can be induced by an immediate subsequent treatment at optimal concentration. When the internal concentration in a larva is not everywhere at a permissive concentration, partially metamorphosed animals are expected to develop. Such animals are found not only after the noted treatment but also spontaneously. When larvae are getting old, some undergo metamorphosis spontaneously, a few producing stolons only, while the larval posterior persists. In old larvae we found the internal concentration of NH_4^+ to be increased (Walther et al., 1996).

III. Concluding Remarks on Metamorphosis

Control of metamorphosis is well studied in the highly organized insects and amphibians. In both, metamorphosis starts when a larva has reached maturity. In most marine invertebrate phyla, where, generally speaking, metamorphosis is poorly understood, it commences in response to externals cues, not when the larva has reached maturity. It appears that this type of metamorphosis control is ancestral. Therefore the work with *Hydractinia*, an animal which is close to the basis of the animal kingdom, can be expected to provide some insight into an ancestral type of metamorphosis control. Further, during metamorphosis the pattern-forming system of the larva loses influence and that of the polyp takes over. Thus, by studying metamorphosis we hope to gain insight into the biochemical basis of pattern control.

Hydractinia has several technical advantages. It is possible to get thousands of larvae several times a week throughout the year. Metamorphosis can be triggered deliberately. Substances applied to the surrounding seawater are easily taken up by the larvae and influence the development. One day after the triggering treatment the larvae either are still larvae or have developed into an easily distinguished polyp. Because several substances are able to induce or antagonize induction, we now have "islands" of knowledge in the net of biochemical pathways controlling metamorphosis. We are aware that some islands will not become part of that pathway by which metamorphosis is controlled naturally. But we hope that finally we will be able to understand the whole net that is somehow linked to that control. One of the most important aims of future research is to fill the gaps between the islands.

IV. Control of Pattern Formation

A. Pattern Formation during Embryogenesis, Larval Development, and Metamorphosis

1. Transmission of Polarity

In *Hydractinia* up to the 10th division (ninth hour of development) cleavage is synchronous. Then the cycling slows down, becomes unsynchronous, and becomes more and more restricted to the endoderm of the central region of the larva (Plickert *et al.*, 1988; Kroiher *et al.*, 1990). A mature larva, consisting of about 10,000 cells, is composed of an ectoderm and an endoderm separated by a basal lamina, the mesogloea. Some cell types, including interstitial stem cells (I-cells), nerve cells, and gland cells, are distributed differentially along the body axis. I-cells are restricted to the endoderm (Van de Vyver 1964; Weis *et al.*, 1985). The decision of an embryonic cell to develop into an ectodermal or an endodermal cell can be influenced experimentally by fragmentation of the embryo (e.g., by treatment with ethanol or dimethylsulfoxide during early cleavage stages). The resulting spheres are covered by ectodermal epithelial cells, even when they are very small and contain only a few endodermal cells (Berking, 1991a). This indicates that the number of ectodermal cells is strongly increased at the expense of the endodermal cells. Thus, it appears that a cell does not become committed to develop into an ectodermal or an endodermal cell by an internal patterning process but rather by an external cue. Cells that happen to cover the outer surface of a sphere develop into ectodermal cells; those that are inside of the embryo become endodermal cells. The signals involved in that control are unknown.

In *Hydractinia* the polarity of the oocyte is transferred to the larva (Freeman, 1981a, 1981b; Schlawney and Pfannenstiel, 1991). The first cleavage usually starts at the animal pole. This pole becomes the larval posterior and, on metamorphosis, the polyp's head. However, cleavage also has a strong influence (Freeman, 1981b). When zygotes of *Phialidium* are treated with cytochalasin B, cleavage is blocked. Following treatment, cleavage often starts simultaneously at two positions. If these sites are not too close to each other, larvae with two posterior ends will develop. If cleavage takes place consecutively, a normal larva develops. Further, treatment of early cleavage stages of *Hydractinia* with a low-molecular-weight factor isolated from *Eudendrium*, termed proportion-altering factor or PAF (Plickert, 1987; Kroiher and Plickert, 1992), or with vanadate (Leitz and Wirth, 1991) also causes development of larvae with additional ends. In most cases additional larval posterior ends are formed.

In *Hydractinia* the polarity of the larva is transferred to the polyp (Teissier, 1933; Freeman, 1981a). Kühn (1914) proposed this to be an ancestral type of development. It is also found in Clavidae, Corynidae, Penariidae, and Eudendriidae. In some animals it has been shown that the cues for a polar development are maintained in cells of all body regions. Planulalarvae including that of *Hydractinia* can regenerate missing parts, as is well known from polyps (Teissier, 1933; Müller *et al.*, 1977; Schwoerer-Böhning *et al.*, 1990). It is even possible to obtain larvae from aggregates of larval cells. Aggregates that have almost the size of a larva develop into normal-shaped larvae, whereas larger aggregates develop into larvae with several ends (Freeman, 1981b). From the outset these aggregates are chaotic with respect to orientation and neighborhood of the different cell types. Nevertheless, a polar organization develops, with all cell types at their usual places and the ability to develop into a polyp. It appears that there exists a pattern-forming system in larvae which is able to generate polarity starting from an almost homogenous distribution of larval cells.

In some species the polarity of the larva is not transmitted to the polyp. Larvae of the athecate *Perigonimus* (Hartlaub, 1895, in Kühn, 1914), *Turilopsis nutricula* (Brook and Rittenhouse, 1907, in Kühn, 1914), *Oceanea armata* (Kühn, 1914) attach laterally to the substrate. Then they transform completely into stolons. The stolon grows and branches and produces eventually one or more polyps at a more or less random position on top of the stolons. Therewith, the polar pattern of the polyp is generated *de novo* during or following metamorphosis. In summary, hydrozoa usually transmit the polar organization of the oocyte via the larva to the polyp, but they do not depend on that transmission; they have strong pattern-forming systems that are able to generate polarity from almost homogenous starting conditions.

2. Differences of the Pattern-Forming Systems of Larvae and Polyps

The existence of partially metamorphosed animals demonstrates that the transition to the polyp state can occur in only a group of cells. The cells that have been committed to acquire the polyp stage are obviously unable to trigger the surrounding cells to do the same. On the other hand, larval cells are unable to hinder the transformation of cells in their neighborhood into polyp cells. Larval and polyp cells coexist adjacent to each other. This is interesting for at least two reasons. (1) Larval cells are suspected to be able to move and find suitable neighbors, as shown by the dissociation-reaggregation experiments (see previous discussion), which result in normal larvae. In partially metamorphosed animals the epithelial cells at the border between larval and polyp tissue do not separate and do not move away. It thus appears that the epithelial cells do not alter their specific contact molecules during metamorphosis. (2) Larvae and polyps regenerate missing parts. Polyps also have been shown to regenerate missing parts by intercalation when cells come into contact that usually are not adjacent (Müller, 1964, 1982). Spontaneous regeneration of missing parts was not observed in partially metamorphosed animals. At the common border, polyp and larval cells behave as if they had contact with their normal neighbor cells. It appears that the positional value of the cells (Wolpert, 1969) is unchanged during metamorphosis. Polyps and larvae are therefore proposed to use the same or very similar substances for signaling positional information. It is the interpretation of the signal, the specific response of the cells to these signals, which has been changed by metamorphosis (Berking, 1991b). Interpretation of positional information is a cellular property. A change of these interpretation rules remains locally restricted, as observed in partially metamorphosed animals.

A mature larva of *Hydractinia* does not grow. Cell multiplication decreases down to almost zero with age. On metamorphosis, cell multiplication is resumed (Plickert *et al.*, 1988). Larvae cut transversely regenerate the missing parts in such a way that the reformed larva has almost the size of the isolated tissue piece. There is enough energy and material for growth, but that takes place only after metamorphosis. It appears that epithelial cell multiplication is blocked early in development and that this block is not overcome in regeneration. The block can be overcome in larvae by treatment with proportion-altering factor (PAF) and vanadate. Young larvae treated with these agents elongate dramatically, and this leads in many cases to larvae with more than one end (Kroiher and Plickert, 1992; Leitz and Wirth, 1991). In partially metamorphosed animals, cell multiplication is resumed in polyp tissue only. On feeding, partially metamorphosed animals with a polyp's head and a larval anterior grow in length and produce polyps from gastric tissue at a lateral position, which has never been observed in colonies, while the larval anterior persists almost unchanged in size. Stolons are produced very rarely. Partially metamorphosed animals with stolons at their base and a larval posterior develop into a colony in which the larval posterior persists (Berking and Walther, 1994). It appears that a transition from the larval to the polyp stage permits unlimited growth; in particular, multiplication of epithelial cells is allowed. Because in partially metamorphosed animals adjacent larval and polyp tissues respond quite differently, this permission of growth is not a matter of presence or absence of diffusible signal substances but a matter of a changed cell property.

3. Proportioning of Polyps during Metamorphosis of *Hydractinia*

Several substances when applied during metamorphosis are able to alter the proportion or the body regions of a primary polyp. PAF causes the hypostome and the tentacles to grow at the expense of the other regions (Plickert, 1987). Vanadate and Cs$^+$ ions display a similar influence but not as pronounced as that of PAF (Müller et al., 1977; Berking, 1987; Leitz and Wirth, 1991). A glycoprotein isolated from supernatants of Hydractinia colonies, termed stolon-inducing factor (SIF), caused the opposite: the stolon grows at the expense of the other regions (Lange and Müller, 1991). A high concentration completely transforms the larva into stolon tissue. As mentioned, in some hydrozoa this is the normal way of metamorphosis. The noted endogenous methyldonors have a similar influence (i.e., the stolon-forming region grows at the expense of the head), but the effect is much less pronounced than with SIF (Berking, 1987). Simultaneous treatment with Cs$^+$ ions and either crude extract of larvae or methyldonors causes the transformation of most larval tissue into gastric tissue without head or stolons (Berking, 1984; Walther et al., 1996). The influence of the methyldonors on proportioning may indicate that SAM is involved in positional signaling in polyps of Hydractinia.

B. Pattern Formation in Polyps

Most contributions to our understanding of pattern formation in hydrozoa have been obtained with Hydra. Various strains and mutants have been used. Almost all of the mutants were isolated and initially characterized by T. Sugiyama and his group.

Hydra has a tube-shaped body which is a few millimeters long (see Fig. 1). The two epithelial layers contain several cell types, including epithelial muscle cells, I-cells, gland cells, nerve cells, and nematocytes, which are differentially distributed along the body axis (Bode et al., 1973). When a Hydra regenerates from tissue pieces this differential pattern of cell types is reestablished. The question of what roles the various cell types play in pattern formation has been analyzed by many groups. Hydra allows a special approach to that question, because one can remove all I-cells, which are the precursors of the germ cells, nerve cells, gland cells, and nematocysts (Campbell and David, 1974; David and Gierer, 1974; David and Murphy, 1977; Bode et al., 1987), by treatment with colchicine, hydroxyurea, or lithium chloride (Campbell, 1976; Bode, 1983; Hassel and Berking, 1988) and in the mutant Sf-1 by heat treatment (Marcum et al., 1980). Such animals, when force fed, consist after some time of epithelial cells only but are still able to produce buds and to regenerate. Thus, the epithelial cells are sufficient for pattern control. However, in these animals the epithelial cells look slightly different from normal ones. They were found to contain vesicles normally found in the nerve cells of Hydra (Schaller et al., 1980). Therefore epithelial cells may be able to take over functions of nerve cells, if nerve cells are missing. Because one can recolonize such epithelial Hydra with I-cells of other strains or mutants, one can analyze the effects of nerve cells on pattern formation in combinations with epithelial cells of different origin.

The various approaches to the study pattern formation in *Hydra* can be roughly subdivided into three groups: analysis by transplantation and regeneration methods, by biochemical methods, and by molecular biology methods. The links between these different approaches are models of pattern formation. They are used as tools. They are expected to allow use of the results obtained in one field for the design of critical experiments in another field.

1. Generation and Maintenance of the Polar Body Pattern

Early models of pattern formation were strongly influenced by experiments with sea urchins. Driesch (1893) proposed that the cells of cleavage-stage embryos were forced to develop according to their relative position in the whole, even if the embryo was experimentally enlarged or reduced. Runnström (1929) explained that behavior by postulating signals that are generated at the animal and the vegetal poles and that reduce in intensity when traveling through the embryo to the opposite pole. The developmental fate of a cell is determined by the local ratio of intensity of animal and vegetal signaling. Wolpert (1969), working with *Hydra*, coined the terms "positional information" and "interpretation," that is, the specific response of cells to the local intensity of a positional signal. Transplantation experiments in sea urchins and other animals including *Hydra* showed that cells have a certain memory about their position. Both the local value of the positional information and the memory of that information were termed "positional value" (Wolpert, 1969); however, in this discussion the term is used to describe only the memory, a rather stable cell property. The fate of an implant was found to be determined by both the old positional value and the new positional information. These concepts also helped to understand the observation that in various hydrozoa a head inhibits the formation of a further head from gastric tissue (see Tardent, 1963). On the other hand, they made it possible to see a problem: in all experiments, boundary regions that were proposed to generate positional information were found to be regenerated from nonboundary tissue. That property of the pattern-forming mechanism could not convincingly be explained by these simple rules (Gierer, 1977).

In particular, the finding that polyps can reform from aggregates of single cells of midgastric body sections (Gierer *et al.*, 1972) showed that problem. Obviously, pattern formation starts from an almost homogenous mixture of the various cell types and leads eventually to a polar-organized animal. General features of a pattern-forming mechanism that are able to account for such a development were proposed by Turing (1952). He proposed the existence of substances, termed "morphogens," which have the property to control their own concentration within the tissue by interaction with each other's synthesis. The resulting concentration profile was proposed to control pattern formation. On the basis of this idea Gierer and Meinhardt (1972) developed a model in which the type of interaction between the morphogens was exemplified in detail. The general idea is that there must be some sort of autocatalysis that remains locally restricted, because there is a coupling to lateral inhibition of that autocatalysis. A concrete version of the general model is that a substance A (activator) stimulates (in a nonlinear manner) its own release from special cells termed sources and also stimulates the release of a substance H, which inhibits the release of A. If H diffuses faster than A and both have a limited lifetime, then the enhanced production of A and H become locally restricted and stable in time. The relatively higher concentration of H in the tissue surrounding the A peak prevents further autocatalytically stimulated release of A, laterally. The mechanism of interaction that controls the release of the morphogens is designed in such a way that a homogenous distribution of the morphogens in the tissue is unstable but a differential one is stable. That occurs only in tissue above a certain size. In small pieces of tissue, where H cannot escape essentially by diffusion from the site of its release, the autocatalytic stimulation of the release of A is prevented. Most interestingly, the first differential concentration pattern of the morphogens that forms when a tissue exceeds a certain size is a polar one. If the local concentration of A or H, or both, is used as positional information, then the tissue can be organized; for instance, a head may be formed at the highest concentration of A and head formation is prevented in the surroundings of a head. It is unclear whether diffusible substances are indeed involved in pattern control, but the observation that in aggregates and transplants a head is able to inhibit head formation in its surroundings fits well into that proposition.

No doubt it would be very helpful to know the morphogens. Unfortunately, not only is the search for such substances time-consuming but there is also the principal problem that localized glands for such morphogens, as they are known for hormones, do not exist. We are looking for substances that are able to organize a pattern out of almost homogenous starting conditions. Every cell or every cell of a special type must be able to generate the morphogens, because in *Hydra* tissue of almost every position is able to form a head or a foot. By interaction in control of production and release, a small part of the tissue eventually generates the morphogens. Thus, a "gland" is created by interaction between cells. This gland cannot be isolated, and the tissue can not be depleted of morphogens and deliberately resupplied with the agents. Because this direct way to find morphogens and therewith to understand pattern formation is impossible, we search for rules of pattern formation which we hope will enable us to reduce the number of possible mechanisms and substances involved in pattern control.

The model of Gierer and Meinhardt (1972) also provides a concrete proposition for the gradient of positional values along the body axis which determines the tissue polarity. Gierer *et al.* (1972) found that polarity in *Hydra* is determined by a scalar tissue property and not by a vectorial one. The positional value of a tissue piece is thus determined by a certain quantity contained in that tissue. It is a cellular property, probably a substance contained in certain cells. In *Hydra* that quantity is distributed in a graded manner, it is fairly stable, and it is maintained even in disaggregated cells. The quantity was proposed to be the density of the sources that contain the morphogens. The sources are proposed to release the morphogens at a low basal rate. Thus, the release is high where the density of sources is high. The source density is proposed to change in the long run by the local positional information (*i.e*, by A and/or H). Following transversal cuttings of *Hydra* into sections, the apical cut surface forms a head because this tissue has the highest density of source and therefore the highest basic unregulated release of A. The autocatalytic release of A wins the race at that position, and in the long run the density of the sources changes to the value found in a head. It has to be kept in mind that the graded distribution of the source density merely orientates the pattern, but a pattern will also form if the source density is homogenous.

The model has been applied successfully to head regeneration (Gierer and Meinhardt, 1972; Meinhardt and Gierer, 1974; Gierer, 1977). Foot formation was proposed to be controlled by a mechanism almost identical to that controlling head formation (MacWilliams *et al.*, 1970; Hicklin and Wolpert, 1973; Mac Williams and Kafatos, 1974). The foot was proposed to be the source of a foot inhibitor, which explains the observation that a new foot is formed when the old one is removed by sectioning. It was proposed that *Hydra* has only one head and one foot because the inherent ability to produce both structures is hindered along the whole body axis by the respective inhibitors generated by the head and the foot. The inhibitors antagonize to some extent the activator release in the opposite system, so that the head and the foot are formed at maximal distance in an isolated piece of tissue (Sinha *et al.*, 1984).

Meinhardt (1993) proposed a model with hierarchical features. The primary system controls structure formation (hypostome and foot) and source density; a secondary system controls tentacle formation. That system is able to function only above a certain threshold density of sources, that is, not close to the foot but in the whole gastric region. The primary system makes use of four morphogens. Following sectioning of an animal, the activators are generated. The head activator is generated at the apical end (i.e., the end with the highest source density). If the head inhibitor generated at the apical end reaches the opposite end in sufficient concentration in time, head activator is not released at the basal end; rather, a foot forms (and vice versa). Therewith the model describes where the hypostome and the foot form in such a tissue piece and where the source density increases and decreases. The inhibitors are not proposed to antagonize activator production in the opposite system.

Müller (1995, 1996) proposed an asymmetric model in which tissue of high positional value absorbs (hypothetical) head-promoting factors. In an isolated tissue ring the competition causes a depletion of these substances at the basal end, where the positional value is comparatively low, leading there to foot formation. However, the model cannot explain the aggregation experiments, that is, the generation of a pattern from almost homogenous starting conditions.

In summary, most models are based on the concept developed by Gierer and

Meinhardt (1972). In these models, first, morphogens are proposed to exist which control their own concentration profile in the tissue by interaction, including autocatalysis and lateral inhibition. This interaction results in all-or-none generation of an activator at a certain site up to a certain maximal concentration, irrespective of the starting condition. Second, there exist two such systems, one for the head (hypostome) and one for the foot. Both make use of an activator and an inhibitor. The respective concentration profiles form countercurrent double gradients. Third, the activators determine the formation of a head (hypostome) and a foot, respectively. Therefore *Hydra* is under the control of two organizing regions located at opposite ends of a field. Fourth, in the long run the local concentration of the morphogens determines the local positional value and therewith the polarity of the tissue. The positional value (source density) does not change essentially during head or foot regeneration. These features explain a great number of experimental results, but groups of experiments remain difficult to explain. These experiments will now be discussed.

Under the proposed type of self-regulation, the generation of the activator is an all-or-none decision. When in head-regenerating tissue the activator release becomes autocatalytically enhanced, a hypostome forms inevitably. When in regenerating tissue that activator is not released, a hypostome does not form and the positional value is not increased. It is not possible to have an even slight increase of the positional value without a hypostome being formed there eventually. Some experiments are difficult to explain by these assumptions. When head-regenerating tissue is brought into competition with a head (Browne, 1909; Webster and Wolpert, 1966; Berking, 1979b), a developing head (Webster, 1971), or a developing bud (Roulon and Child, 1937), it does not forms a complete head: depending on the strength of interaction, either the number of tentacles is reduced to four or three with a hypostome in between, or three or two tentacles form without a hypostome in between, or one single tentacle forms at the site where the hypostome usually forms. It appears that the positional value has increased, but the most distal structure, the hypostome, does not form when the interaction is strong. A further example may be a type of polyp produced by *Hydractinia* (see later discussion). Like other polyps of *Hydractinia*, this polyp starts its formation from stolon tissue (i.e., tissue of low positional value). However, after forming a gastric region it does not form tentacles and a hypostome but only one tentacle instead of a hypostome. A signal to increase the positional value is obviously produced at the beginning of polyp formation, but a hypostome does not form. The proposed properties of the hypothetical head activator appear to be at variance with these observations.

Double-gradient models, which postulate structure-specific processes to start after sectioning, are inadequate to explain the formation of two identical structures at opposite ends of a regenerating body section, in particular when the piece is large. *Hydractinia* polyps, when removed by a cut from the colony, can regenerate either stolons or a second head at the basal end. Pieces from small

polyps preferentially regenerate stolons, pieces from big ones a second head
(Müller *et al.*, 1986). Both small and large tissue pieces of big polyps can
regenerate in this way; in particular, tissue pieces from the lower gastric region
regenerate heads at both ends (Müller, 1969). Body sections of *Tubularia* are able
to form heads at both ends, in particular when the pieces are small or rather large.
However, a head regenerates at the apical end and stolons at the basal end when
the basal end is covered by sand (Morgan and Moszkowski, 1907; Tardent, 1963,
1978). When the basal end is closed by a ligature, no structure forms at that site
(Morgan, 1902; Tardent, 1963). *Pennaria* polyp sections produce stolons at a cut
surface when covered by sand, and a hydranth at the free end. Both the apical and
the basal end behave in that way. If the body section is allowed to regenerate
while freely floating, it produces a head at both ends with high frequency (Loeb,
1891, in Przibram, 1909). In *Hydra* the polarity is well known to be rigidly
controlled, but small tissue pieces were also found to regenerate a head at both
ends of a body section (Weimar, 1928). Isolated polyp tissue of *Eirene* regenerate
a head at the apical end and a stolon at the basal end. When, however, the basal
cut, which isolates the polyp from the colony, is made some few hours ahead of
the apical cut, the apical cut surface regenerates a stolon as well, and not a head
(Plickert, 1987). Bipolar regeneration is not restricted to hydrozoa. Isolated heads
of several planarians regenerate a mirror-imaged head at the basal cut surface,
and isolated tails regenerate a further tail, whereas large tissue pieces regenerate
the normal polar body pattern. (Morgan, 1904; Morgan and Moszkowski, 1907).

Double-gradient models, which postulate head- and foot-specific processes to
start after sectioning, are inadequate to explain the fate of implants made of
regenerating tissue. Non-regenerating tissue (used as control) transplanted to a
host at the very same position usually integrates without forming a structure.
When transplanted closer to the head, it forms a foot; when transplanted closer to
the foot, it forms a head. When the implant is regenerating tissue from the apical
(or basal) end of a body section it should inevitably form a head (or foot) if there
is an indication that the pattern-forming system has been activated before transplantation. The results obtained indicate that indeed head- and foot-regenerating
tissues develop in divergent directions, but the decision to form a hypostome or a
foot is not fixed. Head-regenerating tissue acquires the property of non-regenerating tissue at a position somewhat closer to the head; depending on the site of
implantation, the implant forms a head or it integrates without forming a structure, or even forms a foot. The regenerating tissue behaves as if the positional
value (a quantity) has increased but not like tissue in which head formation (a
quality) has been decided. It appears that structure-specific processes probably
start late in development, when the positional value, which has been committed
to change during regeneration, is changed and has reached a mature or functional
state. Foot-regenerating tissue behaves just the other way around (Berking,
1979b).

2. A Hierarchical Model of Pattern Formation

Based on these results, pattern formation was proposed to be controlled hierarchically. A primary pattern-forming system controls the change of the positional value, and secondary systems function at a certain positional values and control the structure-forming processes. In contrast to current opinions in this model, the head is not regarded as a unit including hypostome and tentacles but rather as a composite structure made by different and largely independent secondary pattern-forming systems. The proper arrangement of the structures is caused by the positional value, which was proposed to have its highest value at the tip of the hypostome. Tentacles were proposed to be formed not at the highest but at a slightly lower positional value. This was indicated by the observation that heads implanted in reverse polarity cause the formation of heads from gastric tissue in such a way that eventually both heads, the implanted and the induced, adhere together with the tips of their hypostomes. Foot formation was also proposed to be controlled by a secondary system, which starts to function when a certain low positional value is passed (Berking, 1979b). This model is exemplified in the following discussion, which is focused on the primary system that controls the positional value. (The term "positional value," it should be remembered, is used in a restricted manner; it describes the rather stable tissue property only and does not include the comparatively fast-changing local value of the positional signal.)

The positional value was shown to be determined by a quantity (Gierer *et al.*, 1972). Therefore, the primary system proposed to control the change of that quantity should be able to increase and to decrease that quantity. Both the increase and the decrease should include self-enhancement and lateral inhibition of that enhancement. A simple concrete proposition is that the positional value is determined by the concentration of a substance A (activator) contained in specific cells. (For reasons of simplicity a potential role of the extracellular matrix in control of pattern formation is ignored.) The positional value decreases when A is released. This release is autocatalytically stimulated by external A and antagonized by an inhibitor of activator release (IAR). The release of that inhibitor also is stimulated by A. The positional value increases by production of A in the specific cells. That production is stimulated by external A and inhibited by an external inhibitor of activator production (IAP), the release of which is stimulated by A as well. The low constitutive and the maximal release of A depend on the positional value. The diffusion constants decrease from IAR over IAP to A (Fig. 3).

This hierarchical model of pattern formation can be summarized as follows:

- A subgroup of the cells, termed sources, contain a substance A (activator). The concentration of A in a cell determines the positional value.
- A is released constantly from the sources at a low rate proportional to the

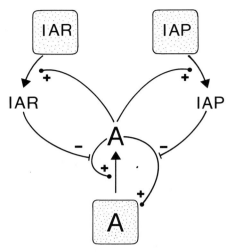

Fig. 3 Scheme of the hierarchical model. Shown is the hypothetic interaction of activator (A), inhibitor of activator release (IAR), and inhibitor of activator production (IAP). For details see text.

internal concentration of A. External A is able to stimulate its own release from the sources (autocatalysis) in a nonlinear manner, as proposed by Gierer and Meinhardt (1972). The maximal rate of stimulated release of A is proportional to the internal concentration of A.

- In the sources A is produced at a low constant rate. External A is able to stimulate the production of A in the sources.
- A also stimulates the release of two inhibitors: a substance IAR (inhibitor of activator release), which antagonizes the stimulating influence of A on the release of A [similar to H in the model of Gierer and Meinhardt (1972)], and a substance IAP (inhibitor of activator production), which antagonizes the stimulating influence of A on the production of A. (In the alternative, IAR and IAP may be degradation products of A, produced externally.)
- The diffusion constants of the morphogens decrease in the order IAR > IAP > A.
- Secondary systems function at a certain positional value (within a certain range). Hypostome formation starts at the highest possible value, foot formation at the lowest possible one. In between, several systems may function. Interaction between the involved morphogens of the secondary systems, including mutual local exclusion (Meinhardt, 1982), is proposed to allow a clear decision for one or the other structure to be formed.
- The change in the external concentration of the morphogens by stimulated release is a fast process (Gierer and Meinhardt, 1972). The change in the

internal concentration is comparatively slow. There is a delay between the initiation of a change in the internal concentration of A and the response of the cell, leading to a change of differentiation and the functioning of secondary pattern-forming systems.

All models must explain how one parameter, the graded distribution of positional values, decides whether a head or a foot is formed at the end of a body section. In the double-gradient models, the interaction between the morphogens and their sources is proposed to result in generation of different morphogens at the apical and basal cut ends. In the hierarchical model, the same morphogens are proposed to be generated at both ends, but a quantitative difference in the concentration ratio of A to IAP causes the production of A to be unequal at both ends. The IAP, released at a constitutive low rate in tissue adjacent to the cut, adds little to the (stimulated) released IAP at the cut end when the positional values decrease with distance from the cut. But it adds more when the positional values increase adjacent to the cut. Therefore, production of A is allowed at the apical end but not at the basal one; however, this occurs only when there is a sufficiently steep gradient of positional values in the tissue.

If the slope of the gradient of positional value is flat close to the basal cut end (but not necessarily in the whole piece), a head forms there as well. This may explain the bipolar regeneration of large body sections of *Hydractinia*, *Tubularia*, and *Pennaria*. In small tissue pieces, head (but not foot) formation may be favored because more of the more highly diffusible IAP than of A is lost in the surrounding water. As mentioned, in *Tubularia* and *Pennaria* a head is formed from the basal end when the section is freely floating in the water but a stolon is formed when the end is covered by sand. That covering is expected to hinder the loss of IAP and IAR from the cut surface into the surrounding water. The increased concentration of IAR at the site reduces the speed of regeneration or even prevents it, but it does not influence the decision to form either a head or a stolon. However, the increased concentration of IAP antagonizes A production.

The apical end of a body section of *Eirene* formed a stolon and not a head when the polyp was isolated from the colony before decapitation. It is suspected that some IAP generated at the basal end (the end with the head start of regeneration) had reached the apical tissue by the time a cut was made there, some hours after the first cut.

Implants can form a head close to a foot and vice versa. It is suggested that isolation of a tissue piece of the gastric column causes considerable loss of IAR into the surrounding water, in accordance with Newman (1974). He found that head regeneration was delayed after head removal by ligation, which causes a much smaller wound than cutting, and in several cases one tentacle was formed instead of a complete head. Therefore, in an isolated piece of tissue an autocatalytically stimulated release of A is proposed to take place. The resulting maximal concentration of A (and thus of IAR and IAP) in the tissue is proportional to its positional value. When tissue of low positional value is implanted close to a head, the autocatalytically stimulated release of A, once started, may not be blocked by IAR imported from the surrounding host tissue. However, the imported IAP adds to the concentration present in the implant. Therefore, in a tissue of low positional value, implanted close to the head, production of A is antagonized.

It has been argued (Meinhardt, 1993) that a positional information model cannot explain the sequence of events taking place in head regeneration, since the highest level (hypostome formation) cannot be reached without passing through the lower level (tentacle formation). Indeed, head regeneration does not result in a tentacle at the tip followed by a hypostome at that place. Therefore it is argued that the positional value must be stable for some time to obtain a stable decision. In this process, an interaction between the secondary systems of hypostome and tentacle formation may be included.

3. Budding

Several hydrozoa, including *Hydra*, can reproduce asexually by producing buds that separate from the parent animal (see Fig. 1). In almost all cases the tip of the bud transforms into the head of the future adult. The scyphozoa *Cassiopea* is one of the very rare examples in which a foot is produced at the bud's tip. In *Hydra* and other hydrozoa (Kühn, 1914), budding starts with the evagination of the two epithelia at a species-specific position at the body. In *Hydra vulgaris* it is about two thirds of the length measured from the head. The position of the new bud is specified in two steps. At first a ring is activated and then the position within the ring is specified (Berking and Gierer, 1977). Subsequently, preparatory steps of visible bud formation start, including commitment of the multipotent I-cells to development as nerve cells. That commitment takes place in the middle of an I-cell's S phase (Berking, 1979a). The head of a *Hydra* has a higher concentration of nerve cells than the gastric region from which the bud forms (Bode *et al.*, 1973), so additional nerve cells have to be formed during bud development. The first of these additional nerve cells become committed 1 day before budding starts visibly. Commitment takes place in presumptive bud tissue, and following commitment the neuroblasts migrate into the bud's tip. These neuroblasts do not finish development. When a certain density of precursors is attained, budding starts visibly and nerve cell development proceeds synchronously (Berking, 1980). The bud grows by recruiting tissue of the parent animal (Otto and Campbell, 1977). Before evagination, about 800 epithelial cells are visibly recruited for the bud to be produced, and this number increases by about 5000 within a day (Graf and Gierer, 1980). Thereafter, the bud size increases mainly by proliferation of the cells within the bud. On recruitment the epithelial cells assume a columnar shape, with a smaller contact area facing the mesogloea, accompanied by a decrease in volume that is mostly accounted for by devacuolization. Evagination proceeds without reorientation of epitheliomuscular fibers, whereas elongation of the bud is accompanied by fiber reorientation (Graf and Gierer, 1980). One day after the onset of budding, the head starts to form visibly. Then at the base a constriction forms and a foot differentiates. Three to four days after the visible onset of budding the new animal, which is about one quarter the size of an adult, separates from the parent animal.

Young animals grow in length and do not form buds. Older ones do not grow in length; they produce buds but only at a certain position. The finding that budding is enhanced when the head of the parent animals is removed (Tardent, 1972; Shostak, 1974) indicates an inhibitory influence of apical tissue. Budding is proposed to start where that inhibitory influence is low enough to allow the release of activator (A) to become autocatalytically stimulated. Meinhardt (1993) showed by simulation experiments that this release can remain locally restricted due to autocatalysis and lateral inhibition, resulting in the formation of a secondary axis with a head at its distal end.

Up to now, the most difficult problem was to explain why a foot forms at the bud's base. Indeed, no model was able to explain that, and it was argued repeatedly that budding is controlled by a pattern-forming system different from that which controls head and foot regeneration. Thus it appears critical for our understanding of pattern control to understand foot formation at the bud's base. A deep insight into that process allows an approach introduced by Mutz (1930). Budding can be initiated experimentally by implantation of the tip of the hypostome or, technically simpler, by implantation of a head in reversed polarity (Fig. 4). (The latter method allows a small contact between tissue of maximal positional value and gastric tissue; the reversion itself is without influence on the results). When the tissue is implanted into a position between the budding region and the foot, a bud (i.e., a secondary axis) is induced to form and separates from the parent after having formed a foot at its base. If the tissue is implanted between budding region and head, a secondary axis forms which neither separates nor forms a foot. The closer the hypostomal tissue is implanted to the host's foot, the smaller is the obtained bud (Berking, 1979b).

Based on these observations, budding is proposed to be controlled not by a separate system but rather by the general one. As proposed by Meinhardt (1993) and also in accordance with the hierarchical model, a local peak of released activator (A) forms in gastric tissue some distance away from the head. The center of the area in which A release takes place becomes the new bud's tip. The reason is that in this center the ratio of the concentration of A to that of IAP is high, causing the positional value to increase. However, due to the difference in diffusion constants, the concentration ratio of A to IAP decreases with distance from the center. Thus, the area in which A is released does not respond uniformly; rather, in the very periphery of that area the ratio is suggested to decrease below the threshold ratio, which balances the loss of A by release with the gain by production. If the positional value decreases in that ring to a sufficiently low

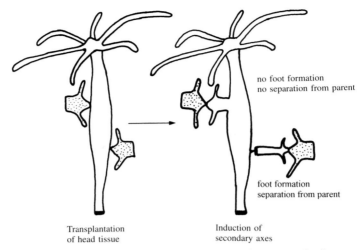

no foot formation
no separation from parent

foot formation
separation from parent

Transplantation   Induction of
of head tissue   secondary axes

Fig. 4 Induction of different types of secondary axes in *Hydra* as a result of contact between hypostomal tissue of a graft and gastric tissue of a host.

value, the foot system starts. If the activator is generated in tissue in which the positional value is too high, a secondary axis forms without a foot at its base. There is a further limitation: if in a certain species the diffusion constants of A, IAR, and IAP are such that within the area of stimulated A release the concentration ratio of A to IAP remains above the threshold in question, the resultant secondary axis does not form a foot at its base. Such a species develops a colony-like pattern in which polyps branch from gastric regions of polyps or polyps emerge from stolons.

The model fits in the finding of Sanyal (1966) that in the gastric region of the parent animal surrounding the very young bud (i.e., in prospective foot tissue of the bud), the positional value is already decreased. Pérez (1996) found by treatment with various chemicals that affect phosphorylation (see later) that the decrease of the positional value at the future bud's base is an early process, taking place some hours before and after onset of evagination and therefore at least 2 days before the foot visibly forms.

The role of the local positional value in foot formation was shown in the following experiment (Fig. 5). Tissue from a position close to the head was transplanted into the budding region. It did not form a structure there but happened to be recruited by a developing bud in its basal part. The developing secondary axis formed a foot in the form of a crescent from parent tissue, but the implant did not form a foot (Berking, 1979b). It appears that in the implant the positional value was too high initially and thus, by reducing the positional value

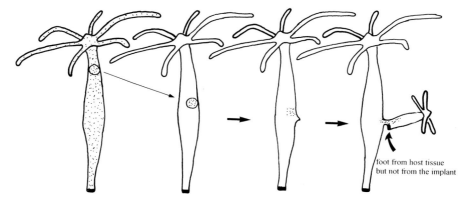

Fig. 5 Transplanted tissue of high positional value into the budding region is not included in foot formation at the bud's base.

during axis formation, the low value necessary for initiation of the foot pattern-forming system was not attained. Finally, a secondary axis formed which was partly separated from the parent but connected to it by the implant.

Hypostomal tissue implanted close to the budding zone causes some induced secondary axes to not separate but to produce a lateral foot patch instead of the ring-shaped foot produced by a normal bud (see Fig. 1). The foot patch points in almost all cases to the parent animal's foot, meaning that the foot is formed from tissue of the lowest positional value of the tissue recruited by the bud. This finding further indicates that the parent animal's foot is unable to hinder foot formation in its vicinity and has led to the proposition that the inhibitory influence of a foot has no long range (Berking, 1979b).

The scyphozoan *Cassiopea* produces buds in a way completely different from *Hydra*. The tip of the bud is the future foot and not the head (Fig. 6). However, tentacles, a mouth opening, and a foot form only after separation from the parent animal and contact with a suitable substrate. A straightforward explanation on the basis of the hierarchical model is that in the tip A is released autocatalytically stimulated, as in buds of all other cnidarians, but that in the budding region of *Cassiopea* the concentration of IAP is so high that the positional value does not increase as usual but decreases. In the future bud's head, at least, the final increase of the positional value necessary for head formation is hindered. Neumann (1979) found that sectioning of a freely swimming bud allows head formation. She proposed that sectioning removes the source of an inhibitor which emerges from the tip. In terms of the hierarchical model, sectioning removes the main source of IAR and IAP.

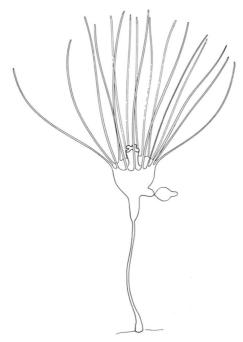

Fig. 6 Budding in *Cassiopea andromeda* (scyphozoa).

4. Growth and Size Control

In *Hydra* (and other cnidarians), cells of both epithelia proliferate, causing an expansion of the tissues, but a mature animal does not grow in length. The most important limitation of growth is caused by budding. After its visible initiation, the bud very quickly recruits tissue of the parent animal. A further limitation is caused by the loss of cells at the body ends (Campbell 1967a). The structures like head and foot apparently persist unchanged, but the cells that form these structures are continuously replaced by new ones. The tissue appears to flow through the structures (Tripp, 1928; Campbell 1967a, 1967b). This may be a unique feature of cnidarians. In *H. vulgaris*, cells of the apical one fifth of the body column eventually reach the head, and cells below that border reach the foot (or the bud). According to the model proposed, the ratio of A to IAP is high at the apical end of the animal and decreases with distance from that end. Above a threshold ratio the positional value is forced to increase; below that value the positional value decreases. Both the increase and the decrease of positional value are slow but inevitable. The decrease is particularly slow close to the foot (Campbell 1967a), which fits in with the proposition that the decrease is mainly caused by the low constitutive release of the proposed activator. The change of the positional values takes place even during starvation, when cell proliferation is low. The animal shrinks, but the apparent flow of tissue through the structures is maintained and the body proportions are generally maintained (see later discussion).

In polyps that do not form lateral buds (e.g., colonial marine hydrozoa, anthozoa), the body size may simply be determined by the increase in cell number due to proliferation and loss of cells, in particular at the body ends. The stability of the body pattern results from the forced slow change of the positional values of the cells along the body axis. That may be sufficient to maintain the integrity of a polyp even if it grows during its life by a factor of 1000.

In *Hydra*, proportion regulation was studied in detail. Experimentally it was found that the tentacle tissue was exactly proportioned over a 20-fold size range, even if in small pieces not one but two heads were formed by regeneration (Bode and Bode, 1980). The basal disc was proportioned exactly over a 40-fold size range. In contrast, with decreasing size, the hypostome and the tissue between the tentacles became an increasing fraction of the animal at the expense of the body tissue, and in the very smallest it regenerated at the expense of the tentacle tissue (Bode and Bode, 1984). The dynamics of pattern regulation is not well understood. When by cutting at the apical end head regeneration was allowed to start in a large tissue piece and some hours later a second cut was made to isolate the piece (including the head-regenerating tissue), the head of the resultant small isolate was relatively large. With increasing time between cuts, the head became increasingly larger. This kinetics was thought to reflect the initiation of the change in positional value at the apical end. Surprisingly, the head also became larger when foot regeneration was started ahead of head regeneration (Berking and Schindler, 1983). This observation excludes some simple double-gradient models as explanations of size regulation. Size regulation, for instance, is not the result of a simple competition between the head and the foot system in recruiting intervening tissue.

In *Hydra*, the diameter of the body tube increases with increasing body size. How this is caused is not understood. One may argue that the hypostome which forms an increasingly larger cone at the apical end of the animal with increasing body size determines with its basal diameter the diameter of the body column. In large animals the absolute distance from the head to the border of divergent fate is large. That forces more cells to become head cells per unit time, which in turn would increase the size of the head. The distance increases with increasing body length because IAP produced at the apical end has increasingly more space to move in. When in basal tissue the concentration of IAP reaches the low level determined by constitutive release, the head does not grow further.

The length of a tentacle depends on the speed of recruitment of new epithelial cells at its base and the lifetime of these cells. Tentacle cells do not proliferate, and they have a limited lifetime (Campbell 1967a, 1967b; Bode *et al.*, 1973; Hobmayer *et al.*, 1990a, 1990b). The number of cells contained in all tentacles of an animal is a constant fraction of the animal's total cell number. This is not true for the number of tentacles or the number of cells in one tentacle (Bode and Bode, 1984). Thus, one may argue that a tentacle is unable to cause commitment autonomously of epithelial cells to tentacle cells in a concentric ring surrounding it. The finding is consistent with the proposition that commitment is possible only at a certain positional value (within a certain range). The process of tentacle formation from these cells appears to be a secondary event which may include local activation for final differentiation and lateral inhibition, as indicated by the regular arrangement of tentacle primordia during budding and in head regeneration.

The tentacles of an adult *Hydra* are usually arranged in a precise ring. Tentacles that happen to form outside of that ring, due to certain treatments, either move up to the tentacle ring or become displaced down the body columns while becoming smaller and finally disappearing. This fits in with the proposition that they are unable to recruit cells and to commit them to tentacle cells. The moving up of ectopic tentacles to the ring may indicate that at the base of a tentacle only cells that are just before final differentiation become integrated. Others are not integrated and thus may hinder the growth of a tentacle and cause its apical displacement. Because of this, the positioning of the tentacles in a narrow concentric ring around the hypostome is dynamically stabilized; it does not reflect a narrow ring of tentacle cell determination. In several marine species including *Hydractinia*, and in particular in Clavidae, tentacles are not formed in a narrow ring but are almost regularly spaced in a broad belt. In these animals the dynamics of tentacle growth is unknown, but if the mechanism is essentially similar to that in *Hydra*, one may suspect that in these animals the inhibitory field surrounding a tentacle is large and that within this field epithelial cells proliferate and do not attain final differentiation. The belt indicates the range of positional values in which the tentacle system can function. Thus, whether a belt of tentacles or a narrow ring is formed appears to depend largely on the size of the inhibitory field of the tentacles, the speed of recruitment of tentacle cells at the base of a tentacle, and the range of positional values in which tentacle cell commitment is possible.

Above the tentacle ring the positional value is proposed to be too high to allow tentacle cell commitment. The few cells that proliferate here replace the existing hypostomal cells. Further, cells between the tentacles within the (hypothetical) inhibitory field of a tentacle may happen to attain to high a positional value before they are forced to start terminal differentiation into tentacle cells. There may be an interaction between the systems that control hypostome formation and tentacle formation, as proposed by Meinhardt (1993), but, for the reasons discussed, it is assumed that the hypostome is not organized by the activator of the primary system.

5. Biochemical Approaches to the Control of Pattern Formation

From *Hydra* tissue morphogenetic active substances have been isolated. The small peptide "head activator" (Schaller and Bodenmüller, 1981), applied externally, slightly increases the number of tentacles, but the general morphology does not change. The substance was also found to slightly stimulate foot formation (Javois and Frazier-Edwards, 1991). It stimulates nerve cell formation and shortens the G_2 phase of epithelial cells (Schaller, 1976a, 1976b; Hobmayer *et al.*, 1990a). Head activator has been shown to affect the cyclic adenosine monophosphate pathway (for review, see Galliot, 1997). Two peptides have been isolated which increase the production of an enzyme that is preferentially but not exclusively found in the foot (Hoffmeister, 1996). Their role in foot formation remains to be established. Inhibitors of regeneration have been isolated, but the structure of none of them is known (Berking, 1977; Schaller *et al.*, 1979; Berking, 1983; Schaller, 1984). One of them hinders both head and foot regeneration by hindering a change of the positional value. Consistently, it also antagonizes bud formation. This substance therefore has features of the hypothetical morphogene IAR of the hierarchical model. The agent prevents the commitment of I-cells to develop into nerve cells (Berking, 1979a) and prevents epithelial cells from leaving the G_2 phase (Herrmann and Berking, 1987). Endogenous and nonendogenous substances affecting the proportioning of larvae and metamorphosing *Hydractinia* were mentioned previously. From tissue of *Tubularia* an activity antagonizing head regeneration has been isolated; the structure is unknown (see Tardent, 1963).

Several nonendogenous substances have been found to affect pattern formation in *Hydra*. The reason to study the effects of these substances is that one can expect to get insights into the biochemical pathways involved in the generation of the morphogens and/or the response of the target cell to the morphogens. Most investigations have been done with substances that affect phosphorylation, including LiCl, TPA, DAG, and cantharidin. Li^+ ions are well known to antagonize dephosphorylation of phosphoinositols, TPA and DAG stimulate protein kinase C, and cantharidin is a phosphatase inhibitor (Berridge *et al.*, 1989; Li and Casida, 1992). The results obtained are not easy to understand, because slight changes in the regimen of treatment give completely different results. Further, species which look similar and display similar properties in regeneration and transplantation experiments respond very differently.

In general, continuous treatment with low (1 m*M*) concentrations of Li^+ ions causes a decrease of the positional value, while a pulse treatment (with higher concentrations) causes an increase of the positional value. After some days of continuous treatment *H. vulgaris* produces foot patches at the basal ⅔ of the body axis. Some animals form a belt of foot tissue that eventually leads to constriction and separation of an animal into two. Budding is suppressed. Sectioning during treatment causes foot formation at both ends of the isolated tissue ring. *Hydra magnipapillata* and some other strains fail to respond (Hassel *et al.*, 1993). In contrast, starting the Li$^+$ treatment after cutting causes head formation at both ends in *Pelmatohydra robusta* (Yasugi, 1974) and in *H. vulgaris* (Hassel *et al.*, 1993). Regenerates of *Hydra littoralis* produce more tentacles when maintained in LiCl after sectioning (Ham *et al.*, 1956). A single treatment for 3 hr of *H. vulgaris* and *H. magnipapillata* bearing a young bud increases the positional value at the future bud's base in such a way that foot formation cannot start 3 days later (Hassel and Berking, 1990; Hassel *et al.*, 1993). The resulting secondary axes do not separate from the parent. When both *H. vulgaris* and *H. magnipapillata* were treated for 2 days with 4 mM LiCl and then shifted to 1 mM LiCl, but not to zero, they developed ectopic tentacles and hypostomes along the body axis. In *H. vulgaris* some foot patches were formed in addition (Hassel *et al.*, 1993; Hassel and Bieller, 1996).

The biochemical basis of the observed Li$^+$ effects is unclear. Li$^+$ ions are known to affect the activity of several enzymes, including enzymes of the PI cycle. In *Hydra* an influence on that pathway was found. The shift from 4 mM to 1 mM LiCl increases the level of inositol phosphates (Hassel and Bieller, 1996). Unfortunately, inositol can not be introduced by injection, and such a large hydrophilic substance can hardly be taken up from surrounding water. Therefore we do not know whether inositol is able to antagonize the influence of LiCl, and it remains unknown how Li$^+$ affects pattern formation. The general problem with the biochemical approach is that is it not known whether one has to look for several separate influences of Li$^+$ leading to head and foot formation or whether it is possible that there is only one influence which causes both head and foot formation. The hierarchical model favors the latter.

A straightforward explanation is that Li$^+$ treatment stimulates the release of A, either directly or by reducing the antagonizing influence of IAR on that release. The increased release of A all along the body axis increases the concentration of both IAR and IAP everywhere. The release of A is stimulated, but the production is antagonized; the positional value decreases, leading eventually to foot formation (in *H. vulgaris* but not in *H. magnipapillata*). Based on that proposition, a pulse treatment with Li$^+$ should cause an increase in the concentration of all morphogens and a subsequent decrease. Following treatment, IAR and IAP level out faster than A, due to the differences in diffusion constant and lifetime. That causes an increase of the positional value, as observed in all species tested. The release of A tends to become locally increased by autocatalysis. Hills and valleys of positional values emerge, leading to head structures and foot patches, respectively. The observation that *H. magnipapillata* does not form foot patches along the body axis following continuous treatment with Li$^+$ indicates a lower release of A during treatment, compared to *H. vulgaris*. This could have several explanations, including a lower stimulating influence of Li$^+$ on A release and a higher concentration of IAR in these animals during treatment.

Continuous treatment of *H. magnipapillata* or *H. vulgaris* with DAG for some days causes disintegration or no dramatic effect on patterning, depending on the concentration applied. The exception is budding. In *H. vulgaris*, continuous treatment with a low concentration of DAG or a single pulse for 3 hr caused young buds to eventually not form a foot at their base. A similar result was obtained with Li$^+$ ions (already discussed). This indicates an increased positional value at that site. *H. magnipapillata* did not respond (Pérez and Berking, 1994). There is a further indication that both treatments have similar effects in *Hydra vulgaris*. During treatment with LiCl or TPA (DAG was not tested), commitment of the multipotent I-cells to nerve cells was prevented, and after treatment commitment is enhanced (Hassel and Berking, 1988; Greger and Berking, 1991). Why buds of *H. magnipapillata* did not respond to DAG or TPA pulse treatments is yet unclear. However, even in *H. vulgaris* the TPA or DAG pulse treatment has moderate influence. Only some animals respond, but almost all respond when treated with DAG or TPA and simultaneously with an inhibitor of a non C–type kinase, such as staurosporin, genistein or H-7 (Pérez and Berking, 1994). Further, a simultaneous pulse treatment of *H. vulgaris* with catharidin and TPA, both applied in subthreshold concentrations, caused all animals to respond (Pérez, 1996). These results indicate that several kinases with different specificity are involved in pattern control in *H. vulgaris*. *H. magnipapillata* also did not respond to these treatments, but other agents had an influence; repeated pulse treatments with K-252a, which inhibits protein kinases including PKC, and xanthate D609, which is suspected to inhibit phospholipase C (phosphoinositidase), an enzyme that liberates DAG, caused buds to not separate from the parent (Müller, 1990). These substances have not yet been applied in a single pulse to *H. magnipapillata* or to *H. vulgaris*. The different response of *H. magnipapillata* and *H. vulgaris* are not understood. It may turn out that in *H. magnipapillata* other kinases have a strong influence on pattern control than in *H. vulgaris*. However, we do not expect the pattern-forming systems in these species to be very different; rather, the various treatments may be suspected to affect processes that are only linked to pattern control but that are not part of the main pathway. Thus, studies on the influence of pattern control in different species by such well known agents is one of the most promising approaches to understanding the biochemical basis of pattern formation. The specificity of action of the various inhibitors mentioned in *Hydra* is unknown. An analysis at the biochemical level is a prerequisite for the understanding of that control.

Daily pulse treatments of *H. magnipapillata* with high concentrations of DAG, combined with (heavy) feeding, caused after 1 wk of treatment the formation of tentacles and complete heads at the body (Müller, 1989). M. Kroiher (unpublished data) found that heavy feeding alone causes almost the same results as DAG treatment but with some additional delay (Fig. 7). Further, repeated pulse treatment with DAG not accompanied by feeding does not cause ectopic structures to form. *H. vulgaris* does not form ectopic head structures along the body axis, but

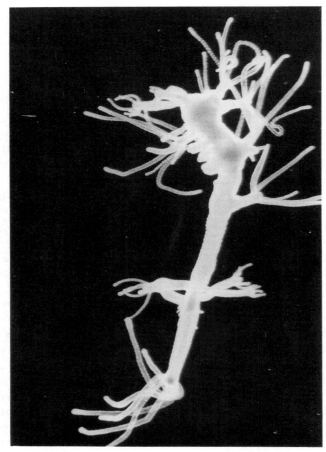

Fig. 7 *Hydra magnipapillata*, after 14 days of feeding with about 25 *Artemia* nauplii once per day, exhibits ectopic formation of tentacles and hypostomes (photograph courtesy of M. Kroiher).

more tentacles are formed in the head. In *H. vulgaris* it has been shown that feeding has some features in common with pulse treatments with TPA (DAG was not tested) or with LiCl (see previous discussion). Also, on feeding a short signal is generated which causes commitment of stem cells to nerve cells (Berking, 1979a).

The most interesting questions appear to be, why do we get ectopic head structures, why are repetitions necessary to have that result, and why do *H. vulgaris* and *H. magnipapillata* respond differently? An access to the answers to these questions appears to be the difference in morphology of the treated animals. With heavy feeding (combined or not combined with DAG treatment), *H. vulgaris* grows almost proportionally: the circumference of the body, the size of the hypostome, and the number of tentacles increase with increasing length. *H. magnipapillata* grows unproportionally: the body length increases, in particular the distance between the head and the budding region increases, but the circumference of the animal, the hypostome, and the number of tentacles only slightly increase (Bode and Bode, 1984; Müller, 1989; M. Kroiher, unpublished data). In *H. magnipapillata* the ectopic heads and tentacles are found in the lower half of the region, between the head and the budding region.

Based on the hierarchical model, pulse treatment with DAG and/or heavy feeding is proposed to cause release of A (and IAR and IAP) in pulses. The concentrations of the morphogens are higher at the apical than at the basal end. Due to differences in diffusion constant, the release and the concentration of A are not homogeneous along the body axis, in particular not after several repetitions. Thus, the positional value changes in patches. Where the concentrations of IAR and IAP are sufficiently low A, if released from such a patch following treatment, eventually becomes autocatalytically stimulated. Consistently, the increase of the positional value in the form of patches takes place in the lower half of the disproportionately enlarged area between the head and the budding region in *H. magnipapillata*. But it does not take place in the quite well proportioned *H. vulgaris*. Thus, the question remains why *H. magnipapillata* and *H. vulgaris* grow differently in response to these treatments. It appears possible that the different morphology is the result of a difference in the kinetics of the recovery from feeding (and pulse treatment with DAG). A fast recovery has little influence on body proportions. Long-lasting, high concentrations of IAR and IAP should prevent budding beyond the usual distance from the head, causing the lower gastric column to increase. Further, it should prevent the proportional growth of the head when the body length increases (see previous discussion).

6. Molecular Approaches to the Control of Pattern Formation

Genes of several families which are known to be important for pattern formation in different animals have been found in coelenterates. At present their significance in pattern formation is mostly deduced from their expression pattern in normal and in experimentally manipulated animals. HOM/HOX genes (CNOX) from various cnidarians have been described including *Eleutheria* (Schierwater et al., 1991), *Hydra* (Schummer et al., 1992; Shenk, Bode, et al., 1993; Shenk, Gee, et al., 1993), and *Hydractinia* (Schierwater et al., 1991). Based on the kinetics of their expression pattern in regeneration, the respective genes of *Hydra* were proposed to play a role in head or foot specification. In several species further homeobox gene homologues have been detected, including in *Hydra* homologues of *msh* (Schummer et al., 1992), *Nk-2*, *PAX*, and *prd* (Grens et al., 1996). In *Hydra* the existence of transcription factors has been determined, including one of the basic helix-loop-helix types (Grens et al., 1995) and from the *forkhead* family (Martinez et al., in preparation). Moreover complementary DNAs have been isolated of genes that are involved in mediating signal transduction, including *Hydra ras* (Bosch *et al.*, 1995), *Hydra creb* (Galliot *et al.*, 1995), PKC homologues (Hassel *et al.*, in preparation), and a whole set of different homologues of receptor tyrosine kinases (Chan *et al.*, 1994; Steele *et al.*, 1996). In *Hydra* genes have been reported to be expressed locally to which homologues are not yet known, including *KS1* (Weinziger *et al.*, 1994) and *Hyp1* (Hermans-Borgmeyer *et al.*, 1996), which are both almost restricted to the head of the animal. The message of both genes becomes turned on shortly after decapitation in regenerating gastric tissue. In *Podocoryne*, a gland cell gene is preferentially expressed during regeneration and transdifferentiation (Baader *et al.*, 1995). By means of monoclonal antibodies obtained against fractionated *Hydra* cells, position-dependent patterns of different antigens were detected. Examples are the tentacle-specific antigen TS19 (Bode *et al.*, 1988) and the hypostome-specific antigen L96 (Technau and Holstein, 1995). A current description of genes and antigens of developmental interest known in *Hydra* (and related species) has been published by Galliot (1997).

In general, the expression patterns are more or less uniformly distributed along the body axis or decrease along the axis in a graded manner. Some are concentrated to the head or to the foot or to the developing bud. Sharp borders of expression are found between tentacle base and gastric region and between basal disc and gastric region, but the epithelial cells do also look different at both sides of the border. Taking all together, the approach to understanding pattern formation by means of molecular biologic methods is a fast-growing field. At present there are correlations between expression patterns and developmental processes. In the future some genes may become valuable tools for a detailed analysis of pattern formation. However, it should be kept in mind that in regeneration gastric tissue of *Hydra* is able to form either a head, a different area of the gastric region, or a foot, depending on the position of the cut. Therefore, the fate of that tissue is determined by interaction between the cells. The type of interaction controls differential gene expression. Thus differential gene expression is the second step in pattern formation. Of particular importance are the so-called early-response genes. They are expected to allow a view upstream and downstream in the cascade of events in pattern control.

C. Pattern Formation in Colonies

Colonies consist of polyps (hydranths) connected by channels of the common gastrovascular cavity. The simplest colonies are the so-called stolon colonies of athecate hydrozoa. In these colonies the polyps look similar to a *Hydra* and they are connected to each other by hollow tubes at their base, the stolons. *Hydractinia* forms such a colony. Late in the development, the colony of *Hydractinia* develops into a further type of colony often found in cnidarians, a mat covered with polyps. Generally, thecate hydrozoa do not form simple polyps on top of their stolons but rather form stems (hydrocaulis) with several hydranths. In athecate hydrozoa the stolons and the mat are covered by a chitinous periderm. In thecate hydrozoa the stem and its branches are covered by a periderm as well. The hydranths can retract into a (hydro)theca, that is, a tube at the end of the branches where the hydranths are formed.

1. Stolon Formation

A stolon elongates exclusively at the tip, where the periderm is thin. In the tip the periderm is produced and just behind that site it hardens. Though the stolon elongates exclusively at its distal end, at that site neither mitotic cells nor cells in S phase were found (Hale, 1964; Wyttenbach, 1965; Braverman, 1971; Plickert et al., 1988). Two days after pulse labeling, tritium thymidine–labeled cells were found in both tissue layers in the stolon tip of *Podocoryne*. It appears that cells from the proliferating region of the stolon migrate into the tip (Braverman, 1974). The tip is a multicellular locomotory organ (Beloussov et al., 1989). Following separation of a tip from the stolon net, it moves slowly over the substrate. Movement occurs by taking up water and ions into the epithelial cells and expelling both rhythmically. In this process the epithelial cells change their form and their position relative to the mesogloea and to the periderm (Beloussov et al., 1989). By this activity the tip stretches the epithelial sheets. In this way, proliferating epithelial cells downstream of the tip can remain close to each other after cytokinesis. Thus, a stolon elongates exclusively at the tip, but the necessary increase in cell number occurs by proliferation along the whole stolon excluding the tip.

Stolons can form lateral branches (see Fig. 2). Where a stolon bud forms, the periderm must be opened. However, the stolon bud visibly forms within the periderm tube. The enzymatic opening of the periderm is a late process. A stolon tip is able to induce the formation of a further stolon tip when approaching another stolon (Müller et al., 1987). When tips of the same colony meet, they fuse and therewith extinguish their activity as tips. This process is observed generally in athecate hydrozoa, where the periderm is thin, but is rarely observed in thecate hydrozoa, where the periderm is thick. When a tip meets a stolon of a different species or of the same species but of a different tissue type, it often induces a tip but does not fuse with it; rather, both prepare to destroy each other (Hauenschild, 1954; Müller, 1964; Buss et al., 1984; Lange et al., 1989). In *Hydractinia* special nematocytes are produced which move to the site of contact and, on firing, dissolve the tissue of the enemy. Lange and Müller (1991) isolated a glycoconjugate, termed "stolon-inducing factor" (SIF), which is able to induce stolon tips when applied locally. In the presence of SIF the primary polyps of *Hydractinia* produce fewer tentacles, but also the length of stolons is reduced. Most tissue forms a basal globe. Stolons may emerge from that globe or from other parts of the gastric column. At high concentrations the primary polyp appears to transform completely into stolon tissue covered with periderm.

In *Eirene*, stolon branching was induced by pressing the stolon for some minutes with a thin thread (Plickert, 1980). Stolon tissue of every position is able to respond in that way except near to (within about 400 μm of) an existing tip. When two such pressings are made at a certain distance from each other, two stolon branches develop. With decreasing distance only one of them is successful, and finally only one stolon tip is formed in the middle of both pressings. These observations are explained by the proposition that pressing stimulates a pattern-forming system including local activation and lateral inhibition. Computer simulations support that view (Meinhardt, 1982). Lateral inhibition is also observed in regenerating stolons. In small isolates in which the cut surfaces of a section are close to each other, only one tip forms; otherwise, the stolon section produces tips at both ends (Müller and Plickert, 1982).

On the basis of the hierarchical mode it is proposed that in the stolon tip A (together with IAR and IAP) is released. The release of A is autocatalytically stimulated. The presence of IAP causes a decrease of the positional value. Therewith, the tip is the tissue with the lowest positional value in the whole colony. That value causes the generation of secondary signals that control the differentiation and the property of the tip as a locomotory organ. Stolon tissue is proposed to have an almost constant low positional value mainly controlled by basic release and basic production of A (see later discussion). The proposition that A is produced in the tip may help to explain why in *Eirene* the tip of a stolon occasionally transforms into a polyp when it has no contact with the substrate. The concentration of IAP in the very tip may be reduced when the stolon tip is surrounded by seawater. That would allow the released A to cause the production of A at that very site.

2. Polyp (Hydranth) Formation

Stolons produce polyp buds that grow like buds of *Hydra* but do not form a foot at their base and do not separate (see Fig. 2). As with budding in *Hydra*, initially tissue of the stolon is recruited to form the polyp bud. When the new polyp (hydranth) has reached a certain size, it does not elongate any further, but proliferation of epithelial cells does not cease. From that time onward the tissue at the hydranth's base is exported into the stolon (Müller, 1964). On the basis of the hierarchical model and as proposed for *Hydra*, it is suggested that the driving force of this movement is the decrease of the positional value caused by the ratio of A to IAP in the respective tissue. When a certain basal value is attained, the polyp tissue transforms into stolon tissue. The current concept is that *Hydra* is under the control of two organizing regions, the head and the foot, located at opposite ends of a field (Meinhardt, 1993). In colonial hydrozoa this concept may lead to problems in explaining hydranth formation from stolons and hydranth branching in thecate hydrozoa, because the organizing influence of the foot homologue has to be maintained in the tissue of the gastric region of the hydranth when the foot homologue is very close and far away, respectively.

In some colony-forming species, the distance between polyps is regular; in others it is quite irregular but a minimal distance is maintained, indicating lateral inhibition of polyps on polyp formation. Transplantation and regeneration experiments have confirmed that proposition. Young polyps were found to have a stronger inhibitory influence than old ones, allowing intercalation of new polyps between old existing ones (Braverman, 1971; Plickert *et al.*, 1987). Factors affecting spacing have been postulated to exist and have been partially purified by several authors (for a review, see Plickert, 1990), but their chemical nature is largely unknown. Exceptions are the noted endogenous methyldonors found to be involved in control of metamorphosis and proportioning of polyps. These substances were found also to affect spacing of polyps. In *Eirene*, 0.1 μM of homarine was found to increase the distance between a polyp and the next polyp bud, while 0.1 μM of sinefungin, an antagonist of transmethylation, reduced that distance (Berking, 1986b). Polyps, in particular in their heads, contain the methyldonors in a stored form (Berking, 1987). Therefore it was argued that polyps release methyldonors from the stores into stolon tissue. At a position where the concentration is sufficiently low, budding can start. Thus, methyldonors have properties of the inhibitor of activator release, IAR, in the model proposed. Consistently, the treatments with homarine and sinefungin did not affect polyp bud formation only, but also affected stolon bud formation in the same way. A problem with this concept is that shrimps, which are used for feeding, contain homarine in a very high concentration. Therefore the substance is expected to be present on feeding in stolon tissue, at least in transiently high concentration.

Partially metamorphosed animals that bear stolons and a larval posterior are unable to feed. But the yolk is sufficient for stolons to grow out and to produce secondary polyps on top of these stolons. With the exception of the structure in the center, this small colony looks identical to one with a polyp in the center. It appears that a larval posterior generates the same or a very similar inhibitory signal as that generated by a polyp (Berking and Walther, 1994). Larvae also contain methyldonors in large amounts in a stored form (Berking, 1987).

In colonies with regularly spaced polyps, a polyp bud is formed on a growing stolon just outside the inhibitory influence of an existing polyp and outside of that of a stolon tip. In the thecate hydrozoa *Dynamena pumila*, both the spacing between polyps (hydrocaulus with several hydranths) and that between stolons is quite regular. On top of a stolon, close to its tip and some distance away from an existing polyp on that stolon, a bud of a polyp starts to grow. This bud is surrounded by several stolon tips, yet it is within the periderm tube of the stolon. However, these tips do not grow out. When the polyp bud grows in length, one stolon tip at each side of the original stolon may start to grow out at almost right angles while the others are reintegrated into stolon tissue. It appears that the same pattern-forming system that causes a local increase of the positional value at the roof of the stolon causes a decrease laterally, leading to stolon tips. A similar mechanism was proposed for bud formation in *Hydra* (see the section on budding).

In several athecate species, such as *Hydractinia*, the polyp spacing pattern is not regular (although a minimal distance is kept) and there is no obvious correlation between the points where a stolon tip is formed and where a polyp bud is formed. An autocatalytically stimulated release of A may start due to some statistical heterogeneity in stolon tissue. The rule that polyps are formed from the stolon roof and stolon tips are formed laterally may reflect differences in the architecture of stolon tissue in the circumference. These differences also include differences in the ability to release morphogens into the surrounding water. In particular, the exchange of agents at the basal surface is strongly hindered by the firm contact to the substrate. In *Hydractinia* a polyp bud does not cause the formation of a stolon tip close to it, as was observed in *D. pumila*. A straightforward explanation is that at the periphery of the area of stimulated A release (the presumptive polyp bud), the ratio of A to IAP is not low enough to antagonize A production (see the section on budding).

The colony of *Hydractinia* starts with a polyp that can feed (gastrozooid), as described (see Fig. 2). When the net of stolons becomes dense a mat forms, and on top of this mat a new type of polyp develops, a so-called gonozooid, which produces the gonads. These gonads are argued to be reduced medusae (Kühn, 1914). For unknown reasons such gonozooids do not form on stolons. *Hydractinia* produces two other types of polyps. In one type, the body is shaped like a spiral while the head strongly resembles that of a gonozooid; the other has a normally shaped body, but the apical end is not a head but rather a long single tentacle. They are found only at the orifice of the shell when a hermit crab inhabits the shell. It has been argued that they are malformations caused by the stress performed by the crab (Hauenschild and Kanellis, 1953; Müller, 1961). The athecate *Thecocodium* produces two types of polyps: one is big and can feed but is unable to catch the prey; the other is very small and can catch the prey but has no mouth opening to take up the food. The small ones are large in number and are scattered between the large ones. It is obvious that there must be an interdependence in the formation of the two types to guarantee optimal feeding. That indeed was found. In isolates, a polyp can not survive without one of the other type, not even for a short time, whereas (unfed) isolates with mixed polyps survive. Furthermore, one type of polyp stimulates the formation of the other and inhibits the formation of an additional polyp of its own type in such isolates. It appears that lateral help (Meinhardt, 1982) between differently structured polyps in combination with homotypic and heterotypic inhibition is involved in control of polyp polymorphisms and patterning of these polyps in colonies (Pfeifer and Berking, 1995). This example may show that the pattern of a colony cannot be understood only on the basis of a primary pattern-forming system as presented here; a successful model should at least explain lateral help as well.

V. Concluding Remarks on Pattern Formation

Various approaches are used to study pattern formation in hydrozoa, including regeneration and transplantation techniques, methods of molecular biology, and biochemical methods. Models play the role of a link between the different fields and the different species. They allow researchers to handle the results obtained and to design critical experiments. A model is never able to explain all data. The aim can only be to show whether or not it is basically correct. Basically correct does not mean that the morphogens predicted in a model do exist in reality; for instance, activation can be obtained by inhibition of an inhibition. But it is much easier in discussing experiments to speak of an activator than to speak of all possible equivalents. A basic principle is the hierarchical feature, which predicts that in the primary system neither head and foot activators nor head and foot inhibitors exist. This conflicts with the propositions of countercurrent gradients of qualitatively different morphogens that directly organize the formation of terminal structures and that are proposed to organize pattern formation in the whole body. How pattern formation is controlled can be discovered by a combination of at least two approaches. One is to further try to explain critical experiments on the basis of the various models. Computer simulations may help greatly in this way. The most important approach, however, is to look at the biochemical and molecular level for the compounds and interactions involved in pattern control. On the other hand, looking at the biochemical level without having a model about what one can expect may be fruitless for a long time, because there are too many possible substances that affect pattern formation. Therefore, the aim of the model presented here is to be a tool in the search for the mechanisms of pattern formation in hydrozoa.

Acknowledgments

I thank M. Hassel, K. Herrmann, M. Kroiher, H. Meinhardt, and the members of the local hydrozoa research group for critical comments; J. Jacobi for help with the figures; and B. Schreiner for help with the text.

References

Baader, C. D., Heiermann, R., Schuchert, P., Schmid, V., and Plickert, G. (1995). Temporally and spatially restricted expression of a gland cell gene during regeneration and in vitro trans-differentiation in the hydrozoan *Podocoryne carnea. Roux's Arch. Dev. Biol.* **204,** 164–171.

Beloussov, L. V., Labas, J. A., Kazakova, N. I., and Zaraisky, A. G. (1989). Cytophysiology of growth pulsations in hydroid polyps. *J. Exp. Zool.* **249,** 258–270.

Berking, S. (1977). Bud formation in *Hydra*: Inhibition by an endogenous morphogen. *Roux's Arch. Dev. Biol.* **181,** 215–225.

Berking, S. (1979a). Control of nerve cell formation from multipotent stem cells in *Hydra. J. Cell Sci.* **40,** 193–205.

Berking, S. (1979b). Analysis of head and foot formation in *Hydra* by means of an endogenous inhibitor. *Roux's Arch. Dev. Biol.* **186,** 189–210.

Berking, S. (1980). Commitment of stem cells to nerve cells and migration of nerve cell precursors in preparatory bud development in *Hydra. J. Embryol. Exp. Morphol.* **60,** 373–387.

Berking, S. (1983). The fractionation of *Hydra*-derived inhibitor into head and foot inhibitors may be an artefact. *Roux's Arch. Dev. Biol.* **192,** 327–332.

Berking, S. (1984). Metamorphosis of *Hydractinia echinata*: Insights into pattern formation in hydroids. *Roux's Arch. Dev. Biol.* **193,** 370–378.

Berking, S. (1986a). Is homarine a morphogen in the marine hydroid *Hydractinia*? *Roux's Arch. Dev. Biol.* **195,** 33–38.

Berking, S. (1986b). Transmethylation and control of pattern formation in *Hydrozoa. Differentiation* **32,** 10–16.

Berking, S. (1987). Homarine (*N*-methylpicolinic acid) and trigonelline (*N*-methylnicotinic acid) appear to be involved in pattern control in a marine hydroid. *Development* **99,** 211–220.

Berking, S. (1988a). Ammonia, tetraetylammonium, barium and amiloride induce metamorphosis in the marine hydroid *Hydractinia. Roux's Arch. Dev. Biol.* **197,** 1–9.

Berking, S. (1988b). Taurine found to stabilize the larval state is released upon induction of metamorphosis in the hydrozoan *Hydractinia. Roux's Arch. Dev. Biol.* **197,** 321–327.

Berking, S. (1991a). Effects of the anticonvulsant drug valproic acid and related substances on developmental processes in hydroids. *Toxic. in Vitro* **5,** 109–118.

Berking, S. (1991b). Control of metamorphosis and pattern formation in *Hydractinia* (Hydrozoa, Cnidaria). *Bioessays* **13,** 323–329.

Berking, S., and Gierer, A. (1977). Analysis of early stages of budding in *Hydra* by means of an endogenous inhibitor. *Roux's Arch. Dev. Biol.* **182,** 117–129.

Berking, S., and Herrmann, K. (1990). Dicapryloylglycerol and ammonium ions induce metamorphosis of ascidian larvae. *Roux's Arch. Dev. Biol.* **198,** 430–432.

Berking, S., and Schindler, D. (1983). Specification of the head-body proportion in *Hydra attenuata* regenerating the head. *Roux's Arch. Dev. Biol.* **192,** 333–336.

Berking, S., and Schüle, T. (1987). Ammonia induces metamorphosis of the oral half of buds into polyp heads in the scyphozoan *Cassiopea. Roux's Arch. Dev. Biol.* **196,** 388–390.

Berking, S., and Walther, M. (1994). Control of metamorphosis in the hydroid *Hydractinia. In* "Perspectives in Comparative Endocrinology" (K. G. Davey, R. E. Peter, and S. S. Tobe, Eds.), pp. 381–388. National Research Council of Canada, Ottawa.

Berridge, M. J., Downes, C. P., and Hanley, M. R. (1989). Neural and developmental action of lithium: A unifying hypothesis. *Cell* **59,** 411–419.

Bode, H. R. (1983). Reducing populations of interstitial cells and nematoblasts with *hydroxyurea. In* "*Hydra* Research Methods" (H. M. Lenhoff, Ed.), pp. 291–294. Plenum Press, New York.

Bode, H. R., Berking, S., David, C. N., Gierer, A., Schaller, H., and Trenkner, E. (1973). Quantitative analysis of cell types during growth and morphogenesis in *Hydra. Roux's Arch. Dev. Biol.* **171,** 269–285.

Bode, H. R., Heimfeld, S., Chowand, M. A., and Huang, L. (1987). Gland cells arise by differentiation from interstitial cells in *Hydra attenuata. Dev. Biol.* **122,** 577–585.

Bode, P. M., Awad, T. A., Koizumi, O., Nakashima, Y., Grimmelikhuijzen, C. J. P., and Bode, H. R. (1988). Development of the two-part pattern during regeneration of the head in *Hydra. Development* **102,** 223–236.

Bode, P. M., and Bode, H. R. (1980). Formation of pattern in regenerating tissue pieces of *Hydra attenuata.* I. Head-body proportion regulation *Dev. Biol.* **78,** 484–496.

Bode, P. M., and Bode, H. R. (1984). Formation of pattern in regenerating tissue pieces of *Hydra attenuata.* II. Degree of proportion regulation is less in the hypostome and tentacle zone than in the tentacles and basal disc. *Dev. Biol.* **103,** 304–312.

Bosch, T. C. G., Benitez, E., Gellner, K., Praetzel, G., and Salgado, L. M. (1995). Cloning of a *ras*-related gene from *Hydra* which responds to head-specific signals. *Gene* **167**, 191–195.

Braverman, M. (1971). Studies on hydroid differentiation. VII. The hydrozoan stolon. *J. Morphol.* **135**, 131–152.

Braverman, M. (1974). The cellular basis of colony form in *Podocoryne carnea*. *Am. Zool.* **14**, 673–698.

Brewer, R. H. (1978). Larval settlement behaviour in the jellyfish *Aurelia aurita* (Linnaeus) (Scyphozoa: Semaeostomae). *Estuaries* **1**, 120–122.

Browne, E. N. (1909). The production of ney hydranths in *Hydra* by insertion of small grafts. *J. Exp. Zool.* **7**, 1–23.

Bruland, K. W. (1983). Trace elements in seawater. *In* "Chemical Oceanography" (J. P. Riley and G. Skirrow, Eds.), pp. 157–220. Academic Press, New York.

Buss, L. W., McFadden, C. S., and Keene, D. R. (1984). Biology of hydractiniid hydroids. 2. Histocompatibility effector system/competitive mechanism mediated by nematocyst discharge. *Biol. Bull. Mar. Biol. Lab. Woods Hole* **167**, 139–158.

Campbell, R. D. (1967a). Tissue dynamics of steady-state growth in *Hydra littorlis*. I. Patterns of tissue movements. *Dev. Biol.* **15**, 487–502.

Campbell, R. D. (1967b). Tissue dynamics of steady-state growth in *Hydra littorlis*. II. Patterns of tissue movements. *J. Morphol.* **121**, 19–28.

Campbell, R. D. (1976). Elimination of *Hydra* interstitial and nerve cells by means of colchicine. *J. Cell Sci.* **21**, 1–13.

Campbell, R. D., and David, C. N. (1974). Cell cycle kinetics and development of *Hydra attenuata*. II. Interstitial cells. *J. Cell Sci.* **16**, 349–358.

Chan, T. A., Chu, C. A., Rauen, K. A., Kroiher, K., Tatarewicz, S., and Steele, R. E. (1994). Identification of a gene encoding a novel protein-tyrosine kinase containing SH2 domains and ancyrin-like repeats. *Oncogene* **9**, 1253–1259.

Chicu, S. A., and Berking, S. (1997). Interference with metamorphosis induction in the marine cnidaria *Hydractinia echinata* (Hydrozoa): A structure-activity relationship analysis of lower alcohols, aliphatic and aromatic hydrocarbons, thiophenes, tributyltin and crude oil. *Chemosphere* **34**, 1851–1866.

Coon, S. L., Walch, M., Fitt, W. K., Weiner, R. M., and Bonar, D. B. (1990). Ammonia induces settlement behavior in oyster larvae. *Biol. Bull.* **179**, 297–303.

David, C. N., and Gierer, A. (1974). Cell cycle kinetics and development of *Hydra attenuata*. III. Nerve and nematocyte differentiation. *J. Cell Sci.* **16**, 339–375.

David, C. N., and Murphy, S. (1977). Characterization of interstitial stem cells in *Hydra* by cloning. *Dev. Biol.* **57**, 372–385.

Driesch, H. (1893). Entwicklungsmechanische Studien X. Über einige allgemeine entwicklungsmechanische Ergebnisse. *Mitt. Zool. Stat. Neapel.* **2**, 221–253.

Edwards, C. N., Thomas, M. B., Long, B. A., and Amyotte, S. J. (1987). Catecholamine induce metamorphosis in the hydrozoan *Halochordyle disticha* but not in *Hydractinia echinata*. *Roux's Arch. Dev. Biol.* **196**, 381–384.

Fitt, W. K., and Coon, S. L. (1992). Evidence for ammonia as a natural cue for recruitment of oyster larvae to oyster beds in a Georgia salt marsh. *Biol. Bull.* **182**, 401–408.

Freeman, G. (1981a). The cleavage initiation site establishes the posterior pole of the hydrozoan embryo. *Roux's Arch. Dev. Biol.* **190**, 123–135.

Freeman, G. (1981b). The role or polarity in the development of the hydrozoan planula larva. *Roux's Arch. Dev. Biol.* **190**, 168–184.

Freeman, G. (1993). Metamorphosis in the brachiopod *Terebratalia*: Evidence for a role of calcium channel function and the dissociation of shell formation from settlement. *Biol. Bull.* **184**, 15–24.

126   Stefan Berking

Gajewski, M., Leitz, T., Schloßherr, J., and Plickert, G. (1996). LWamides from Cnidaria constitute a novel family of neuropeptides with morphogenetic activity. *Roux's Arch. Dev. Biol.* **205**,
232–242.
Galliot, B. (1997). Signalling molecules in regenerating *Hydra*. *Bioessays* **19**, 37–46.
Galliot, B., Welschof, M., Schuckert, O., Hoffmeister, S. H., and Schaller, C. (1995). The cAMP
response element binding protein is involved in *Hydra* regeneration. *Development* **121**, 1205–
1216.
Gierer, A. (1977). Biological features and physical concepts of pattern formation exemplified by
Hydra. *Curr. Top. Dev. Biol.* **11**, 17–59.
Gierer, A., Berking, S., Bode, H., David, C. N., Flick, K., Hansmann, G., Schaller, H., and
Trenkner, E. (1972). Regeneration of *Hydra* from reaggregated cells. *Nat. New Biol.* **239**, 98–
101.
Gierer, A., and Meinhardt, H. (1972). A theory of biological pattern formation. *Kybernetik* **12**,
30–39.
Gilmour, T. H. J. (1991). Induction of metamorphosis of echinoid larvae. *Am. Zool.* **31**, 105A.
Graf, L., and Gierer, A. (1980). Size, shape, and orientation of cells in budding *Hydra* and regula-
tion of regeneration in cell aggregates. *Roux's Arch. Dev. Biol.* **188**, 141–151.
Greger, V., and Berking, S. (1991). Nerve cell production in *Hydra* is deregulated by tumor-pro-
moting phorbol ester. *Roux's Arch. Dev. Biol.* **200**, 234–236.
Grens, A., Gee, L., Fisher, D. A., and Bode, H. R. (1996). *CnNK-2*, a NK-2 homeobox gene, has
a role in patterning the basal end of the axis in *Hydra*. *Dev. Biol.* **180**, 473–488.
Grens, A., Mason, E., Marsh, J. L., and Bode, H. R. (1995). Evolutionary conservation of a cell
fate specification gene: the *Hydra achaete-scute* homolog has proneural activity in *Drosophila*.
Development **121**, 4027–4035.
Grimmelikhuijzen, C. J. P., and Wesfall, J. A. (1995). The nervous system of cnidarians. *In* "The
Nervous System of Invertebrates: An Evolutionary and Comparative Approach" (O. Breiden-
bach and W. Kutsch., Eds.), pp. 7–24. Birkhäuser, Basel.
Hale, L. J. (1964). Cell movement, cell division and growth in the hydroid *Clytia johnstoni*.
J. Embryol. Exp. Morphol. **12**, 517–538.
Ham, R. G., Fitzgerald, D. C., Jr., and Eakin, R. E. (1956). Effects of lithium ion on regeneration
of *Hydra* in a chemically defined environment. *J. Exp. Zool.* **133**, 559–572.
Hassel, M., Albert, K., and Hofheinz, S. (1993). Pattern formation in *Hydra vulgaris* is controlled
by lithium sensitive processes. *Dev. Biol.* **156**, 362–371.
Hassel, M., and Berking S. (1988). Nerve cell and nematocyte production in *Hydra* is deregulated
by lithium ions. *Roux's Arch. Dev. Biol.* **197**, 471–475.
Hassel, M., and Berking S. (1990). Lithium ions interfere with pattern control in *Hydra vulgaris*.
Roux's Arch. Dev. Biol. **198**, 382–388.
Hassel, M., and Bieller, A. (1996). Stepwise transfer from high to low lithium concentrations in-
crease the head forming potential in *Hydra vulgaris* and possibly activates the PI-cycle. *Dev.
Biol.* **177**, 439–448.
Hauenschild, C. (1954). Genetische und entwicklungsphysiologische Untersuchungen über Inter-
sexualität und Gewebeverträglichkeit bei *Hydractinia echinata* Flemm. (Hydroz. Bougainvill.).
Roux's Arch. Dev. Biol, **147**, 1–41.
Hauenschild, C., and Kanellis, A. (1953). Experimentelle Untersuchungen an Kulturen von *Hy-
dractinia echinata*, Flemm. zur Frage der Sexualität und Stockdifferenzierung. *Zool. Jb. Abt.
allg. Zool. u. Physiol.* **64**, 1–13.
Hermans-Borgmeyer, I., Schinke, B., Schaller, H. C., and Hoffmeister-Ullerich, S. A. H. (1996).
Isolation of a marker for head-specific cell differentiation in *Hydra*. *Differentiation* **61**, 95–
101.
Herrmann, K., and Berking, S. (1987). The length of S-phase and G_2-phase of epithelial cells is
regulated during growth and morphogenesis of *Hydra attenuata*. *Development* **99**, 33–39.

Hicklin, J., and Wolpert, L. (1973). Positional information and pattern regulation in *Hydra*: Formation of the foot end. *J. Embryol. Exp. Morphol.* **30**, 727–740.

Hobmayer, E., Holstein, T. W., and David, C. N. (1990a). Tentacle morphogenesis in *Hydra*. 1. The role of head activator. *Development* **109**, 887–895.

Hobmayer, E., Holstein, T. W., and David C. N. (1990b). Tentacle morphogenesis in *Hydra*. 2. Formation of a complex between a sensory nerve cell and a battery cell. *Development* **109**, 897–904.

Hoffmeister, S. A. H. (1996). Isolation and characterization of two new morphogenetically active peptides from *Hydra vulgaris*. *Development* **122**, 1941–1948.

Javois, L. C., and Frazier-Edwards, A. (1991). Simultaneous effects of head activator on the dynamics of apical and basal regeneration in *Hydra vulgaris* (formerly *Hydra attenuata*). *Dev. Biol.* **144**, 78–85.

Kroiher, M., and Plickert, G. (1992). Analysis of pattern formation during embryonic development of *Hydractinia echinata*. *Roux's Arch. Dev. Biol.* **201**, 95–104.

Kroiher, M., Plickert, G., and Müller, W. A. (1990). Pattern of cell proliferation in embryogenesis and planula development of *Hydractinia echinata* predicts the postmetamorphic body pattern. *Roux's Arch. Dev. Biol.* **199**, 156–163.

Kroiher, M., Walther, M., and Berking, S. (1991). Necessity of protein synthesis for metamorphosis in the marine hydroid *Hydractinia echinata*. *Roux's Arch. Dev. Biol.* **200**, 336–341.

Kroiher, M., Walther, M., and Berking, S. (1992). Heat shock as inducer of metamorphosis in marine invertebrates. *Roux's Arch. Dev. Biol.* **201**, 169–172.

Kühn, A. (1914). Entwicklungsgeschichte und Verwandschaftsbeziehungen der Hydrozoen. 1. Teil: Die Hydroiden. *Erg. Fortschr. Zool.* **4**, 1–284.

Lange, R. G., and Müller, W. A. (1991). SIF, a novel morphogenetic inducer in *Hydrozoa*. *Dev. Biol.* **147**, 121–132.

Lange, R. G., Plickert, G., and Müller, W. A. (1989). Histoincompatibility in a low invertebrate, *Hydractinia echinata*: Analysis of the mechanism of rejection. *J. Exp. Zool.* **249**, 284–292.

Latorre, R., Coronado, R., and Vergara, C. (1984). K$^+$-channels gated by voltage and ions. *Annu. Rev. Physiol.* **46**, 485–495.

Leitz, T., and Klingermann, G. (1990). Metamorphosis in *Hydractinia*: Studies with activators and inhibitors aiming at protein kinase C and potassium channels. *Roux's Arch. Dev. Biol.* **199**, 107–113.

Leitz, T., and Lay, M. (1995). Metamorphosin A is a neuropeptide. *Roux's Arch. Dev. Biol.* **204**, 276–297.

Leitz, T., Morand, K., and Mann, M. (1994). Metamorphosin A: A novel peptide controlling development of the lower metazoan *Hydractinia echinata* (Coelenterata, hydrozoa). *Dev. Biol.* **163**, 440–446.

Leitz, T., and Müller, W. A. (1987). Evidence for the involvement of PI-signalling and diacylglycerol second messengers in the initiation of metamorphosis in the hydroid *Hydractinia echinata* Fleming. *Dev. Biol.* **121**, 82–89.

Leitz, T., and Wagner, T. (1993). The marine bacterium *Alteromonas espejiana* induces metamorphosis of the hydroid *Hydractinia echinata*. *Mar. Biol.* **115**, 173–178.

Leitz, T., and Wirth, A. (1991). Vanadate, known to interfere with signal transduction, induces metamorphosis in *Hydractinia* (Coelenterata; Hydrozoa) and causes profound alterations of the larval and postmetamorphic body pattern. *Differentiation* **47**, 119–127.

Leviev, I., and Grimmelikhuijzen, C. J. P. (1995). Molecular cloning of a preprohormone from sea anemones containing numerous copies of a metamorphosis-inducing neuropeptide: A likely role for dipeptidyl aminopeptidase in neuropeptide precursor processing. *Proc. Natl. Acad. Sci. U.S.A.* **92**, 11647–11651.

Li, Y., and Casida, J. E. (1992). Cantharidin-binding protein: Identification as protein phosphatase 2A. *Proc. Natl. Acad. Sci. U.S.A.* **89**, 11867–11870.

MacWilliams, H. K., and Kafatos, F. C. (1974). The basal inhibition in *Hydra* may be mediated by a diffusing substance. *Am. Zool.* **14**, 633–645.

MacWilliams, H. K., Kafatos, F. C., and Bossert, W. H. (1970). The feedback inhibition of basal disk regeneration in *Hydra* has a continuously variable intensity. *Dev. Biol.* **23**, 380–398.

Marcum, B. A., Fujisawa, T., and Sugiyama, T. (1980). A mutant *Hydra* strain (SF-1) containing temperature-sensitive interstitial cells. *In* "Developmental and Cellular Biology of Coelenterates" (P. Tardent and R. Tardent, Eds.), pp. 429–434. Elseview North-Holland Biochemical Press, Amsterdam.

McCauley, D. W. (1996). Serotonin induces metamorphosis in planulae of *Phialidium gregarium* (Cnidaria: Hydrozoa). *In* "Settlement and Metamorphosis of Marine Invertebrate Larvae." Programme and Abstracts of an International Symposium (A. S. Clare, N. Fusetani, H. Hirota, B. Jones, Organizing Committee). University of Plymouth, Plymouth.

Meinhardt, H. (1982). "Models of Biological Pattern Formation." Academic Press, London.

Meinhardt, H. (1993). A model of biological pattern formation of hypostome, tentacles and foot in *Hydra*: How to form structures close to each other, how to form them at a distance. *Dev. Biol.* **157**, 321–333.

Meinhardt, H., and Gierer, A., (1974). Application of a theory of biological pattern formation based on lateral inhibition. *J. Cell Sci.* **15** 321–346.

Morgan, T. H. (1902). Further experiments on the regeneration of *Tubularia*. *Roux's Arch. Dev. Biol.* **13**, 528–544.

Morgan, T. H. (1904). Regeneration of heteromorphic tails in posterior pieces of *Planaria simplicissima*. *J. Exp. Zool.* **1**, 385–393.

Morgan, T. H., and Moszkowski, M. (1907). "Regeneration." Verlag Wilhelm Engelmann, Leipzig.

Müller, W. A. (1961). Untersuchungen zur Stockdifferenzierung von *Hydractinia echinata*. *Zool. Jb. Abt. allg. Zool. u. Physiol.* **69**, 317–324.

Müller, W. A. (1964). Experimentelle Untersuchungen über Stockentwicklung, Polypendifferenzierung und Sexualchimären bei *Hydractinia echinata*. *Roux's Arch. Dev. Biol.* **155**, 181–268.

Müller, W. A. (1969). Die Steuerung des morphogenetischen Fließgleichgewichts in den Polypen von *Hydractinia echinata*. 1. Biologisch-experimentelle Untersuchungen. *Roux's Arch. Dev. Biol.* **163**, 334–356.

Müller, W. A. (1973). Metamorphose-Induktion bei Planulalarven: I. Der bakterielle Induktor. *Roux's Arch. Dev. Biol.* **173**, 107–121.

Müller, W. A. (1982). Intercalation and pattern regulation in hydroids. *Differentiation* **22**, 141–150.

Müller, W. A. (1985). Tumor-promoting phorbol esters induce metamorphosis and multiple head formation in the hydroid *Hydractinia*. *Differentiation* **29**, 216–222.

Müller, W. A. (1989). Diacylglycerol induces multiple head formation in *Hydra*. *Development* **105**, 306–316.

Müller, W. A. (1990). Ectopic head and foot formation in *Hydra*: Diacylglycerol-induced increase in positional value and assistance of the head in foot formation. *Differentiation* **42**, 131–143.

Müller, W. A. (1995). Competition for factors and cellular resources as a principle of pattern formation in *Hydra*. II. Assistance of foot formation by heads and buds and a new model of pattern control. *Dev. Biol.* **167**, 175–189.

Müller, W. A. (1996). Competition-based head versus foot decision in chimeric hydras. *Int. J. Dev. Biol.* **40**, 1133–1139.

Müller, W. A., and Buchal, G. (1973). Metamorphose-Induktion bei Planulalarven: II. Induktion durch monovalente Kationen: Die Bedeutung des Gibbs-Donnan-Verhältnisses und der Na^+/K^+-ATPase. *Roux's Arch. Dev. Biol.* **173**, 122–135.

Müller, W. A., Hauch A., and Plickert, G. (1987). Morphogenetic factors in hydroids. I. Stolon tip activation and inhibition. *J. Exp. Zool.* **243**, 111–124.

Müller, W. A., Mitze, A., Wickhorst, J. P., and Meier-Menge, H. N. (1977). Polar morphogenesis in early hydroid development: Action of caesium, of neurotransmitters and of an intrinsic head activator on pattern formation. *Roux's Arch. Dev. Biol.* **182**, 311–328.

Müller, W. A., and Plickert, G. (1982). Quantitative analysis of an inhibitory gradient field in the hydrozoan stolon. *Roux's Arch. Dev. Biol.* **191**, 56–63.

Müller, W. A., Plickert, G., Berking, S. (1986). Regeneration in hydrozoa: Distal versus proximal transformation in *Hydractinia*. *Roux's Arch. Dev. Biol.* **195**, 513–518.

Mutz, E. (1930). Transplantationsversuche an *Hydra* mit besonderer Berücksichtigung der Induktion, Regionalität und Polarität. *Roux's Arch. Dev. Biol.* **121**, 210–271.

Netherton, J. C., and Gurin, S. (1982). "Biosynthesis and Physiological Role of Homarine in Marine Shrimp." *J. Biol. Chem.* **257**, 11971–11975.

Neumann, R. (1979). Bacterial induction of settlement and metamorphosis in the planula larva of *Cassiopea andromeda* (Cnidaria; Scyphozoa, Rhizostomae). *Mar. Ecol. Prog. Ser.* **1**, 21–28.

Newman, S. A. (1974). The interaction of the organizing regions in *Hydra* and its possible relation to the role of the cut end in regeneration. *J. Embryol. Exp. Morphol.* **31**, 541–555.

Otto, J. J., and Campbell, R. D. (1977). Budding in *Hydra attenuata*: Bud stages and fate map. *J. Exp. Zool.* **200**, 417–428.

Pechenik, J. A., and Heyman, W. D. (1987). Using KCl to determine the size at competence for larvae of the marine gastropod, *Crepidula fornica* (L). *J. Exp. Mar. Biol. Ecol.* **112**, 27–38.

Pérez, F. (1996). Effects of cantharidin and a phorbol ester on bud formation in *Hydra vulgaris*. *J. Dev. Biol. Suppl.* **1**, 273.

Pérez, F., and Berking, S. (1994). Protein kinase modulators interfere with bud formation in *Hydra vulgaris*. *Roux's Arch. Dev. Biol.* **203**, 284–289.

Pfeifer, R., and Berking, S. (1995). Control of formation of the two types of polyps in *Thecocodium quadratum* (Hydrozoa, Cnidaria). *Int. J. Dev. Biol.* **39**, 395–400.

Plickert, G. (1980). Mechanically induced stolon branching in *Eirene viridula* (Thecata, Campanulinidae). *In* "Developmental and Cellular Biology of Coelenterates" (P. Tardent and R. Tardent, Eds.), pp. 185–190. Elsevier, Amsterdam.

Plickert, G. (1987). Low-molecular-weight factors from colonial hydroids affect pattern formation. *Roux's Arch. Dev. Biol.* **196**, 248–256.

Plickert, G. (1989). Proportion-altering factor (PAF) stimulates nerve cell formation in *Hydractinia echinata*. *Cell Diff. Develop.* **26**, 19–28.

Plickert, G. (1990). Experimental analysis of developmental processes in marine hydroids. *In* "Experimental Embryology in Aquatic Plants and Animals" (H.-J. Marthy, Ed.), pp. 59–81. Plenum Press, New York.

Plickert, G., Heringer, A., and Hiller, B. (1987). Analysis of spacing in a periodic pattern. *Dev. Biol.* **120**, 399–411.

Plickert, G., Kroiher, M., and Munck, A. (1988). Cell proliferation and early differentiation during embryonic development and metamorphosis of *Hydractinia echinata*. *Development* **103**, 795–803.

Przibram, H. (1909). Experimental-Zoologie 2. Regeneration. Franz Denticke Verlag, Leipzig.

Rulon, O., and Child, C. M. (1937). Observations and experiments on developmental pattern in *Pelmatohydra oligactis*. *Physiol. Zool.* **10**, 1–13.

Runnström, J. (1929). Über Selbstdifferenzierung und Induktion bei dem Seeigelkeim. *Roux's Arch. Dev. Biol.* **117**, 129–145.

Sanyal, S. (1966). Bud determination in *Hydra*. *Indian J. Exp. Biol.* **4**, 88–92.

Schaller, H. C. (1976a). Action of the head activator as a growth hormone in *Hydra*. *Cell Diff.* **5**, 1–11.

Schaller, H. C. (1976b). Action of the head activator on the determination of interstitial cells in *Hydra*. *Cell Diff.* **5**, 13–20.

Schaller, H. C. (1984). The head and the foot inhibitors are not Dowex artefacts. *Roux's Arch. Dev. Biol.* **193**, 117–118.

Schaller, H. C., and Bodenmüller, H. (1981). Isolation and amino acid sequence of a morphogenetic peptide from *Hydra*. *Proc. Natl. Acad. Sci. U.S.A.* **78**, 7000–7004.

Schaller, H. C., Rau, T., and Bode, H. (1980). Epithelial cells in nerve-free *Hydra* produce morphogenetic substances. *Nature* **283**, 589–591.

Schaller, H. C., Schmidt, T., and Grimmelikhuijzen, C. J. P. (1979). Separation and specificity of action of four morphogenes from *Hydra*. *Roux's Arch. Dev. Biol.* **186**, 139–149.

Schierwater, B., Murtha, M., Dick, M., Ruddle, F. H., and Buss, L. W. (1991). Homeoboxes in cnidarians. *J. Exp. Zool.* **260**, 413–416.

Schlawny, A., and Pfannenstiel, H. D. (1991). Prospective fate of early blastomeres in *Hydractinia echinata* (Cnidaria, Hydrozoa). *Roux's Arch. Dev. Biol.* **200**, 143–148.

Schneider, T., and Leitz, T. (1994). Protein kinase C in hydrozoans: Involvement in metamorphosis of *Hydractinia* and in pattern formation of *Hydra*, *Roux's Arch. Dev. Biol.* **203**, 422–428.

Schummer, M., Scheurlen, I., Schaller, C., and Galliot, B. (1992). HOM/HOX homeobox genes are present in *Hydra* (*Chlorohydra viridissima*) and are differently expressed during regeneration. *EMBO J.* **11**, 1815–1823.

Schwoerer-Böhning, B., Kroiher, M., and Müller, W. A. (1990). Signal transmission and covert prepattern in the metamorphosis of *Hydractinia echinata* (Hydrozoa). *Roux's Arch. Dev. Biol.* **198**, 245–251.

Shenk, M. A., Bode, H. R., and Steel, R. E. (1993). Expression of *Cnox-2*, a HOM/HOX homeobox gene in *Hydra*, is correlated with axial pattern formation. *Development* **117**, 657–667.

Shenk, M. A., Gee, L., Steele, R. E., and Bode, H. R. (1993). Expression of *Cnox-2*, a HOM/HOX gene, is suppressed during head formation in *Hydra*. *Dev. Biol.* **160**, 108–118.

Shostak, S. (1974). Bipolar inhibitory gradients' influence on the budding region of *Hydra viridis*. *In* "The Developmental Biology of the Cnidaria" (R. L. Miller and C. R. Wyttenbach, Eds.). *Am. Zool.* **14**, 619–632.

Sinha, S. N. V., Joshi, J., Rao, S., and Mookerjee, S. (1984). A 4 variable model for the pattern forming mechanism in *Hydra*. *Biosystems* **17**, 15–22.

Spindler, K.-D., and Müller, W. A. (1972). Induction of metamorphosis by bacteria and by a lithium-pulse in the larvae of *Hydractinia echinata* (Hydrozoa). *Roux's Arch. Dev. Biol.* **169**, 271–280.

Stanfield, P. R. (1983). Tetraethylammonium ions and the potassium permeability of excitable cells. *Rev. Physiol. Biochem. Pharmacol.* **97**, 1–67.

Steele, R. E., Lieu, P., Mai, N. H., Shenk, M. A., and Sarras Jr., M. P. (1996). Response to insulin and the expression pattern of a gene encoding an insulin receptor homologue suggest a role for an insulin-like molecul in regulating growth and patterning in *Hydra*. *Dev. Genes Evol.* **206**, 247–259.

Tabor, C. W., and Tabor, H. (1984). Polyamines. *Annu. Rev. Biochem.* **53**, 749–790.

Takahashi, T., Muneoka, T., Lohmann, J., de Haro, M., Solleder, G., Bosch, T. C. G., David, C., Bode, H. R., Koizumi, O., Shimizu, H., Hatta, M., Fujisawa, T., and Sugiyama, T. (1997). Systematic isolation of peptide signal molecules from *Hydra*. I. Strategy for isolation and characterization of the LWamide and PW families. *Proc. Natl. Acad. U.S.A.* **94**, 1241–1246.

Tardent, P. (1963). Regeneration in the hydrozoa. *Biol. Rev.* **38**, 293–333.

Tardent, P. (1972). Experimente zum Knospenbildungsprozess von *Hydra attenuata* (PALL.). *Rev. suisse. Zool.* **79**, 355–375.

Tardent, P. (1978). Coelenterata, Cnidaria. In "Morphogenese der Tiere" (F. Seidel, Ed.). Gustav Fischer Verlag. Stuttgart.

Technau, U., and Holstein, T. W. (1995). Head formation in *Hydra* is different in apical and basal fragments. *Development* **121**, 1273–1282.

Teissier, G. (1933). Recherches sur les potentialités de l'oeuf des Hydraires: Polarité des larves complexes produites par greffe embryonnaire. *C. r. Séanc. Soc. Biol. Paris* **113**, 26–27.

Thomas, M. B., Freeman, G., and Martin, V. J. (1987). The embryonic origin of neurosensory cells and the role of nerve cells in metamorphosis in *Phialidium gregarium* (Cnidaria, Hydrozoa). *Int. J. Invert. Reprod. Dev.* **11**, 265–287.

Tripp, K. (1928). Die Regenerationsfähigkeit von *Hydren* in verschiedenen Körperregionen nach Regenerations- und Transplantationsversuchen. *Z. wiss. Zool.* **132**, 470–525.

Turing, A. (1952). The chemical basis of morphogenesis. *Philos. Trans. R. Soc. Lond.* **237**, 37–72.

Van de Vyver, G. (1964). Étude histologique de développement d'*Hydractinia echinata* (Flemm.). *Cah. Biol. Mar.* **5**, 295–310.

Walther, M., and Berking, S. (in press). Polyamine metabolism has an impact on metamorphosis induction of the planula larva of the marine hydroid *Hydractinia echinata*.

Walther, M., Ulrich, R., Kroiher, M., and Berking, S. (1996). Metamorphosis and pattern formation in *Hydractinia echinata*, a colonial hydroid. *Intern. J. Dev. Biol.* **40**, 313–322.

Webster, G. (1971). Morphogenesis and pattern formation in hydroids. *Biol. Rev.* **46**, 1–46.

Webster, G., and Wolpert, L. (1966). Studies on pattern regulation in *Hydra*. 1. Regional differences in time required for hypostome determination. *J. Embryol. Exp. Morphol.* **16**, 91–104.

Weimar, B. (1928). The physiological gradients in *Hydra*. I. Reconstitution and budding in relation to length of piece and body level in *Pelmatohydra oligactis*. *Physiol. Zool.* **1**, 183–230.

Weinziger, R., Salgado, L. M., David, C. N., and Bosch, T. C. G. (1994). *KS1*, an epithelial cell-specific gene, responds to early head signals of head formation in *Hydra*. *Development* **120**, 2511–2517.

Weis, V. M., Keene, D. R., and Buss, L. W. (1985). Biology of the hydractiniid hydroids. 4: Ultrastructure of the planula of *Hydractinia echinata*. *Biol. Bull.* **168**, 403–418.

Wolpert, L. (1969). Positional information and the spatial pattern of cellular differentiation. *J. Theor. Biol.* **25**, 1–47.

Wyttenbach, C. R. (1965). Sites of mitotic activity in the colonial hydroid, *Campanularia flexuosa*. *Anat. Rec.* **151**, 483.

Yasugi, S. (1974). Observations on supernumerary head formation induced by lithium chloride treatment in the regenerating *Hydra*, *Pelmatohydra robusta*. *Rev. Growth Differ.* **16**, 171–180.

Yool, A. J., Grau, S. M., Hadfield, M., Jensen, R. A., Markell, D. A., and Morse, D. E. (1986). Excess potassium induces larval metamorphosis in four marine invertebrate species. *Biol. Bull.* **170**, 255–266.

4

Primate Embryonic Stem Cells

James A. Thomson and Vivienne S. Marshall
Wisconsin Regional Primate Research Center
University of Wisconsin
Madison, Wisconsin 53715-1299

Primate embryonic stem (ES) cells are derived from preimplantation embryos, have a normal karyotype, and are capable of indefinite, undifferentiated proliferation. Even after culture for more than a year, primate ES cells maintain the potential to differentiate to trophoblast and derivatives of embryonic endoderm, mesoderm, and ectoderm. In this review, we compare the characteristics of ES cell lines from two primate species, the rhesus monkey (*Macaca mulatta*) and the common marmoset (*Callithrix jacchus*), with the characteristics of mouse ES cells and human embryonal carcinoma cells. We also discuss the implications of using primate ES cells to understand early human development and discuss the practical and ethical implications for the understanding and treatment of human disease. Copyright © 1998 by Academic Press.

The Simiadae then branched off into two great stems, the New World and Old World monkeys; and from the latter at a remote period, Man, the wonder and the glory of the universe, proceeded.

CHARLES DARWIN, *The Descent of Man*, 1871

I. What Are Embryonic Stem Cells?

In the adult mammal, cells with a high turnover rate are replaced by a highly regulated process of proliferation, differentiation, and apoptosis from undifferentiated "stem cells." Tissue from which stem cells have been extensively studied include the hematopoietic system, the epidermis, and the intestinal epithelium (Hall and Watt, 1989). In the human small intestine, for example, approximately 10^{11} cells are shed and must be replaced daily (Potten and Loeffler, 1990).

Current Topics in Developmental Biology, Vol. 38

Although various definitions for stem cells have been proposed, the characteristics proposed by Reynolds and Weiss (1996) are representative and include (1) prolonged proliferation, (2) self-maintenance, (3) generation of large numbers of progeny with the principal phenotypes of the tissue, (4) maintenance of developmental potential over time, and (5) generation of new cells in response to injury. Thus, stem cells in the adult sustain a relatively constant number of cells and cell types, although adjustments can occur in response to environmental or physiologic changes.

In the embryo, cell proliferation and differentiation elaborate an increasing number of cells and cell types. As cells become committed to particular lineages, a progressive decrease in developmental potential occurs. In the mouse, each cell of the cleavage-stage embryo has the developmental potential to contribute to any embryonic or extraembryonic cell type, but by the blastocyst stage cells of the trophectoderm are irreversibly committed to the trophectoderm lineages (Winkel and Pedersen, 1988). Soon afterward, the inner cell mass (ICM) is separated into a layer of primitive endoderm, which gives rise to extraembryonic endoderm, and a layer of primitive ectoderm, which gives rise to the embryo proper and to some extraembryonic derivatives (Gardner, 1982). Although the cells of the ICM and, later, of the primitive ectoderm are the precursor cells to all adult tissues, these embryonic cells proliferate and replace themselves in the intact embryo for only a limited time before they become committed to specific lineages; therefore, they do not satisfy the criteria for stem cells that are usually applied to adult tissues.

In contrast, if the mouse ICM is removed from the normal embryonic environment and dissociated, under appropriate conditions, the cells will remain undifferentiated, replace themselves indefinitely, and maintain the developmental potential to contribute to all adult cell types. These ICM-derived cells satisfy the criteria for stem cells outlined above. The initial mouse ICM-derived cell lines were termed either EK or ESC cells, to distinguish them from embryonal carcinoma (EC) cells, the pluripotent cells derived from teratocarcinomas (Evans and Kaufman, 1981; Martin, 1981; Axelrod, 1984). Later, the term embryo-derived stem cell (ES cell) was proposed to distinguish them from both EC cells and the pluripotent cells within intact embryos (Rossant and Papaioannou, 1984). Subsequent authors shortened this to embryonic stem (ES) cells (Doetschman et al., 1985), the term now in general use.

Although mouse ES and EC cells have similar properties, their different origins are often reflected in different developmental potentials. In general, EC cells have a more restricted developmental potential than ES cells (Rossant and Papaioannou, 1984). Mouse EC cell lines are more likely to have karyotypic abnormalities, and they have developmental potentials that vary widely between cell lines. These differences between ES and EC cells are thought to arise from the selective pressures of the teratocarcinoma environment, which are avoided by the *in vitro* derivation of ES cells; however, some mouse EC cell lines participate in normal development when formed into chimeras with intact embryos, and in some cases EC cells have been reported to contribute to the germ line in chimeras (Brinster, 1974; Mintz and Illmensee, 1975; Illmensee and Mintz, 1976; Mintz and Cronmiller, 1978, 1981; Stewart and Mintz, 1982). Mouse ES cells more consistently contribute to chimeras and particularly to the germ line (Bradley *et al.*, 1984). When combined in chimeras with tetraploid embryos, a few select mouse ES cell lines can form entire, viable fetuses (Nagy *et al.*, 1993). The ability of mouse ES cells to enter the germ line in chimeras allows the introduction of specific genetic changes into the mouse genome and thus offers a direct approach to understanding gene function in the intact animal (Rossant *et al.*, 1993). Yet the majority of ES cells used to form successful germ-line chimeras have been from a single mouse strain (129) in combination with a single host embryo strain (C57BL/6). Although other strain combinations have been used successfully, germ-line transmission is highly strain dependent (Hogan *et al.*, 1994).

What criteria are appropriate to apply the term "ES cell" to a cell line derived from a primate? Given the criteria used for stem cells in adult tissues, the properties of mouse ES cells, and the historical introduction of the term ES cell primarily to distinguish its origin from an EC cell, we propose the following criteria for primate ES cells: (1) derivation from the preimplantation or peri-implantation embryo, (2) prolonged undifferentiated proliferation, and (3) stable developmental potential after prolonged culture to form derivatives of all three embryonic germ layers. The primate cell lines we recently derived meet these criteria (Thomson *et al.*, 1995, 1996). In addition, at least some of these primate ES cell lines maintain a normal karyotype through undifferentiated culture for at least 2 years, maintain a stable developmental potential throughout this culture period that greatly exceeds that of human EC cell lines, and maintain the potential to form trophoblast *in vitro*, a property that may distinguish them from mouse ES cells. The potential to enter the germ line in chimeras may also be a property of primate ES cells, but for practical or ethical reasons, in many primate species including humans, this is impossible to test.

II. Species Choice for Experimental Primate Embryology

Primate ES cell lines offer exciting possibilities for establishing a robust experimental primate embryology and provide a powerful new model for understanding human development and disease. Historical use, well-defined genetics, and favorable reproductive characteristics have made the mouse the mainstay of mammalian experimental embryology. However, early mouse development and early human development differ significantly. For example, human and mouse embryos differ in the timing of embryonic genome expression (Braude *et al.*, 1988); in the formation, structure, and function of the fetal membranes and placenta

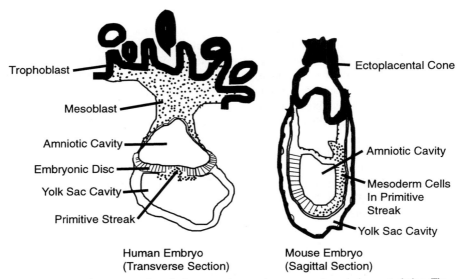

Fig. 1   Schematic representation of a human embryo and a mouse embryo during gastrulation. The human embryo depicted is at approximately 16 postovulatory days (O'Rahilly and Muller, 1987) and the mouse embryo is at approximately 7 postovulatory days (Kaufman, 1992).

(Luckett, 1975, 1978; Benirschke and Kaufmann, 1990); and in formation of an embryonic disc instead of an egg cylinder (Fig. 1) (O'Rahilly and Muller, 1987; Kaufman, 1992). Although there is some variation between species, nonhuman primate development and human development are extremely similar. Thus, if one is interested in the development of a human tissue that differs significantly from the corresponding mouse tissue, such as the yolk sac or the placenta, a primate model is desirable. Although there is active experimental work on primate embryos, including work on oocyte maturation, fertilization, and preimplantation development (Lopata *et al.*, 1988; Wolf *et al.*, 1990, 1996; Seshagiri *et al.*, 1993, 1994; Seshagiri and Hearn, 1993; Wilton *et al.*, 1993; Zhang *et al.*, 1994; Schramm and Bavister, 1995; Weston *et al.*, 1996), the study of early developmental events that occur after implantation is largely limited to descriptive histologic and ultrastructural work. Techniques that form the backbone of mouse experimental embryology, such as chimera formation, lineage analysis, and transgenesis, have never been developed in primates.

   Because most mammalian embryologists are primarily familiar with the mouse, a brief introduction to some practical implications of primate reproductive biology is useful. There are approximately 200 species of primates in the world, of which only a small handful are used routinely in biomedical research (King *et al.*, 1988). The major division among "higher" primates is between Old World and New World species. Humans are more closely related to Old World

Fig. 2   Adult female rhesus monkey (*Macaca mulatta*) with infant.

species. Macaques, such as the rhesus monkey (*Macaca mulatta*) and the cynomolgus monkey (*Macaca fascicularis*), are the most widely used Old World primates in biomedical research (Fig. 2). The evolutionary relatedness between humans and macaques is reflected by close similarities in reproduction and development; it is precisely these similarities that make the study of experimental embryology in macaques both very desirable and very difficult. Reproductive characteristics of macaques include the ovulation of a single oocyte during each monthly menstrual cycle and thus the birth of single young, a relatively long gestation period (about 5 mo), and an age at sexual maturity of 4–5 years. In addition, it is not routinely possible to synchronize the reproductive cycle of female macaques, so achieving synchrony between a donor and a recipient female for embryo transfer requires either that the embryo be frozen or that a large pool of naturally cycling females be maintained. A nonsurgical uterine-stage embryo recovery technique has been developed for macaques, but given that only a single oocyte is ovulated each month and that donor animals cannot be synchronized, it is difficult to recover groups of developmentally matched embryos for controlled experiments (Goodeaux *et al.*, 1990; Seshagiri *et al.*, 1993; Seshagiri and Hearn, 1993). Improvements in ovarian hyperstimulation protocols and *in vitro* fertilization have increased the availability of oocytes and early cleavage–stage macaque embryos (Wolf *et al.*, 1990; Alak and Wolf, 1994;

138   James A. Thomson and Vivienne S. Marshall

Zhang *et al.*, 1994; Weston *et al.*, 1996; Wolf *et al.*, 1996). However, rhesus monkeys become refractory to the human hormones used for stimulation because of the development of antibodies. Recombinant macaque follicle-stimulating hormone (FSH) and luteinizing hormone (LH) for ovarian hyperstimulation would allow more hyperstimulation cycles to be performed, but the macaque hormones are not yet available. Even with improvements in preimplantation embryo availability, because of the difficulties in performing significant numbers of embryo transfers, postimplantation events *in vivo* will continue to be very difficult to examine experimentally in macaques.

Some New World primate species, such as the common marmoset, have reproductive characteristics that make them well suited for experimental primate embryology (Fig. 3). New World primates are more distantly related to humans than are Old World primates, and this distance is reflected by some differences in development, such as the structure of the placenta (Phillips, 1976; Chambers and Hearn, 1985; Moore *et al.*, 1985). However, these differences are minor compared with the differences between human and mouse development. Common marmosets are small (300–400 g), routinely have twins or triplets, and have a relatively short time to sexual maturity (about 18 mo) (Hearn *et al.*, 1978). In

Fig. 3   Adult female marmoset monkey (*Callithrix jacchus*).

addition, marmoset ovarian cycles can be synchronized by prostaglandin administration, which makes the collection of groups of developmentally matched embryos possible and allows the efficient transfer of embryos to synchronized recipients (Summers *et al.*, 1985, 1987; Lopata *et al.*, 1988). Marmoset embryos recovered by surgical techniques have provided an important model for understanding primate implantation and maternal recognition of pregnancy, processes that differ significantly between primates and mice (Hearn, Gidley-Baird, *et al.*, 1988; Hearn, Hodges, *et al.*, 1988; Hearn and Webley, 1993). However, because repeated surgery is stressful to the animal and is severely restricted in many countries including the United States, marmoset embryo availability has been limited. Recently developed nonsurgical procedures allow both the collection and transfer of *in vivo*–produced, uterine-stage preimplantation marmoset embryos without trauma to the donor (Thomson *et al.*, 1994), with the embryo transfer procedure allowing a pregnancy rate of 44% (Marshall *et al.*, 1997). An effective ovarian hyperstimulation protocol for marmosets has not yet been developed.

Even with improved ovarian hyperstimulation, *in vitro* fertilization, and embryo recovery procedures, primate embryos will continue to be in very limited supply for experimental studies. However, primate ES cells can be expanded indefinitely *in vitro* and still retain the potential to differentiate to derivatives of all three embryonic germ layers and to extraembryonic tissues, so the mechanisms controlling the differentiation of specific early primate lineages now can be studied experimentally in detail for the first time. Rhesus ES cells will be particularly useful as an accurate *in vitro* model to help us understand the differentiation of specific human tissues because of the close evolutionary relatedness between humans and rhesus monkeys. Marmoset ES cells will be particularly useful for experiments that require embryo transfer, such as chimera formation and lineage analysis, because of the reproductive characteristics of marmosets.

III. Isolation and Propagation of Primate ES Cell Lines

The derivation and propagation of primate ES cells is similar to that for mouse ES cells, yet they differ significantly enough to merit a detailed description of the culture of primate ES cell lines (Thomson *et al.*, 1995, 1996). Mouse embryonic fibroblasts mitotically inactivated with 3000 rads γ-radiation are prepared at 5×10^4 cells/cm^2 on tissue culture plastic previously treated by overnight incubation with 0.1% gelatin (Robertson, 1987). The fibroblasts, prepared the day before ICM isolation and cultured in 80% Dulbecco's modified Eagle medium (DMEM; 4500 mg of glucose per liter, without sodium pyruvate), 20% fetal bovine serum (FBS), 0.1 mM 2-mercaptoethanol, and 1% nonessential amino acid stock (GIBCO), are allowed to equilibrate in a mixture of 5% carbon dioxide and 95% air at 37°C overnight. Expanded *in vivo*–produced rhesus blastocysts collected 6 days after ovulation (Seshagiri and Hearn, 1993) or marmoset blastocysts collected 8

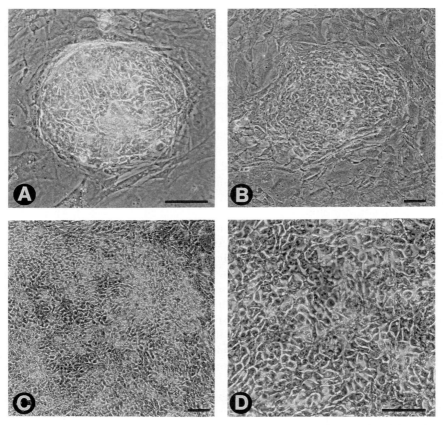

Fig. 4   Progressive stages of rhesus ES cell derivation and growth. (A) Rhesus ICM 6 days after isolation, immediately prior to dissociation. (B) Rhesus ES cell colony 1 day after passage. (C, D) Rhesus ES cell colony 7 days after passage. Bar, 100 μm.

days after ovulation (Thomson *et al.*, 1994) are incubated in 0.5% pronase-DMEM while being observed under a dissecting microscope. Immediately after zona pellucida dissolution, the blastocysts are washed twice in DMEM. To remove the trophectoderm by immunosurgery (Solter and Knowles, 1975), the blastocysts are exposed to rabbit anti-rhesus or anti-marmoset spleen cell antiserum in DMEM for 30 min, washed three times in DMEM, and then incubated in a 1:10 dilution of Guinea pig complement for 30 min. After two further washes in DMEM, lysed trophectoderm cells are removed from the intact ICM by gentle pipetting, and the ICM is plated on mouse inactivated embryonic fibroblasts. After 7–10 days (Fig. 4A), ICM-derived masses are removed from endoderm outgrowths, exposed to 0.05% trypsin–ethylenediamine tetraacetic acid (EDTA)

Table I. Karyotypes of Current Rhesus ES Cell Lines and Their Ability to Form Teratomas with Unambiguous Derivatives from All Three Germ Layers in Severe Combined Immunodeficient Mice

Cell Line	Karotype	Teratomas
R278	42 XY	Yes
R366	42 XY	Yes
R367	42 XY	Yes
R394	42 XX	Yes
R420	42 XX	Yes
R456	42 XX	Not tested
R460	42 XY	Not tested

or Ca^{2+}/Mg^{2+}-free phosphate-buffered saline (PBS) with 0.5 mM EDTA and 1% serum for 3–5 min, and dissociated by gentle pipetting through a micropipette. Dissociated cells are replated on embryonic feeder layers in fresh medium and observed for colony formation. Colonies composed of closely packed cells with high nuclear/cytoplasmic ratios, resembling human EC cells, are individually selected with a micropipette and passaged again onto fresh fibroblasts (see Fig. 4). Early-passage cells are karyotyped, and aliquots are frozen and stored in liquid nitrogen (Robertson, 1987). To date, we have isolated eight marmoset (Thomson et al., 1996) and seven rhesus ES cell lines (Table I).

With current culture conditions, some background differentiation occurs at each passage for both the rhesus and marmoset ES cell lines. For long-term culture, it is necessary to periodically separate undifferentiated colonies from differentiated cells, or the culture is eventually entirely lost to differentiation. Separation must be done prior to extensive contact between the undifferentiated and differentiated cells, or rapid differentiation ensues. Undifferentiated colonies can be separated by incubating the culture in Ca^{2+}/Mg^{2+}-free PBS with 0.5 mM EDTA and 1% FBS until the cells individualize. The undifferentiated colonies can then be identified under a dissecting microscope, picked off by micropipette, and plated on new fibroblast feeder layers. If there is a reasonably pure population of undifferentiated cells, the entire culture can be passaged with Ca^{2+}/Mg^{2+}-free PBS with 0.5 mM EDTA by pipetting and gently scraping the cells. The time during which the cells are not attached to feeder layers during passaging must be minimized. The cells fragment and die rapidly when removed from feeders, and even if they are passaged gently and rapidly the cloning efficiency is poor. Dispersal of the cells to 5–10 cells per clump appears best. If they are dispersed to single cells, the cloning efficiency is very poor (about 1–2%); if the clumps are too large, the cells tend to differentiate. It is difficult to set up a regular passaging schedule for the primate ES cells, but they usually need to be passaged every 7–10 days to prevent differentiation.

Water and serum quality are both critical, and, in particular, primate ES cells are extremely sensitive to endotoxin. Endotoxin levels that would not affect the growth of most other cell lines result in the death or differentiation of primate ES cell lines. We initially screened batches of serum by examining the cloning efficiency of mouse ES cells grown with leukemia inhibitory factor (LIF) in the absence of feeder layers; we now test new batches directly on rhesus ES cells. The quality of the embryonic fibroblasts is critical; we use only low-passage cells (up to passage 10) and try to keep them actively dividing before formation of the feeder layer. Improperly maintained fibroblasts fail to adequately inhibit the differentiation of primate ES cells, resulting in mixed, differentiated populations of cells. Immortal mouse fibroblasts such as STO and 3T3 cells also prevent the differentiation of primate ES cells (for approximately 1 wk), but, in our experience, primary mouse embryonic fibroblasts support active, undifferentiated proliferation for a much longer time (approximately 2 wk). In the absence of fibroblasts, primate ES cells uniformly differentiate or die within 7–10 days.

IV. Comparison of Primate ES Cells, Mouse ES Cells, and Human EC Cells

The morphologies of mouse ES/EC cells, primate ES cells, and human EC cells are similar but distinct. Mouse ES cells have a high nuclear/cytoplasmic ratio, have prominent nucleoli, and form very compact, piled-up colonies, with cells so closely packed that it is difficult to discern individual cells within colonies (Fig. 5A). Rhesus ES cells form much flatter colonies, initially with irregular borders (Fig. 5D,F). Individual cells can be discerned easily in the small colonies formed within the first few days after splitting. After rhesus ES cells have been cultured for 1 wk or longer, the large colonies consist of very tightly packed cells. Rhesus ES cells closely resemble human EC cells (Fig. 5B) by phase contrast or electron microscopy (Fig. 6). In general, human EC cells form more piled-up, irregular colonies, and undifferentiated growth is promoted at high cell densities. In contrast, rhesus ES cells differentiate rapidly if they are allowed to pile up after the cultures grow to confluence. Marmoset ES cells (Fig. 5C,E) have a morphology intermediate between those of mouse ES cells and human EC cells but again form much flatter colonies than mouse ES cells. These differences in colony morphology may reflect the different structures of the rounded mouse egg cylinder and the flat primate embryonic disc, although the developmental potential of primate ES cells suggests an earlier developmental origin (see later discussion).

The surface antigens that characterize human EC cells are the same as those that characterize primate ES cells but differ from those that characterize mouse EC and ES cells. Undifferentiated human EC, marmoset ES, and rhesus ES cells express specific globoseries glycolipids, which carry stage-specific embryonic antigens 3 and 4 (SSEA-3 and SSEA-4), high-molecular-weight glycoproteins

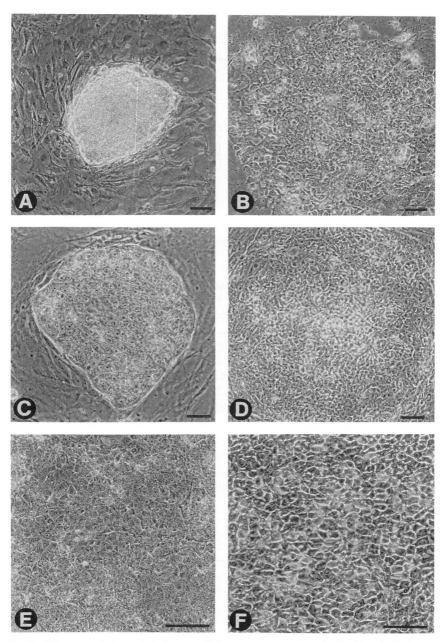

Fig. 5 Colony morphology of pluripotent cell lines on mouse embryonic fibroblast feeder layers. (A) Mouse ES cell colony. (B) Human EC cell colony (GCT 27; gift of Marin Pera). (C, E) Marmoset ES cell colony. Note the large nuclei, prominent nucleoli, and similar colony morphology to the mouse ES cell colony shown in A. (D, F) Rhesus ES cell colony. Note the large nuclei, prominent nucleoli, and the similar colony morphology to the human EC cell colony shown in B. Bar, 100 μm.

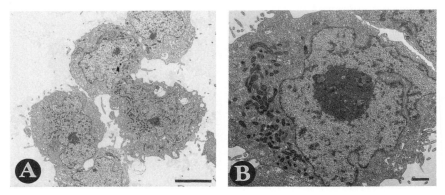

Fig. 6 Electron micrographs of rhesus ES cells. (A) Bar, 5 μm. (B) Bar, 1 μm. Note the prominent nucleoli, irregularly shaped nucleus, and sparse microvilli. Lead citrate and uranyl acetate.

(e.g., TRA-1-60 and TRA-1-81), and alkaline phosphatase (Solter and Knowles, 1978; Kannagi *et al.*, 1983; Andrews, Banting, *et al.*, 1984; Andrews, 1988; Wenk *et al.*, 1994; Thomson *et al.*, 1995, 1996) (Table II). Human choriocarcinoma and yolk sac carcinoma cell lines, which represent the earliest lineages to differentiate from the ICM, lack this combination of markers (Andrews, 1988). Mouse cleavage-stage embryos express SSEA-3 and SSEA-4, but mouse ICM, ES, and EC cells express the lactoseries glycolipid SSEA-1 and do not express SSEA-3, SSEA-4, TRA-1-60, or TRA-1-81 (Solter and Knowles, 1978; Kannagi *et al.*, 1983). Differentiation of human EC cells such as NTERA2 cl.Dl results in the loss of SSEA-3, SSEA-4, TRA-1-60, and TRA-1-81 expression and an increase in SSEA-1 expression (Andrews, Damjanov, *et al.*, 1984; Andrews, 1988). Because no one has studied the expression of these antigens on early intact human or nonhuman primate embryo cells, it is unknown whether the markers expressed by human EC cells reflect fundamental embryologic differences be-

Table II. Comparison of Cell Surface Markers of Pluripotent Cell Lines Derived from Humans, Nonhuman Primates, and Mice

Marker	Human EC (NT2/D1)	Marmoset ES (CJ11)	Rhesus ES (R278.5)	Mouse ES (D3)
SSEA-1	−	−	−	+
SSEA-3	+	+	+	−
SSEA-4	+	+	+	−
TRA-1-60	+	+	+	−
TRA-1-81	+	+	+	−
Alkaline phosphatase	+	+	+	+

tween mice and humans or the malignant origin of human EC cells. Shared characteristic cell surface markers suggest close embryologic similarities between primate ES and human EC cells and also suggest significant embryologic differences from mouse ES cells. Determining the distribution of these antigens in the early primate embryo may provide clues to their functions and should identify the embryonic cells most closely related to primate ES cells.

The factors fibroblasts produce that prevent the differentiation of primate ES cells are unknown. Mouse ES cells remain undifferentiated and proliferate in the absence of fibroblasts when grown in the presence of LIF, ciliary neurotropic factor (CNF), or oncostatin M (Williams *et al.*, 1988; Conover *et al.*, 1993; Rose *et al.*, 1994). Each of these factors acts through the LIF receptor complex, a heterodimer of the LIF receptor and the interleukin-6 (IL-6) signal transducer, gp130; (Wolf *et al.*, 1994; Yoshida *et al.*, 1994). A combination of IL-6 and soluble IL-6 receptor (sIL-6R) can bypass the LIF receptor and directly activate the gp130 signaling pathway in mouse ES cells, supporting undifferentiated proliferation (Yoshida *et al.*, 1994). Both marmoset and rhesus ES cells cultured in the presence of LIF and in the absence of fibroblasts uniformly differentiate or die within 7–10 days (Thomson *et al.*, 1995, 1996). We have found no effect of exogenous LIF on cloning efficiency, growth rate, or reduction of background differentiation of primate ES cells grown in the presence of fibroblast feeder layers (Thomson and Marshall, unpublished data). Indeed, rhesus ES cells grown on primary fibroblasts derived from homozygous LIF-knockout mice (gift of Colin Stewart; Stewart *et al.*, 1992) continue undifferentiated proliferation for at least three passages (Fig. 7A). Similarly, oncostatin M fails to prevent the differentiation of rhesus ES cells (Thomson and Marshall, unpublished data). Some human EC cell lines are feeder dependent, and LIF, oncostatin M, and CNF also fail to prevent their differentiation (Roach *et al.*, 1993). Further work is required to clarify whether the gp130 signaling pathway is activated by other molecules in primate ES or human EC cells, or whether an entirely different signaling pathway maintains undifferentiated proliferation.

None of the factors we have tested to date allows the undifferentiated proliferation of primate ES cells in the absence of fibroblasts. Because of the combined importance of LIF, steel factor (SF), and basic fibroblast growth factor (bFGF) in the isolation of mouse embryonic germ (EG) cell lines (Matsui *et al.*, 1992; Resnick *et al.*, 1992), we have tested each of these factors on rhesus ES cells. None of these factors, alone or in combination, prevents the differentiation of rhesus ES cells in the absence of fibroblast feeder layers (Fig. 7D). Rhesus ES cells continue undifferentiated proliferation for at least three passages when grown on fibroblasts (Sl/Sl[4]) derived from mice with a homozygous deletion for SF (gift of David Williams; Toksoz *et al.*, 1992) (Fig. 7B). We have found no effect of exogenous SF on rhesus ES cells grown in the presence of fibroblasts. A trypsin-sensitive factor from a human yolk sac carcinoma cell line (GCT 44) supports the growth of feeder-dependent human EC cells in the absence of

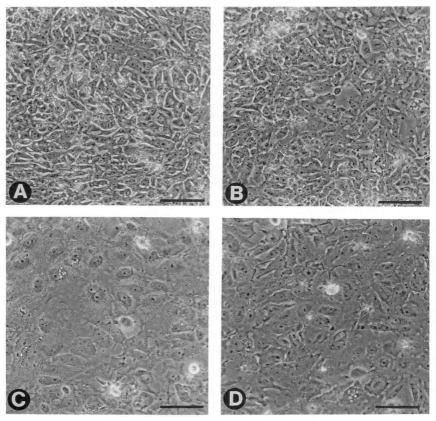

Fig. 7 (A) Rhesus ES cells grown on primary embryonic fibroblasts derived from homozygous leukemia inhibitory factor (LIF)–knockout mice. (B) Rhesus ES cells grown on embryonic fibroblasts derived from mice with a homozygous deletion for steel factor (S1/S1⁴). (C) Differentiated rhesus ES cells grown on extracellular matrix derived from mouse embryonic fibroblasts. (D) Differentiated rhesus ES cells grown on extracellular matrix derived from mouse embryonic fibroblasts and in the presence of LIF (20 ng/ml), steel factor (100 ng/ml), and basic fibroblast growth factor (20 ng/ml). Bar, 100 μm.

fibroblasts (Roach *et al.*, 1993), but conditioned medium from GCT 44 cells (gift of Martin Pera) fails to prevent the differentiation of rhesus ES cells. Indeed, conditioned medium from embryonic fibroblasts also fails to prevent the differentiation of rhesus ES cells, suggesting that whatever is produced by the fibroblasts is unstable, is present in limiting quantities, or is attached to the fibroblast cell membrane or extracellular matrix. Fibroblast matrix produced by lysing of attached fibroblasts by exposure to nonionic detergent (Abbondanzo *et al.*, 1993) improves the attachment and survival of rhesus ES cells but fails to prevent their differentiation (Fig. 7C). To develop a strategy to identify the factors that prevent the differentiation of primate ES cells, additional experiments are required to determine whether fibroblasts are producing soluble, cell surface, or matrix-associated factors.

Primate ES cells and human EC cells share the ability to differentiate to extraembryonic lineages, including yolk sac and trophoblast (Andrews, 1988; Thomson et al., 1995, 1996). Mouse germ cell tumors sometimes include yolk sac, but they do not include trophoblast (Andrews et al., 1980). Mouse ES cells rarely contribute to trophoblast in chimeras (Beddington and Robertson, 1989), and convincing evidence of in vitro differentiation to trophoblast by mouse ES cells is lacking. Both rhesus and marmoset ES cell lines spontaneously differentiate in vitro to endoderm (probable extraembryonic endoderm) and trophoblast, as evidenced by α-fetoprotein, chorionic gonadotropin (CG) α-subunit, and CG β-subunit mRNA synthesis, and by the secretion of bioactive CG into the culture medium (Thomson et al., 1995, 1996). In the mouse embryo, the last cells capable of contributing to derivatives of both trophectoderm and late ICM are early ICM cells of the expanding blastocyst (Winkel and Pedersen, 1988). The timing of commitment to ICM or trophectoderm has not been established for any primate species, but because human EC cells differentiate to trophoblast and to derivatives of all three germ layers, and because they express SSEA-3 and SSEA-4, it has been suggested that human EC cells resemble an earlier stage of embryogenesis than mouse EC/ES cells (Andrews et al., 1980). Similarly, the potential of primate ES cells to contribute to derivatives of trophoblast, endoderm, and all three germ layers suggests that they most closely resemble early totipotent embryonic cells (Fig. 8). However, the apparent difference in the ability of mouse and primate ES cells to differentiate to trophoblast should be interpreted cautiously. The fact that mouse ES cells contribute to the trophoblast even rarely in chimeras suggests that they indeed have the developmental potential to form trophoblast but that the environmental conditions in chimeras do not allow efficient differentiation to trophoblast by ES cells. It is possible that mouse ES cells in chimeras are simply physically excluded from contributing to the trophoblast because of differences in cell–cell adhesion or in the cell cycle between ES cells and host embryo cells. The failure to document in vitro differentiation of mouse ES cells to trophoblast may simply reflect the lack of a sensitive trophoblast marker, such as CG, for the mouse. Indeed, mouse ES cells spontaneously differentiate to large unidentified cells similar in appearance to trophoblast giant cell outgrowths (Beddington and Robertson, 1989). Therefore, although it is clear that primate ES cells can differentiate to trophoblast in vitro, it is unclear whether this represents a fundamental difference from mouse ES cells.

Like their mouse ES/EC counterparts, rhesus ES cell lines have both more advanced and more consistent developmental potentials than human EC cell lines. For example, the most commonly studied human EC cell line, NTERA2 cl.Dl, when injected into nude mice, forms foci of EC cells, simple tubular

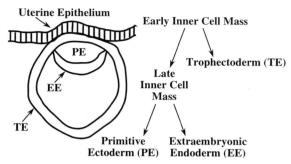

Fig. 8   Schematic representation of a stage 4 human embryo showing the derivation of trophectoderm (TE), primitive ectoderm (PE), and extraembryonic endoderm (EE).

structures resembling primitive gut, neural rosettes, and tissue resembling neuropile (Andrews, Damjanov, *et al.*, 1984). The developmental potential of other human EC cell lines varies but in general is rather limited, possibly reflecting the malignant origin and the severe karyotypic abnormalities of human EC cells (Roach *et al.*, 1993). Rhesus ES cells injected into severe combined immunodeficient (SCID) mice form advanced differentiated structures of all three embryonic germ layers (Thomson *et al.*, 1995). To date, we have tested the developmental potential of five independent rhesus ES cell lines, three with a normal XY karyotype and two with a normal XX karyotype. All form teratomas when injected into SCID mice, with a consistent timing and range of differentiation (see Table I). Although undifferentiated EC-like cells are present soon after injection, advanced rhesus ES cell tumors in SCID mice are characterized by a consistent lack of an EC cell component and a failure to metastasize, at least up to 3 mo of growth. Easily identified differentiated cells in rhesus ES cell teratomas include smooth muscle, striated muscle, bone, cartilage, gut epithelium, respiratory epithelium, keratinizing squamous epithelium, neurons, and ganglia (Figs. 9 and 10). Many other unidentified cells occur, including clusters of cells resembling hepatocytes, but unambiguous identification of these other cells awaits the analysis of cell-specific marker expression.

What is remarkable about the primate ES cell teratomas is not only the range of individual cell types but the formation of complex structures requiring coordinated interactions between different cell types. For example, neural tube–like structures form stratified layers closely resembling ventricular and mantle layers (see Fig. 10A) (Thomson *et al.*, 1995). Entire gut-like structures develop, complete with regional differentiation of small and large intestine (see Fig. 9A,B). Some segments, for example, demonstrate a close resemblance to small intestine, with villi composed of goblet cells, absorptive enterocytes, and an underlying lamina propria, and are surrounded by a well-organized smooth muscle wall (see Fig. 9B). The smooth muscle wall of the gut-like structures is sometimes subdi-

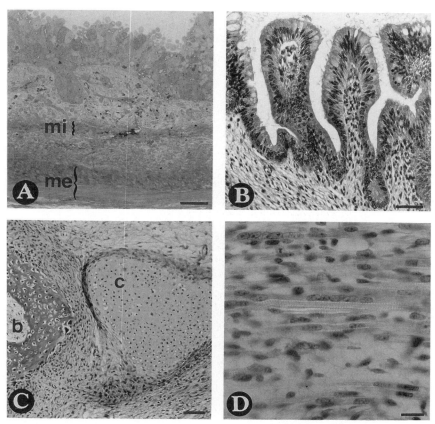

Fig. 9 Differentiated cells in rhesus ES cell teratomas grown in severe combined immunodeficient (SCID) mice. (A) Gut-like structure with well-organized muscular wall. me, smooth muscle bilayer resembling muscularis externa; mi, smooth muscle layer resembling muscularis interna. Toluidine blue staining; bar, 50 μm. (B) Gut-like structure resembling small intestine, showing villi composed of goblet cells, absorptive enterocytes, and an underlying lamina propria. Hematoxylin and eosin (H&E) stain; bar, 50 μm. (C) Bone (b) and cartilage (c). H&E stain; bar, 200 μm. (D) Skeletal muscle. Note presence of multinucleated cells and striations. H&E stain; bar, 20 μm.

vided into distinct inner and outer layers, with the outer layers further subdivided into two perpendicular layers (see Fig. 9A). This arrangement of the smooth muscle wall has an uncanny resemblance to the muscularis interna and externa in the normal gut wall and suggests highly coordinated interactions between endodermal epithelium and adjacent mesenchyme. Similarly, the development of hair follicles, complete with hair shafts, requires coordinated inductive interactions between epithelium and mesenchyme (see Fig. 10B). And finally, perhaps the most impressive example of epithelial–mesenchymal interactions in these tumors

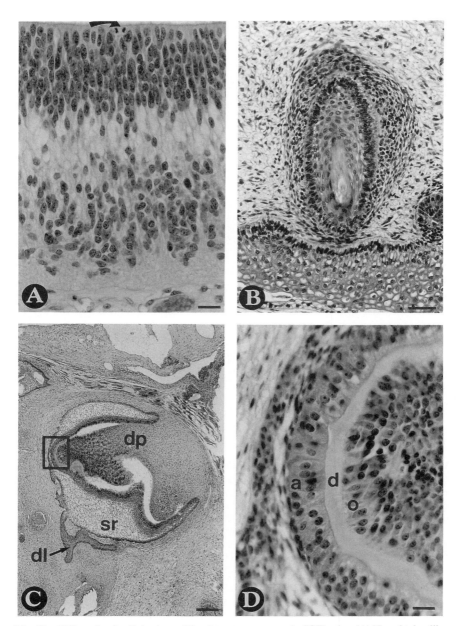

Fig. 10 Differentiated cells in rhesus ES cell teratomas grown in SCID mice. (A) Neural tube–like structure showing stratified layers closely resembling ventricular (*top*) and mantle (*bottom*) layers. Mitotic figures (*arrow*) are restricted to the ventricular layer. H&E stain; bar, 20 μm. (B) Hair follicle. H&E stain; bar, 50 μm. (C) Tooth bud showing dental papilla (dp), stellate reticulum (sr), and dental lamina (dl). H&E stain; bar, 200 μm. (D) Magnification of area marked with box in (C) showing ameloblasts (a), odontoblasts (o), and dentin (d). H&E stain; bar, 20 μm.

is the formation of well-organized tooth buds, complete with ameloblasts, odontoblasts, and intervening dentin (see Fig. 10C,D). Because of the consistent development of specific complex structures in teratomas, genetic manipulations, such as targeted disruption or overexpression of specific genes, can be applied to rhesus ES cells and the developmental effects examined in teratomas.

When grown in the absence of intact fibroblasts, primate ES cell colonies differentiate to simple, flattened epithelial colonies (see Fig. 7C). When allowed to grow to confluence on fibroblasts and pile up, both marmoset and rhesus ES cells spontaneously differentiate into more complex structures. The differentiation of rhesus ES cells *in vitro* is consistently disorganized but includes cardiac muscle, neurons, endoderm, trophoblast, and numerous unidentified cell types (Thomson and Marshall, unpublished data). Marmoset cells, on the other hand, often differentiate to vesicular structures, and some of these vesicular structures demonstrate an organized resemblance to the early postimplantation embryo. Occasionally these marmoset embryoid bodies include a bilaterally symmetric pyriform embryonic disc with early primitive streak formation demonstrating early separation of embryonic endoderm, mesoderm, and ectoderm (Fig. 11). Despite the organized development of the embryoid body in Fig. 11, embryoid body formation by marmoset ES cells to date has been inconsistent and asynchronous, and we have not observed development of marmoset embryoid bodies beyond the initiation of primitive streak formation. If culture conditions can be established that allow efficient, synchronous development of organized primate embryoid bodies *in vitro*, then it will be possible to dissect genetically the mechanisms controlling the initiation of primitive streak formation in primates.

V. Primate ES Cells as an *in Vitro* Model of the Differentiation of Human Tissue

Primate ES cells will be particularly useful for *in vitro* developmental studies of any lineage that differs substantially between humans and mice. Here we consider just one example among many: the differentiation of the trophoblast and the placenta. Commitment and differentiation of the trophectoderm/trophoblast lineage occurs before implantation in all mammals, testifying to a crucial role in the very early events of pregnancy. In humans and nonhuman primates, proliferating trophectoderm cells penetrate the uterine epithelium, invade the endometrium, and differentiate to distinct populations of trophoblast, ultimately forming the outer layers of the definitive placenta (Benirschke and Kaufmann, 1990). In addition to physically connecting the embryo to its mother, the trophectoderm and its trophoblast derivatives produce hormones that modulate maternal physiology. In humans, this critically timed dialogue between the embryo and the mother is initiated by the trophectoderm at about the time of implantation and is essential both for implantation and for the maintenance of pregnancy (Flint *et al.*,

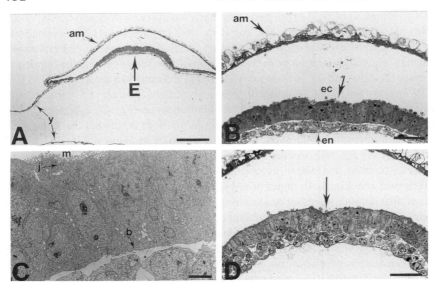

Fig. 11 Embryoid body formed from marmoset ES cells after 4 wk of spontaneous differentiation.
(A) Structures resembling yolk sac (y), amnion (am), and embryonic disc (E). Toluidine blue staining;
bar, 200 μm. (B) Section in cranial third of embryonic disc. Note that the primitive ectoderm (ec)
forms a distinct cell layer from the underlying primitive endoderm (en). The amnion (am) is composed of two layers; the inner layer is continuous with primitive ectoderm at the margins. Toluidine
blue staining; bar, 50 μm. (C) Electron micrograph of embryonic disc. Apical microvilli (m) and
junctional complexes (j) are present in the polarized ectoderm layer, and a basement membrane (b)
separates the ectoderm from the underlying endoderm. Lead citrate and uranyl acetate; bar, 5 μm.
(D) Section in the caudal third of the embryonic disc. Note the central groove (*arrow*) and the mixing
of primitive ectoderm and endoderm cells (early primitive streak). Toluidine blue staining; bar, 50
μm. Reprinted with permission from Thomson *et al.* (1996, Fig. 5, p. 257).

1990). Although accurate estimates are difficult to obtain for women, up to 50%
of embryos may be lost during the first few weeks of natural pregnancy, and
losses after embryo transfer from embryos produced by *in vitro* fertilization
(IVF) are even higher (Edwards, 1988; Hearn and Webley, 1993).

The mouse provides an inadequate model for study of the early events of
human pregnancy. Even the basic mechanisms of maternal recognition and maintenance of pregnancy differ between mice and humans. For example, CG secretion by the trophectoderm at about the time of implantation is an essential signal
for the maternal recognition of pregnancy in all primates studied to date, including humans (Knobil, 1973; Flint *et al.*, 1990), yet the mouse does not even
produce a CG. In primates, if the maternal corpus luteum is exposed to CG,
progesterone secretion and pregnancy are maintained; in the absence of CG, the
corpus luteum regresses and a new ovarian cycle is initiated. Bioactive CG
secretion begins at about the time of implantation in primate embryos, and it can

be measured in the culture medium of IVF-produced human blastocysts by 7–8 days, or in serum of pregnant women by 9–12 days, after fertilization (Dokras *et al.*, 1991; Hearn and Webley, 1993). CG is composed of an α- and a β-subunit. The initiation of CGα and CGβ mRNA synthesis in the preimplantation human embryo remains incompletely defined. In a single published report, CGβ-subunit mRNA was detected by *in situ* hybridization in six- to eight-cell triploid IVF-produced human embryos, which is before the differentiation of the trophectoderm. The significance of this finding for normal diploid embryos is unclear. Later in human development, CG is differentially regulated in the distinct populations of trophoblast cells. The villous cytotrophoblast represents a relatively undifferentiated, mononuclear, replicating cell type which undergoes fusion and terminal differentiation to form the syncytiotrophoblast, a multinuclear cell layer that represents the primary fetal interface with the maternal blood space (King and Blankenship, 1993). Replicating villous cytotrophoblast expresses small amounts of the CGα-subunit, and essentially no CGβ; the syncytiotrophoblast expresses high levels of CGα and CGβ, but this expression is also subject to developmental regulation during the course of pregnancy.

Although CG is an essential embryonic signal to the mother, little is known about its regulation in the early human embryo, largely because of the limited experimental access to peri-implantation human embryos. Nonhuman primate embryos have provided some insights into CG regulation and function, but the limited numbers of embryos available severely restricts the nature and scope of the experiments that can be performed. Because of the lack of a CG, mouse embryos do not provide a model for understanding CG regulation and function. The transcriptional regulation of CG has been extensively characterized in human choriocarcinoma cell lines, but these malignant cell lines may not accurately reflect regulation in the intact embryo, and they provide no information about the ontogeny and function of CG expression. CG regulation has also been studied in primary cultures of human and nonhuman primate trophoblasts but the cytotrophoblast in these studies spontaneously differentiates to syncytial trophoblast and ceases cell division (Golos *et al.*, 1992). Therefore, any genetic technique requiring stable transfection and expansion of clones is impossible with these primary cytotrophoblast preparations.

Primate ES cells provide unique normal embryonic material for delineating the genetic mechanisms controlling differentiation and function of the trophoblast lineage and the regulation of trophoblast-specific genes such as that for CG. As mentioned, spontaneously differentiating rhesus and marmoset ES cells form trophoblast *in vitro*. We have begun to study the regulation of CG secretion in early trophoblast and find that CG secretion increases dramatically in differentiating ES cells of both rhesus and marmoset in the presence of gonadotropin-releasing hormone (GnRH) (Fig. 12). Because primate ES cells can be expanded indefinitely, the same genetic and cultural manipulations now standard for mouse ES cells can be applied to the differentiation and function of primate ES cell–

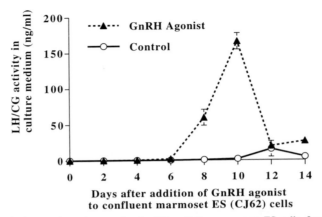

Fig. 12 Chorionic gonadotropin secretion in differentiating marmoset ES cells. Luteinizing hormone/chorionic gonadotropin (LH/CG) bioactivity was measured by Leydig cell bioassay in culture medium conditioned by differentiating ES cells cultured with or without gonadotropin-releasing hormone (GnRH) agonist D-Trp[6]-Pro[9]-NHEt (0.3 nM). Bars represent standard error of the mean. Adapted from Thomson *et al.* (1996, Fig. 4, p. 257).

derived trophoblast. By transfecting undifferentiated primate ES cells with expression constructs, identification of the *cis*-regulatory elements that allow faithful developmental expression of the CGα- and CGβ-subunit mRNA will be possible for the first time. A variety of techniques, including subtractive hybridization (Rothstein *et al.*, 1993), differential display (Liang and Pardee, 1992; Liang *et al.*, 1993), serial analysis of gene expression (Velculescu *et al.*, 1995), and promoter/enhancer trapping (Hill and Wurst, 1993), can now be used to identify genes uniquely expressed in ES cell–derived early trophoblast. Because differentiation of ES cells to trophoblast can be followed *in vitro*, newly identified genes can be assessed for their functional ability to direct trophoblast differentiation. The same general approaches could be applied to any cell type that differentiates from primate ES cells *in vitro*.

VI. Primate ES Cells and Primate Transgenic Models of Human Disease

One of the seminal achievements of mouse embryology is the use of ES cells to alter specific genes in the intact mouse. This has created a link between the *in vitro* manipulations of molecular biology and an *in vivo* understanding of gene function throughout development (Rossant *et al.*, 1993; Hogan *et al.*, 1994). With appropriate transfection and selection strategies, homologous recombination can be used to derive mouse ES cell lines with planned alterations of specific genes, and these genetically altered cells can be used to form chimeras with normal embryos. If the ES cells contribute to the germ line in the chimera, then, in the next generation, a progenitor for the planned mutation is established. Because homologous recombination can be used to "knock out" or modify endogenous genes in ES cells, very specific models of human genetic diseases now can be generated routinely in transgenic mice. However, given the differences between human and mouse anatomy and physiology in specific systems, transgenic mice can provide only a limited understanding of some human diseases. Transgenic primates would offer extremely accurate models of human disease. However, the possibility of producing transgenic primates raises important questions of scientific merit, practicality, and ethics.

Transgenic primates would have considerable scientific and medical merit if used as models of human diseases for which there currently exist no cure, no effective treatment, and no other accurate animal model. For example, transgenic mouse models of Alzheimer's disease have been produced by the expression of mutant β-amyloid precursor protein (βAPP) or of mutant presenilin, but to date transgenic mice have not accurately replicated the human disease (Games et al., 1995; Hsiao et al., 1996; Citron et al., 1997; Selkoe, 1997). The expression of a mutant human βAPP in mice, for example, has led to the development of diffuse and neuritic plaques, with accompanying memory deficit, but not the development of the neurofibrillary tangles also characteristic of the human disease (Hsiao et al., 1996). Future transgenic mice will probably lead to better models of Alzheimer's disease, but because of the basic differences in structure and function of the human and mouse brains, it is not yet clear whether a truly accurate mouse model is possible. In addition, mouse size, lifespan, physiology, brain complexity, and social behavior all limit therapeutic testing in any mouse model of human degenerative neural diseases. Based on an increased understanding of the molecular pathogenesis of Alzheimer's disease, broad categories of drugs are already being envisioned for potential treatments (Selkoe, 1997). These drugs are likely to be most effective early in the disease, and an accurate animal model is essential to test efficacy and safety, because direct testing on early clinical cases in humans could result in unexpected and permanent neurologic side effects. Transgenic primate models of degenerative diseases of the central nervous system would thus provide an increased understanding of pathogenesis and allow testing of the efficacy and safety of specific therapies. In primates, therapeutic efficacy could be assessed not only by morphologic and biochemical changes in the brain but by measures of improvement of specific complex behaviors.

Primate reproductive biology places practical restrictions on the generation of transgenic primates, and the use of primate ES cells in chimeras may prove particularly difficult. Typically, when generating transgenic mice with ES cells using conventional blastocyst injection, chimeras are formed, returned to foster mothers to develop to term (3 wk), allowed to develop to sexual maturity (6–8 wk), and then test bred to confirm germ line transmission (another 3-wk gestation); therefore, a minimum of about 4 mo is required to confirm germ line transmission of the genetic change of interest, and the process typically takes much longer. Such an approach for rhesus monkeys, with the birth of single young and sexual maturity occurring at 4–5 years, would be impractical. Even for the common marmoset, the process would be very time-consuming. With a gestation of 144 days and an age at sexual maturity of about 18 mo, well over 2 years would ultimately be required to confirm germ line transmission if the same techniques were used for the common marmoset. Also, the success of germ line transmission using mouse ES cells is highly strain-dependent and varies among research groups. As an example of the success rate, using a blastocyst injection technique, authors at a mouse embryology laboratory with extensive mouse ES cell experience reported that 40% of mouse chimeric embryos transferred resulted in live young, 11.4% were chimeric males (the ES cell line was XY), and 3.8% produced germ-line transmitters (Wood *et al.*, 1993). If similar success rates were obtained with marmoset ES cell chimeras, then about 1 in 26 chimeras transferred would result in germ line transmission, and the transfer of 60 chimeras would be required to have a greater than 90% chance ($1 - 0.962^{60} = 0.902$) of obtaining at least one chimera with a contribution to the germ line. Because only certain mouse strain combinations produce high-frequency germ line transmission, and because inbred strains of primates do not exist, it is possible that the frequency of germ line transmission even with proven germ line–competent marmoset ES cells could be substantially lower than for mouse ES cells.

The lack of inbred strains of primates poses another obstacle to the analysis of transgenic primates. The strength of ES cells for generating transgenic animals lies in the ability to modify specific endogenous genes, and usually the phenotypic effects of these targeted modifications must be studied in homozygotes. Outbred primates harbor deleterious recessive genes, and breeding of related individuals results in serious inbreeding depression (Ralls and Ballou, 1982). If a single founder animal harboring a recessive targeted mutation were to be bred through successive generations to produce homozygous offspring, it would be very difficult to determine whether phenotypic changes were caused by the disruption of the gene of interest or by endogenous deleterious genes bred to homozygosity. Inbreeding depression would also mean that routine maintenance of lines of primates homozygous for a transgene would be difficult or impossible, even if the transgene were innocuous. To study recessive targeted disruptions and avoid the effects of inbreeding, it would be necessary to produce independent (i.e., derived initially from genetically unrelated ES cell lines), heterozygous transgenic primate lines and to study the F1 homozygotes resulting from crosses between the two lines. This is a time-consuming proposition for any primate species. Methods to produce entire fetuses in one generation from primate ES cells, such as chimera formation with tetraploid embryos (Nagy *et al.*, 1993) or nuclear transfer to enucleated oocytes (Sims and First, 1994), may prevent some of these difficulties, but not enough is known about primate ES cells to predict whether these techniques are biologically possible.

Beyond the potential benefits to human health and the practical difficulties of primate transgenesis, is genetic manipulation of the nonhuman primate germ line ethically acceptable? Transgenic primates raise legitimate concerns, both for the welfare of individual monkeys and for the broader implications to human society. Because of nonhuman primate intelligence and evolutionary relatedness to humans, nonhuman primates are generally afforded special respect by society. In the United States this special respect among vertebrates is reflected by requirements in the U.S. Department of Agriculture Animal Welfare Act for environmental enhancements adequate to promote the psychological well-being of captive primates (Johnson et al., 1995). Additional requirements of the Animal Welfare Act include that animal care, treatment, and procedures ensure minimal animal pain and distress. Careful application of the current requirements of the Animal Welfare Act by Institutions' Animal Care and Use Committees (IACUC) should continue to protect the welfare of individual animals, even for any use of transgenic primates. In the authors' opinion, heightened scrutiny should be applied to any proposed use of transgenic primates, and experiments that are likely to produce severe pain or distress should be unacceptable even though they may provide important insights into human disease. Beyond the welfare of the individual animals, the broader implications of transgenic primates to society also emerge from the evolutionary kinship of human and nonhuman primates. Among multiple concerns is the fear that techniques that are developed for nonhuman primates may one day be applied to humans. At present, human reproductive biology provides a robust practical block to the use of human ES cell–intact embryo chimeras to genetically manipulate the human germ line. Human generation times are simply too long. However, future techniques could someday overcome this block, and other methods of introducing transgenes, such as direct DNA microinjection or retroviral vectors, would probably work for human embryos. As molecular biology and human gene therapy advance over the coming decades and it becomes increasingly possible to genetically modify the human germ line, strong international agreements will be needed to address these concerns.

VII. Implications of Human Embryonic Stem Cells

In the only published attempt to isolate human ES cells, ICM cells derived from donated, excess, IVF-produced embryos were grown in the presence of LIF but in the absence of fibroblast feeder layers (Bongso et al., 1994). Under these culture conditions, permanent cell lines were not established, suggesting that LIF is insufficient to prevent the differentiation of human ES cells, just as it is insufficient to prevent the differentiation of other primate ES cell lines. Because the same culture conditions allowed the derivation of ES cell lines from both an Old World and a New World primate species, it is very likely that similar conditions would allow the derivation of human ES cell lines. The derivation of human ES cell lines would have important ethical consequences and far-ranging potential medical benefits.

It is not known whether human ES cells could form a complete, viable embryo by any method, but this possibility has raised the greatest concern about derivation of human ES cells. If it were possible to form a complete embryo from human ES cells, it would be possible to form multiple complete embryos. Although ES cells have the potential to differentiate to any cell in the body, ES cells are not the equivalent of an intact embryo. If a clump of ES cells were transferred to a uterus, the ES cells would not form a viable fetus. Chimeras formed from mouse ES cells and tetraploid embryos allow the formation of a complete, viable fetus by the ES cell component in some cases (Nagy *et al.*, 1993), but it is not known whether this would be biologically possible with human ES cells. Nuclear transfer from mouse ES cells to enucleated oocytes or zygotes has never resulted in a live birth. However, in some domestic species, births have been reported after nuclear transfer from embryonic cells, early-passage embryonic cells, or an established embryonic cell line (First and Prather, 1991; Collas and Barnes, 1994; Sims and First, 1994; Campbell *et al.*, 1995). Births have also resulted from nuclear transfer of blastomere nuclei to enucleated oocytes in the rhesus monkey (Don Wolf, personal communication). The cloning of a sheep by transfer of an adult mammary epithelial cell nucleus to an enucleated oocyte (Wilmut *et al.*, 1997) surprised the developmental biology community and altered the concerns about cloning embryos from human ES cells. If nuclear transfer from adult cells to enucleated oocytes would allow development to term in humans, then the potential for cloning already exists, and the derivation of human ES cells would not significantly increase the potential for abuse. Attempted nuclear transfer from established cell lines or from primary adult cells to enucleated oocytes has not been reported for any primate species, so it is not known whether the biology of primates, including humans, would allow cloning by these methods.

In 1994 the National Institutes of Health (NIH) convened a panel of medical doctors, scientists, lawyers, sociologists, and ethicists to examine areas of potential research on *ex utero* preimplantation human embryos (Ad Hoc Group, 1994). Among the areas the panel recommended as acceptable for federal funding was "research involving the development of embryonic stem cells, but only with embryos resulting from IVF for infertility treatment or clinical research that have been donated with the consent of the progenitors." Currently in the United States, federal funds cannot be used for research on human preimplantation embryos. However, research is conducted privately in IVF clinics, and in the absence of other federal guidelines the recommendations of the NIH Human Embryo Research Panel should be followed; these include recommendations against funding work involving nuclear transfer to enucleated human oocytes or zygotes followed by embryo transfer and against any formation of chimeras with human embryos. As of this writing, legislation dealing with human cloning is being discussed in the United States.

Human ES cells would have profound implications for the treatment of human disease. All disease ultimately involves the death or dysfunction of cells. When disease results from abnormalities in specific cell types, as in Parkinson's disease (dopaminergic neurons) or juvenile-onset diabetes mellitus (pancreatic β-islet cells), then the replacement of those specific cell types can be envisioned as a potentially lifelong treatment. Because of the potential to differentiate to any cell type of the adult body, *in vitro*–grown and -differentiated human ES cells could provide a limitless source of specific cell types for transplantation. Furthermore, since undifferentiated ES cells can proliferate indefinitely in culture, individual cell lines could be extensively characterized, either for pathogens or for karyotypic abnormalities, before their use as a source for transplantation. Multiple human ES cell lines could be typed for major histocompatability complex (MHC) characteristics, so that close matching between patient and donor cells could be obtained. Alternatively, again because undifferentiated ES cells can proliferate indefinitely, genetic techniques could be used to specifically alter the MHC of ES cells or to alter other components of the cells and thus reduce or actively prevent an immune response. Finally, ES cells could be genetically modified so that their differentiated derivatives would actively combat specific disease.

The possibility of transplanting differentiated derivatives of human ES cells to treat specific disease raises questions of safety and efficacy. Malignant transformation of transplanted cells becomes a particular concern because of the probable long-term culture of ES cells before differentiation and transplantation. Juvenile-onset diabetes, for example, is a serious disease, but pancreatic cancer is usually rapidly fatal (Slack, 1995). Long-term culture could well introduce subtle genetic changes, in spite of the karyotypic stability generally observed in primate ES cells. Therefore, testing of the transplantation of specific cell types in an animal model before human clinical use would be absolutely essential, and rhesus ES cells and rhesus monkeys would provide an accurate animal model for this testing. Indeed, in the example mentioned, Parkinson's disease and diabetes mellitus, extremely accurate models are available in the rhesus monkey (Jones *et al.*, 1980; Burns *et al.*, 1983). Because of the potential for malignant transformation, the introduction of "suicide genes" that would allow the selective killing of transplanted cells by administration of specific drugs would provide an extra level of safety. Fail-safe methods, such as providing the transplanted cells with an immunologically privileged site, could allow the host's immune system to destroy non–MHC-matched cells that escape that site if malignant transformation occurred. For many diseases, more sophisticated methods of preventing immune rejection must be developed; many diseases that could be treated by the transplantation of ES cell–derived differentiated cells are less severe than are the sequelae of the current methods of immunosuppression that may be required to prevent rejection by the immune system.

Primate ES cells offer a new window for understanding the basic processes controlling human development, one that complements and extends other developmental models. Mammalian developmental biology and the Human Genome Project are advancing at a breathtaking pace, and genes regulating specific developmental events are rapidly being identified. Given the pace of these advancements, the ability to direct ES cells to any cell type is not far off. That this basic knowledge could very quickly lead to lifelong treatment of serious human disease makes this an exciting time for developmental biology.

Acknowledgments

The authors thank Martin Pera for the gift of the GCT 27, Peter Andrews for the gift of NTERA2 cells, Colin Stewart for the LIF knockout fibroblasts, and David Williams for the S1/S1⁴ fibroblasts. We also thank Jennifer Kalishman, Michelle Waknitz, and Robert Becker for technical assistance. This research was supported by NIH grants RR00167 and RR11571-01 (to J.A.T.).

References

Abbondanzo, S. J., Gadi, I., and Stewart, C. L. (1993). Derivation of embryonic stem cells. *Methods Enzymol.* **225,** 803–822.

Ad Hoc Group of Consultants to the Advisory Committee to the Director, NIH. (1994). "Report of the Human Embryo Research Panel." National Institutes of Health, Bethesda, Maryland.

Alak, B. M., and Wolf, D. P. (1994). Rhesus monkey oocyte maturation and fertilization *in vitro*: Role of the menstrual cycle phase and of exogenous gonadotropins. *Biol. Reprod.* **51,** 879–887.

Andrews, P. W. (1988). Human teratocarcinomas. *Biochim. Biophys. Acta* **948,** 17–36.

Andrews, P. W., Banting, G., Damjanov, I., Arnaud, D., and Avner, P. (1984). Three monoclonal antibodies defining distinct differentiation antigens associated with different high molecular weight polypeptides on the surface of human embryonal carcinoma cells. *Hybridoma* **3,** 347–361.

Andrews, P. W., Bronson, D. L., Benham, F., Strickland, S., and Knowles, B. B. (1980). A comparative study of eight cell lines derived from human testicular teratocarcinoma. *J. Cancer* **26,** 269–280.

Andrews, P. W., Damjanov, I., Simon, D., Banting, G., Carlin, C., Dracopoli, N., and Fogh, J. (1984). Pluripotent embryonal carcinoma clones derived from the human teratocarcinoma cell line Tera-2. *Lab. Invest.* **50,** 147–162.

Axelrod, H. R. (1984). Embryonic stem cell lines derived from blastocysts by a simplified technique. *Dev. Biol.* **101,** 225–228.

Beddington, R. S. P., and Robertson, E. J. (1989). An assessment of the developmental potential of embryonic stem cells in the midgestation mouse embryo. *Development* **105,** 733–737.

Benirschke, K., and Kaufmann, P. (1990). "Pathology of the Human Placenta." Springer-Verlag, New York.

Bongso, A., Fong, C. Y., Ng, S. C., and Ratnam, S. (1994). Isolation and culture of inner cell mass cells from human blastocysts. *Hum. Reprod.* **9,** 2110–2117.

Bradley, A., Evans, M., Kaufmann, M., and Robertson, E. (1984). Formation of germ-line chimaeras from embryo-derived teratocarcinoma cell lines. *Nature* **309,** 255–256.

Braude, P., Bolton, V., and Moore, S. (1988). Human gene expression first occurs between the four- and eight-cell stages of preimplantation development. *Nature* **332,** 459–461.

Brinster, R. L. (1974). The effect of cells transferred into the mouse blastocyst on subsequent development. *J. Exp. Med.* **140,** 1049–1056.

Burns, R. S., Chiueh, C. C., Markey, S. P., Ebert, M. H., Jacobowitz, D. M., and Kopin, I. J. (1983). A primate model of parkinsonism: selective destruction of dopaminergic neurons in the pars compacta of the substantia nigra by N-methyl-4-phenyl-1,2,3,6-tetrahydropyridine. *Proc. Natl. Acad. Sci. U.S.A.* **80,** 4546–4550.

Campbell, K., McWhir, J., Ritchie, B., and Wilmut, I. (1995). Production of live lambs following nuclear transfer of cultured embryonic disc cells. *Theriogenology* **43,** 181–190.

Chambers, P. L., and Hearn, J. P. (1985). Embryonic, foetal and placental development in the common marmoset (*Callithrix jacchus*). *J. Zool. Lond.* **207,** 545–561.

Citron, M., Westaway, D., Xia, W., Carlson, G., Diehl, T., Levesque, G., Johnson-Wood, K., Lee, M., Seubert, P., Davis, A., Kholodenko, D., Motter, R., Sherrington, R., Perry, B., Yao, H., Strome, R., Lieburg, I., Rommens, J., Kim, S., Schenk, D., Fraser, P., Hyslop, P., and Selkoe, D. J. (1997). Mutant presenilins of Alzheimer's disease increase production of 42-residue amyloid β-protein in both transfected cells and transgenic mice. *Nat. Med.* **3,** 67–72.

Collas, P., and Barnes, F. L. (1994). Nuclear transplantation by microinjection of inner cell mass and granulosa cell nuclei. *Mol. Reprod. Dev.* **38,** 264–267.

Conover, J. C., Ip, N. Y., Poueymirou, W. T., Bates, B., Goldfarb, M. P., DeChiara, T. M., and Yancopoulos, G. D. (1993). Ciliary neurotrophic factor maintains the pluripotentiality of embryonic stem cells. *Development* **119,** 559–565.

Doetschman, T., Eistetter, H., Katz, M., Schmidt, W., and Kemler, R. (1985). The *in vitro* development of blastocyst-derived embryonic stem cell lines: Formation of visceral yolk sac, blood islands and myocardium. *J. Embryol. Exp. Morph.* **87,** 27–45.

Dokras, A., Sargent, I. L., Ross, C., Gardner, R. L., and Barlow, D. H. (1991). The human blastocyst: Morphology and human chorionic gonadotrophin secretion *in vitro*. *Hum. Reprod.* **6,** 1143–1151.

Edwards, R. G. (1988). Human uterine endocrinology and the implantation window. *Ann. N.Y. Acad. Sci.* **541,** 445–454.

Evans, M., and Kaufman, M., (1981). Establishment in culture of pluripotential cells from mouse embryos. *Nature* **292,** 154–156.

First, N. L., and Prather, R. S. (1991). Genomic potential in mammals. *Differentiation* **48,** 1–8.

Flint, A. P. F., Hearn, J. P., and Michael, A. E. (1990). The maternal recognition of pregnancy in mammals. *J. Zool. Lond.* **221,** 327–341.

Games, D., Adams, D., Alessandrini, R., Barbour, R., Berthelette, P., Blackwell, C., Carr, T., Clemens, J., Donaldson, T., Gillespie, F., Guido, T., Hagoplan, S., Johnson-Wood, K., Khan, K., Lee, M., Leibowitz, P., Lieberburg, I., Little, S., Masliah, E., McConlogue, L., Montoya-Zavala, M., Mucke, L., Paganini, L., Penniman, E., Power, M., Schenk, D., Seubert, P., Snyder, B., Soriano, F., Tan, H., Vitale, J., Wadsworth, S., Wolozin, B., and Zhao, J. (1995). Alzheimer-type neuropathology in transgenic mice overexpressing V717F beta-amyloid precursor protein. *Nature* **373,** 523–527.

Gardner, R. L. (1982). Investigation of cell lineage and differentiation in the extraembryonic endoderm of the mouse embryo. *J. Embryol. Exp. Morphol.* **68,** 175–198.

Golos, T., Handrow, R., Durning, M., Fisher, J., and Rilling, J. (1992). Regulation of chorionic gonadotropin and chorionic somatomammotropin messenger ribonucleic acid expression by 8-bromo-adenosine 3′,5′-monophosphate and dexamethazone in cultured rhesus monkey syncytiotrophoblast. *Endocrinology* **131,** 89–100.

Goodeaux, L. L., Anzalone, C. A., Thibodeaux, J. K., Menezo, Y., Roussel, J. D., and Voelkel,

S. A. (1990). Successful nonsurgical collection of *Macaca mulatta* embryos. *Theriogenology* **34,** 1159–1167.

Hall, P. A., and Watt, F. M. (1989). Stem cells: The generation and maintenance of cellular diversity. *Development* **106,** 619–633.

Hearn, J. P., Abbott, D. H., Chambers, P. C., Hodges, J. K., and Lunn, S. F. (1978). Use of the common marmoset, *Callithrix jacchus,* in reproductive research. *Primates in Medicine* **10,** 40–49.

Hearn, J. P., Gidley-Baird, A. A., Hodges, J. K., Summers, P. M., and Webley, G. E. (1988). Embryonic signals during the peri-implantation period in primates. *J. Reprod. Fertil.* (Suppl.) **36,** 49–58.

Hearn, J. P., Hodges, J. K., and Gems, S. (1988). Early secretion of chorionic gonadotrophin by marmoset embryos *in vivo* and *in vitro. J. Endocrinol.* **119** 249–255.

Hearn, J. P., and Webley, G. E. (1993). The regulation of chorionic gonadotropin during embryo-maternal attachment and implantation in primates. *In* "Implantation in Mammals" (L. Gianaroli, A. Campana, and A. O. Trounson, Eds.), pp. 37–47. Raven Press, New York.

Hill, D. P., and Wurst W. (1993). Screening for novel pattern formation genes using gene trap approaches. *Methods Enzymol.* **225,** 664–680.

Hogan, B., Beddington, R., Costantini, F., and Lacey, E. (1994). "Manipulating the Mouse Embryo: A Laboratory Manual." Cold Spring Harbor Laboratory Press, Plainview, New York.

Hsiao, K., Chapman, P., Nilsen, S., Eckman, C., Harigaya, Y., Younkin, S., Yang, F., and Cole, G. (1996). Correlative memory deficits, Aβ elevation, and amyloid plaques in transgenic mice. *Science* **274,** 99–102.

Illmensee, K., and Mintz, B. (1976). Totipotency and normal differentiation of single teratocarcinoma cells cloned by injection into blastocysts. *Proc. Natl. Acad. Sci. U.S.A.* **73,** 549–553.

Johnson, D. K., Morin, M. L., Bayne, K. A. L., and Wolfe, T. L. (1995). Laws, Regulations, and Policies. *In* "Nonhuman Primates in Biomedical Research: Biology and Management" (T. B. Bennett, C. R. Abee, and R. Henrickson, Eds.). Academic Press, San Diego.

Jones, C. W., Reynolds, W. A., and Hoganson, G. E. (1980). Streptozotocin diabetes in the monkey: Plasma levels of glucose, insulin, glucagon, and somatostatin, with corresponding morphometric analysis of islet endocrine cells. *Diabetes* **29,** 536–546.

Kannagi, R., Cochran, N. A., Ishigami, F., Hakomori, S., Andrews, P. W., Knowles, B. B., and Solter, D. (1983). Stage-specific embryonic antigens (SSEA-3 and -4) are epitopes of a unique globo-series ganglioside isolated from human teratocarcinoma cells. *EMBO J.* **2,** 2355–2361.

Kaufman, M. H. (1992). "The Atlas of Mouse Development." Academic Press, London.

King, B. F., and Blankenship, T. N. (1993). Development and organization of primate trophoblast cells. *In* "Trophoblast Cells" (M. J. Soares, S. Handwerger, and F. Talamantes, Eds.), pp. 13–30. Springer-Verlag, New York.

King, F. A., Yarbrough, C. J., Andersen, D. C., Gordon, T. P., and Gould, K. G. (1988). Primates. *Science* **240,** 1475–1482.

Knobil, E. (1973). On the regulation of the primate corpus luteum. *Biol. Reprod.* **8,** 246–258.

Liang, P., Averboukh, L., and Pardee, A. B. (1993). Distribution and cloning of eukaryotic mRNAs by means of differential display: Refinements and optimization. *Nucleic Acids Res.* **21,** 3269–3275.

Liang, P., and Pardee, A. B. (1992). Differential display of eukaryotic messenger RNA by means of the polymerase chain reaction. *Science* **257,** 967–971.

Lopata, A., Summers, P., and Hearn, J. (1988). Births following the transfer of cultured embryos obtained by in vitro and in vivo fertilization in the marmoset monkey (*Callithrix jacchus*). *Fertil. Steril.* **50,** 503–509.

Luckett, W. P. (1975). The development of primordial and definitive amniotic cavities in early rhesus monkey and human embryos. *Am. J. Anat.* **144,** 149–168.

Luckett, W. P. (1978). Origin and differentiation of the yolk sac extraembryonic mesoderm in pre-somite human and rhesus monkey embryos. *Am. J. Anat.* **152,** 59–98.

Marshall, V. S., Kalishman, J., and Thomson, J. A. (1997). Nonsurgical embryo transfer in the common marmoset monkey. *J. Med. Primat.,* in press.

Martin, G. (1981). Isolation of a pluripotent cell line from early mouse embryos cultured in medium conditioned by teratocarcinoma stem cells. *Proc. Natl. Acad. Sci. U.S.A.* **78,** 7634–7638.

Matsui, Y., Zsebo, K., and Hogan, B. L. (1992). Derivation of pluripotential embryonic stem cells from murine primordial germ cells in culture. *Cell* **70,** 841–847.

Mintz, B., and Cronmiller, C. (1978). Normal blood cells of anemic genotype in teratocarcinoma-derived mosaic mice. *Proc. Natl. Acad. Sci. U.S.A.* **75,** 6247–6251.

Mintz, B., and Cronmiller, C. (1981). METT-1: A karyotypically normal *in vitro* line of developmentally totipotent mouse teratocarcinoma cells. *Somat. Cell Genet.* **7,** 489–505.

Mintz, B., and Illmensee, K. (1975). Normal genetically mosaic mice produced from malignant teratocarcinoma cells. *Proc. Natl. Acad. Sci. U.S.A.* **72,** 3585–3589.

Moore, H. D., Gems, S., and Hearn, J. P. (1985). Early implantation stages in the marmoset monkey (*Callithrix jacchus*). *Am. J. Anat.* **172,** 265–278.

Nagy, A., Rossant, J., Nagy, R., Abramow-Newerly, W., and Roder, J. C. (1993). Derivation of completely cell culture-derived mice from early-passage embryonic stem cells. *Proc. Natl. Acad. Sci. U.S.A.* **90,** 8424–8428.

O'Rahilly, R., and Muller, F. (1987). "Developmental Stages in Human Embryos." Carnegie Institution of Washington, Washington.

Phillips, I. (1976). The embryology of the common marmoset. *Adv. Anat. Embryol. Cell Biol.* **52,** 3–47.

Potten, C. S., and Loeffler, M. (1990). Stem cells: Attributes, cycles, spirals, pitfalls, and uncertainties. Lessons for and from the crypt. *Development* **110,** 1001–1020.

Ralls, K., and Ballou, J. (1982). Effects of inbreeding on infant mortality in captive primates. *Int. J. Primatol.* **3,** 491–505.

Resnick, J. L., Bixler, L. S., Cheng, L., and Donovan, P. J. (1992). Long-term proliferation of mouse primordial germ cells in culture. *Nature* **359,** 550–551.

Reynolds, B. A., and Weiss, S. (1996). Clonal and population analyses demonstrate that an EGF-responsive mammalian embryonic CNS precursor is a stem cell. *Dev. Biol.* **175,** 1–13.

Roach, S., Cooper, S., Bennett, W., and Pera, M. F. (1993). Cultured cell lines from human teratomas: Windows into tumour growth and differentiation and early human development. *Eur. Urol.* **23,** 82–88.

Robertson, E. J. (1987). Embryo-derived stem cell lines. *In* "Teratocarcinomas and Embryonic Stem Cells: A Practical Approach" (E. J. Robertson, Ed.), pp. 71–112. IRL Press, Washington, DC.

Rose, T. M., Weiford, D. M., Gunderson, N. L., and Bruce, A. G. (1994). Oncostatin M (OSM) inhibits the differentiation of pluripotent embryonic stem cells in vitro. *Cytokine* **6,** 48–54.

Rossant, J., Bernelot-Moens, C., and Nagy, A. (1993). Genome manipulation in embryonic stem cells. *Philos. Trans. R. Soc. Lond. B Biol. Sci.* **339,** 207–215.

Rossant, J., and Papaioannou, V. (1984). The relationship between embryonic, embryonal carcinoma and embryo-derived stem cells. *Cell Differ.* **15,** 155–161.

Rothstein J. L., Johnson, D., Jessee, J., Skowronski, J., DeLoia, J. A., Solter, D., and Knowles, B. B. (1993). Construction of primary and subtracted cDNA libraries from early embryos. *Methods Enzymol.* **225,** 587–610.

Schramm, R. D., and Bavister, B. D. (1995). Effects of granulosa cells and gonadotrophins on meiotic and developmental competence of oocytes *in vitro* in non-stimulated rhesus monkeys. *Hum. Reprod.* **10,** 887–895.

Selkoe, D. J. (1997). Alzheimer's disease: Genotypes, phenotypes, and treatments. *Science* **275,** 630–631.

Seshagiri, P. B., Bridson, W. E., Dierschke, D. J., Eisele, S. G., and Hearn, J. P. (1993). Non-surgical uterine flushing for the recovery of preimplantation embryos in rhesus monkeys: Lack of seasonal infertility. *Am. J. Primatol.* **29,** 81–91.

Seshagiri, P. B., and Hearn, J. P. (1993). *In vitro* development of *in vivo*–produced rhesus monkey morulae and blastocysts to hatched, attached, and post-attached blastocyst stages: Morphology and early secretion of chorionic gonadotrophin. *Hum. Reprod.* **8,** 279–287.

Seshagiri, P. B., Terasawa, E., and Hearn, J. P. (1994). The secretion of gonadotrophin-releasing hormone by peri-implantation embryos of the rhesus monkey: Comparison with the secretion of chorionic gonadotrophin. *Hum. Reprod.* **9,** 1300–1307.

Sims, M., and First, N. L. (1994). Production of calves by transfer of nuclei from cultured inner cell mass cells. *Proc. Natl. Acad. Sci. U.S.A.* **91,** 6143–6173.

Slack, J. M. W. (1995). Developmental biology of the pancreas. *Development* **121,** 1569–1580.

Solter, D., and Knowles, B. (1975). Immunosurgery of mouse blastocysts. *Proc. Natl. Acad. Sci. U.S.A.* **72,** 5099–5102.

Solter, D., and Knowles, B. B. (1978). Monoclonal antibody defining a stage-specific mouse embryonic antigen (SSEA-1). *Proc. Natl. Acad. Sci. U.S.A.* **75,** 5565–5569.

Stewart, C. L., Kaspar, P., Brunet, L. J., Bhatt, H. Gadi, I., Kontgen, F., Abbondanzo, S. J. (1992). Blastocyst implantation depends on maternal expression of leukaemia inhibitory factor. *Nature* **359,** 76–79.

Stewart, T. A., and Mintz, B. (1982). Recurrent germ-line transmission of the teratocarcinoma genome from the METT-1 culture line to progeny *in vivo*. *J. Exp. Zool.* **224,** 465–469.

Summers, P. M., Shephard, A. M., Taylor, C. T., and Hearn, J. P. (1987). The effects of cryopreservation and transfer on embryonic development in the common marmoset monkey, *Callithrix jacchus*. *J. Reprod. Fertil.* **79,** 241–250.

Summers, P. M., Wennink, C., J., and Hodges, J. K. (1985). Cloprostenol-induced luteolysis in the marmoset monkey (*Callithrix jacchus*). *J. Reprod. Fertil.* **73,** 133–138.

Thomson, J. A., Kalishman, J., Golos, T. G., Durning, M., Harris, C. P., Becker, R. A., and Hearn, J. P. (1995). Isolation of a primate embryonic stem cell line. *Proc. Natl. Acad. Sci. U.S.A.* **92,** 7844–7848.

Thomson, J. A., Kalishman, J., Golos, T. G., Durning, M., Harris, C. P., and Hearn, J. P. (1996). Pluripotent cell lines derived from common marmoset (*Callithrix jacchus*) blastocysts. *Biol. Reprod.* **55,** 254–259.

Thomson, J. A., Kalishman, J., and Hearn, J. P. (1994). Non-surgical uterine stage preimplantation embryo collection from the common marmoset. *J. Med. Primatol.* **23,** 333–336.

Toksoz, D., Zsebo, K. M., Smith, K. A., Hu, S., Brankow, D., Suggs, S. V., Martin, F. H., Williams, D. A. (1992). Support of human hematopoiesis in long-term bone marrow cultures by murine stromal cells selectively expressing the membrane-bound and secreted forms of the human homolog of the steel gene product, stem cell factor. *Proc. Natl. Acad. Sci. U.S.A.* **89,** 7350–7354.

Velculescu, V., Zhang, L., Vogelstein, B., and Kinzler, K. W. (1995). Serial analysis of gene expression. *Science* **270,** 484–487.

Wenk, J., Andrews, P. W., Casper, J., Hata, J., Pera, M. F., von Keitz, A., Damjanov, I., and Fenderson, B. A. (1994). Glycolipids of germ cell tumors: Extended globo-series glycolipids are a hallmark of human embryonal carcinoma cells. *Int. J. Cancer* **58,** 108–115.

Weston, A. M., Zelinski-Wooten, M. B., Hutchison, J. S., Stouffer, R. L., and Wolf, D. P. (1996). Developmental potential of embryos produced by *in-vitro* fertilization from gonadotrophin-releasing hormone antagonist–treated macaques stimulated with recombinant human follicle stimulating hormone alone or in combination with luteinizing hormone. *Hum. Reprod.* **11,** 608–613.

Williams, R., Hilton, D., Pease, S., Wilson, T., Stewart, C., Gearing, D., Wagner, E., Metcalf, D.,

Nicola, N., and Gough, N. (1988). Myeloid leukaemia inhibitory factor maintains the developmental potential of embryonic stem cells. *Nature* **336,** 684–687.

Wilmut, I., Schnieke, A. E., McWhir, J., Kind, A. J., and Campbell, K. H. S. (1997). Viable offspring derived from fetal and adult mammalian cells. *Nature* **385,** 810–813.

Wilton, L. J., Marshall, V. S., Piercy, E. C., and Moore, H. D. M. (1993). *In vitro* fertilization and embryo development in the marmoset monkey (*Callithrix jacchus*). *J. Reprod. Fertil.* **97,** 481–486.

Winkel, G. K., and Pedersen, R. A. (1988). Fate of the inner cell mass in mouse embryos as studied by microinjection of lineage tracers. *Dev. Biol.* **127,** 143–156.

Wolf, D. P., Alexander, M., Zelinski-Wooten, M., and Stouffer, R. L. (1996). Maturity and fertility of rhesus monkey oocytes collected at different intervals after an ovulatory stimulus (human chorionic gonadotropin) in *in vitro* fertilization cycles. *Mol. Reprod. Dev.* **43,** 76–81.

Wolf, D. P., Thomson, J. A., Zelinski-Wooten, M. B., and Stouffer, R. L. (1990). *In vitro* fertilization embryo transfer in nonhuman primates: The technique and its applications. *Mol. Reprod. Dev.* **27,** 261–280.

Wolf, E., Kramer, R., Polejaeva, I., Thoenen, H., and Brem, G. (1994). Efficient generation of chimaeric mice using embryonic stem cells after long-term culture in the presence of ciliary neurotrophic factor. *Transgenic Res.* **3,** 152–158.

Wood, S. A., Allen, N. D., Rossant, J., Auerbach, A., and Nagy, A. (1993). Non-injection methods for the production of embryonic stem cell-embryo chimaeras. *Nature* **365,** 87–89.

Yoshida, K., Chambers, I., Nichols, J., Smith, A., Saito, M., Yasukawa, K., Shoyab, M., Taga, T., and Kishimoto, T. (1994). Maintenance of the pluripotential phenotype of embryonic stem cells through direct activation of gp130 signalling pathways. *Mech. Dev.* **45,** 163–171.

Zhang, L., Weston, A. M., Denniston, R. S., Goodeaux, L. L., Godke, R. A., and Wolf, D. P. (1994). Developmental potential of rhesus monkey embryos produced by *in vitro* fertilization. *Biol. Reprod.* **51,** 433–440.

5

Sex Determination in Plants

Charles Ainsworth,[1] John Parker,[2] and Vicky Buchanan-Wollaston[1]
Plant Molecular Biology Laboratory[1]
Wye College, University of London
Ashford, Kent TN25 5AH, United Kingdom

University Botanic Garden[2]
Cambridge University
Cambridge CB2 1JF, United Kingdom

The majority of flowering plants produce flowers that are "perfect." These flowers are both staminate (with stamens) and pistillate (with one or more carpels). In a small number of species, there is spatial separation of the sexual organs either as monoecy, where the male and female organs are carried on separate flowers on the same plant, or dioecy, where male and female flowers are carried on separate male (staminate) or female (pistillate) individuals. Sex determination systems in plants, leading to unisexuality as monoecy or dioecy, have evolved independently many times. In dioecious plant species, the point of divergence from the hermaphrodite pattern shows wide variation between species, implying that the genetic bases are very different. This review considers monoecious and dioecious flowering plants and focuses on the underlying genetic and molecular mechanisms. We propose that dioecy arises either from monoecy as an environmentally unstable system controlled by plant growth substances or from hermaphroditism where the underlying mechanisms are highly stable and control does not involve plant growth substances. Copyright © 1998 by Academic Press.

Current Topics in Developmental Biology, Vol. 38
Copyright © 1998 by Academic Press. All rights of reproduction in any form reserved.
0070-2153/98 $25.00

I. Introduction

Most animal species are unisexual, with male and female gametes produced directly by meiosis in different individuals. By contrast, the great majority of flowering plants produce flowers that are "perfect." These flowers are both staminate (with stamens) and pistillate (with one or more carpels). The stamens produce microspores by meiosis; within the microspores, the sperm develop by mitosis. Megaspores are produced by meiosis within ovules carried in carpels, and mitotic divisions give rise to an embryo sac containing an egg cell. Thus the stamens can be regarded as the "male" organs of a flower and the carpels as the "female" organs, although these terms technically should be reserved for gametes or the direct producers of gametes (male and female animals). For simplicity, we will use male and female in this paper to refer to stamens and carpels and also to the staminate and pistillate individuals in dioecious species.

Sexual reproduction is the major mechanism by which genetic variation derived from mutation is resorted and redistributed, and in plants the basic flower pattern has been modified such that outcrossing is favored (Darwin, 1876). The different mechanisms adopted can be divided into three classes. First, there can be a temporal separation of the maturation of the male and female organs within an otherwise perfect flower. This phenomenon, termed dichogamy, includes protogyny (in which the stigma of the female is receptive before the release of the pollen from the anthers) and protandry (in which the pollen is shed before the stigma is receptive) (reviewed in Bertin and Newman, 1993). These systems do not preclude self-fertilization, which can occur by crossing between different flowers on the same plant. Second, there can be self-incompatibility mechanisms, both sporophytic (reviewed by Nasrullah and Nasrullah, 1993) and gametophytic (reviewed by Newbegin *et al.*, 1993), in which there is genetic control over the fertilization events that are possible. Third, there can be spatial separation of the sexual organs, either as monoecy, where the male and female organs are carried on separate flowers on the same plant (again allowing selfing), or dioecy, where male and female flowers are carried on separate male (staminate) or female (pistillate) individuals (see Figs. 1 and 3). These unisexual states are not always discrete, and intermediate states exist that are discussed later. Although floral unisexuality among higher plants is much rarer than hermaphroditism, a number of agronomically important plants are monoecious or dioecious. The monoecious species include cucumber (*Cucumis sativus*), melon (*Cucumis melo*), oil palm (*Elaeis guineensis*), hazelnut (*Corylus* spp.), fig (*Ficus carica*), castor bean (*Ricinus communis*), walnut (*Juglans regia*), and maize (*Zea mays*). The dioecious species include asparagus (*Asparagus officinalis*), date palm (*Phoenix dactylifera*), kiwifruit (*Actinidia deliciosa*), papaya (*Carica papaya*), hemp (*Cannabis sativa*), yam (*Dioscorea* spp.), hop (*Humulus lupulus*), pistachio (*Pistacia vera*), cloudberry (*Rubus chamaemorus*), poplar (*Populus* spp.), willow (*Salix* spp.), spinach (*Spinacia oleracea*), and mistletoe (*Viscum album*).

During the last decade, the increasingly sophisticated techniques of molecular biology have been applied to plant developmental biology. One result is that our knowledge of the genetic control of the development of perfect (hermaphrodite) flowers has expanded from a base of almost no information to a situation in which many of the genes that control meristem identity (inflorescence and floral meristems), floral organ identity, and flower shape have been isolated and characterized. Much of this knowledge has resulted from the research of two groups working on two quite different flowering plants—that led by Enrico Coen working on *Antirrhinum majus*, Scrophulariaceae (the snapdragon), and that of Elliot Meyerowitz working on *Arabidopsis thaliana*, Cruciferae—or from the catalytic effects of their findings on research on other plant species. Such fundamental information on the genes themselves and the genetic interactions that take place during the development of hermaphrodite flowers, along with the enabling molecular biology techniques, have allowed the area of sex determination and control of flower development in plants with unisexual flowers to be addressed in a way that was not possible previously.

This review focuses primarily on sex determination in monoecious and dioecious angiosperm (flowering plant) species and, in particular, addresses the molecular biology experiments that have been or are being undertaken in the monoecious plant species maize and cucumber and in a group of contrasting dioecious plant species (Table I).

II. Breeding Systems Used in Flowering Plants

A. Flowering and Inflorescences

In animals, the body plan and organ systems are established during development of the embryo. Flowering plants, however, have adopted a very different developmental strategy, with the various organs arising from meristems, collections of undifferentiated cells set aside during embryogenesis. In the initial stages of plant development, the apical vegetative meristem gives rise to the vegetative organ primordia on its flanks, which develop into leaves or shoots. The transition from the vegetative phase to the phase of floral development causes the vegetative meristem to give rise to either an inflorescence or a flower. In the former case, the apical vegetative meristem changes to an inflorescence meristem and generates a series of lateral floral meristems on its flanks. This is an indeterminate developmental pattern. This pattern is modified in determinate inflorescences, in which either the shoot apex terminates in a solitary flower [as, for example, in most tulips (*Tulipa* species) and single narcissi (*Narcissus* species)] or the inflorescence branches each terminate in a flower.

A. majus produces indeterminate inflorescences, but the *centroradialis* mutation causes the inflorescence to terminate in a flower. The *CEN* gene has been

Table I. Monoecious and Dioecious Plants That Have Been Studied in Detail

Common name	Species	Family	Breeding system	Heteromorphic sex chromosomes	Step blocked in male and female development
Maize	*Zea mays*	Poaceae	monoecious	—	Organ development
Cucumber	*Cucumis sativus*	Cucurbitaceae	monoecious	—	Organ development
Annual mercury	*Mercurialis annua*	Euphorbiaceae	dioecious	None	Organ initiation
Hop	*Humulus lupulus*	Cannabidaceae	dioecious	X:A	Organ initiation
Sorrel	*Rumex acetosa*	Polygonaceae	dioecious	X:A	Organ development
White campion	*Silene latifolia*	Caryophyllaceae	dioecious	Active Y	Organ maturation
Asparagus	*Asparagus officinalis*	Liliaceae	dioecious	?	Late organ maturation
Kiwifruit	*Actinidia deliciosa*	Actinidiaceae	dioecious	None	Pollen/ovule development

cloned and has been shown to be expressed in the inflorescence apex a few days after floral induction. It interacts with the floral meristem identity gene *FLO-RICAULA* to regulate flower position and morphology (Bradley *et al.*, 1996). The *CEN* gene is one of a number of genes that have been isolated that affect meristem identity and are involved in inflorescence and flower production. Because meristem identity does not have a direct impact on sex determination in plants with unisexual flowers, the reader is directed to the reviews by Yanofsky (1995), Weigel (1995), and Weigel and Clark (1996) for further information. The process of floral induction itself is not well understood (see Bernier *et al.*, 1993, and Weigel, 1995, for reviews).

B. Floral Structure in Hermaphrodite Plant Species

Regardless of the type of inflorescence, a basic flower pattern exists in angiosperm species for the major class of plants (i.e., those that carry hermaphrodite flowers). It seems likely (see later discussion) that the unisexual types of flower have evolved from this basic pattern.

From initial research on *Antirrhinum* and *Arabidopsis* it has become clear that the hermaphrodite flower can be divided into four concentric whorls, each whorl containing a different set of organs (Fig. 1). Two whorls comprise the perianth (sepals and petals) and two whorls are reproductive (stamens and carpels). The outermost whorl (whorl 1) contains the sepals; whorl 2 contains the petals; whorl 3 contains the male reproductive organs, the stamens, which are collectively termed the androecium; and the central whorl (whorl 4) contains the female reproductive organs, the carpels, collectively termed the gynoecium (or pistil). The numbers of organs within each whorl varies considerably between different plant species. For example, in *Antirrhinum* there are five sepals, five petals, four stamens (the fifth, dorsal stamen aborts during development), and two carpels. In *Arabidopsis*, there are four sepals, four petals, six stamens, and two carpels. In contrast to these simple patterns, some plants show variation in whorl number or in number of organs between whorls. Members of the family Rosaceae, with a perianth part number based on five, can have 100 or even more stamens. Some plants clearly are more complicated, such as those of *Passiflora* spp., which carry one or several petaloid or staminoid outgrowths from the central column (androgynophore) of the flower, forming a feature known as a corona (Hickey and King, 1988). The homologies of this structure with the organs of more standard flowers are unknown. Other plants, such as *Clematis* spp., have only three whorls, there being only one whorl of perianth segments. Genetic control of whorl and organ number is not well understood, but *FON1*, a gene that controls floral organ number in *Arabidopsis*, has been identified; it functions to terminate floral meristem activity after the floral organ primordia have been formed (Huang and Ma, 1997).

Pioneering genetic analysis and molecular biology work on *Antirrhinum* and *Arabidopsis* homeotic mutants in which the organs of one whorl are replaced by organs of a different type (Coen and Meyerowitz, 1991), now supported by information from a wide variety of other plant species including tomato, tobacco, and petunia, has enabled a model for the control of organ determination in the generic angiosperm flower to be proposed (Coen and Meyerowitz, 1991). In this model, organ position and identity within the flower are controlled by the action of homeotic genes that are active in three overlapping domains (termed A, B, and C) during early development of the floral meristem (Coen and Meyerowitz, 1991). Each domain of gene expression encompasses two adjacent floral whorls, so that the A function includes whorls 1 and 2, the B function includes whorls 2 and 3, and the C function includes whorls 3 and 4 (Fig. 2). The model is combinatorial in that petals and stamens are specified by A plus B and B plus C, respectively, whereas sepals and carpels are each specified by a single function, A and C, respectively. The primary evidence for the ABC model derives from analysis of homeotic mutations affecting these organ whorls; in all mutants, organs in two adjacent whorls are affected (see Fig. 2). In the A function mutants *ovulata* (*Antirrhinum*) and *apetala2* (*Arabidopsis*), the third and fourth whorls are normal but the first and second whorl organs are converted to carpels and stamens, respectively. In the B function mutants *deficiens* and *globosa* (*Antirrhinum*) and *apetala3* and *pistillata* (*Arabidopsis*), the first and fourth whorls are normal but the second and third whorl organs are converted to sepals and carpels, respectively. In the C function mutants *plena* (*Antirrhinum*) and *agamous* (*Arabidopsis*), the first and second whorls are normal but the third and fourth whorl organs are converted to petals and sepals, respectively. In order to explain the phenotypes of the A and C function mutants, it was proposed that A and C are antagonistic, the presence of the A function preventing C expression in those whorls where A is present, and vice versa. Cloning of the A, B, and C function genes from *Antirrhinum* and *Arabidopsis* and analysis of their expression patterns in wild-type and mutant plants, together with genetic data from crosses between the different homeotic mutants and transgenic plants, have elegantly confirmed the predictions of the model (Bradley *et al.*, 1993; Jack *et al.*, 1992; Goto and Meyerowitz, 1994; Sommer *et al.*, 1990; Schwarz-Sommer *et al.*, 1992; Tröbner *et al.*, 1992; Yanofsky *et al.*, 1990; Mandel *et al.*, 1992).

To date, five organ identity genes have been cloned from *Arabidopsis* and three from *Antirrhinum*. These are two A function genes of *Arabidopsis, APETALA1* or *AP1* and *APETALA2* or *AP2* (no *Antirrhinum* homologues have been isolated); the B function genes of *Arabidopsis* (*APETALA3* and *PISTILLATA*) and *Antirrhinum* (*DEFICIENS* and *GLOBOSA*); and the C function genes of *Arabidopsis* and *Antirrhinum, AGAMOUS* and *PLENA*, respectively. All but one of the ABC organ identity genes, *AP2*, belong to a class of regulatory genes encoding transcription factors, the MADS box family (reviewed by Shore and Sharrocks, 1995; and Thiessen and Seidler, 1995), so termed from the first four genes of this type cloned (*MCM1* from yeast, Passmore *et al.*, 1988; Herskowitz, 1989; *AGAMOUS* from *Arabidopsis*, Yanofsky *et al.*, 1990; *DEFICIENS* from *Antirrhinum*, Sommer *et al.*, 1990; and *SRF* from humans, Norman *et al.*, 1988). The MADS box gene family in *Arabidopsis* includes 17 cloned genes, many of which are involved in the control of the development of floral meristems and organs, and there are likely to be more genes as yet uncharacterized (Rounsley *et al.*, 1995). The characteristic feature of the proteins encoded by the MADS box genes is the conserved region of 56 amino acids at the N-terminus, which functions in DNA binding and interactions with other proteins (Schwartz-Sommer *et al.*, 1990; Shore and Sharrocks, 1995). A second conserved region of MADS proteins is the K box, which may be involved in protein–protein interactions (Ma *et al.*, 1991), and it is clear that the two different B function genes found in *Antirrhinum* (*DEFICIENS* and *GLOBOSA*) and in *Arabidopsis* (*APETALA3* and *PISTILLATA*) act as heterodimers (Tröbner *et al.*, 1992; Schwartz-Sommer *et al.*, 1992). A number of plant MADS box proteins can bind to DNA either as homodimers or as heterodimers, suggesting that the number of different regulators could be much greater than the number of genes (Huang *et al.*, 1996).

AP2 differs from the previously described organ identity genes in that it is not a member of the MADS box family and, in addition, the expression of *AP2* is not domain specific (Jofuku *et al.*, 1994). The duplication of A function genes in *Arabidopsis* is puzzling (given that the active protein is not an AP1/AP2 heterodimer, but it may be that *AP1* acts downstream of *AP2* (Weigel and Meyerowitz, 1994).

It is clear that the ABC genes are an absolute requirement for floral organ determination of both perianth parts and sex organs. In the absence of all three functions (in whorls 1 and 4 of AC double mutants and in ABC triple mutants), the floral organs have the characteristics of leaves (Bowman *et al.*, 1991). What is less clear is the way in which the organ identity genes are regulated and the nature of the target genes.

Genes that are known to act upstream of the organ identity genes are meristem identity genes and cadastral genes. The meristem identity gene *LEAFY* (Weigel *et al.*, 1992), in combination with *AP1* (Mandel *et al.*, 1992), is responsible for the initiation of floral meristem development in *Arabidopsis*. A similar combination of genes exists in *Antirrhinum*, the counterpart genes being *FLORICAULA* (Coen *et al.*, 1990) and *SQUAMOSA* (Huijser *et al.*, 1992). It is clear that there is a link between these genes and the organ identity genes, but is not clear whether the link is direct or involves several steps.

The cadastral genes set the boundaries for the expression domains of the organ identity genes. The organ identity genes themselves may have cadastral functions, as shown in the antagonistic effects of the A and C function genes, which limit each other's expression. A number of genes whose primary function is cadastral have been identified. These include the *Arabidopsis* genes *SUPERMAN* (Bowman *et al.*, 1992) and *LEUNIG* (Liu and Meyerowitz, 1995), which are required for suppression of the B function genes in whorl 4 and the C function genes in whorls 1 and 2, respectively. Another gene, *UNUSUAL FLORAL OR-GANS* (*UFO*), from *Arabidopsis* (Wilkinson and Haughn, 1995), and its *Antirrhinum* counterpart, *FIMBRIATA* or *FIM* (Simon *et al.*, 1994), are thought to act at the transcriptional level to regulate the expression of the organ identity genes, thereby setting the boundaries for expression (Levin and Meyerowitz, 1995; Simon *et al.*, 1994; Ingram *et al.*, 1995). The recently cloned *CURLYLEAF* gene of *Arabidopsis*, which has sequence similarity to the *Drosophila* Polycomb-group gene, *Enhancer of zeste*, has been shown to regulate the expression of *AGAMOUS*, suggesting a role for Polycomb-group genes in the determination of cell fate in plants (Goodrich *et al.*, 1997). It has been proposed that boundary formation for the domains of organ identity gene expression may involve inter-cellular trafficking through plasmodesmata, with the products of the organ identity genes themselves acting as supracellular control proteins (SCPs) through the formation of new types of heterodimers; alternatively, SCP cofactors may be involved, with UFO and FIM being suggested as candidate proteins (Mezitt and Lucas, 1996). For a review on flower and inflorescence development in *Antirrhinum*, the reader is directed to Coen (1996).

The MADS box organ identity genes are clearly of ancient origin and are not confined to higher (flowering) plants; homologues have been identified in con-ifers (Tandre *et al.*, 1995). It is possible, therefore, that the notion that all flower-ing plants have arisen from a common ancestor (Cronquist, 1988) can be taken still further in evolutionary terms.

The organ identity genes clearly provide a means of affecting the sex of an individual flower and are an obvious target in the search for the genes that determine sex in plant species with unisexual flowers. Organ identity genes have been cloned from a number of dioecious and monoecious plant species, including maize (Schmidt *et al.*, 1993; Mena *et al.*, 1995), sorrel (Ainsworth *et al.*, 1995), white campion (Hardenack *et al.*, 1994), asparagus (Miller *et al.*, 1995) and cucumber (Rosenmann *et al.*, 1996).

C. Unisexual Flowers: Monoecy and Dioecy

The three main sexual conditions in plants are hermaphroditism, in which perfect flowers are formed (already described), monoecy, in which male and female flowers are carried on the same individual; and dioecy in which individuals carry either male or female flowers throughout their lifespan (Fig. 3). In addition to these major classes, plant species exhibit a range of other sexual states:

- *Gynodioecy*, in which populations are composed of female and hermaphrodite plants; for example, *Plantago coronopus*, Plantaginaceae (Koelewijn and van Damme, 1996)
- *Androdioecy*, in which populations are composed of male and hermaphrodite plants; for example, *Datisca glomerata*, Datiscaceae (Liston *et al.*, 1990)
- *Trioecy* (sometimes called subdioecy), in which populations are composed of male, female, and hermaphrodite plants; for example, *Pachycereus pringlei*, Cactaceae (Fleming *et al.*, 1994)
- *Gynomonoecy*, in which plants carry female and hermaphrodite flowers; for example, *Poa* spp., Poaceae (Anton and Connor, 1995)
- *Andromonoecy*, in which plants carry male and hermaphrodite flowers; for example, *Cucumis melo*, Cucurbitaceae (Rosa, 1928)
- *Trimonoecy*, in which plants carry male, female, and hermaphrodite flowers; for example, *Dimorphotheca pluvialis*, Asteraceae (Correns, 1906).

Of these "mixed" conditions, androdioecy, trioecy, and trimonoecy are rare, the others being relatively common. The existence of these types of sex systems has evolutionary implications that are discussed later.

The survey of floral types found in flowering plants by Yampolsky and Yampolsky (1922) indicated that monoecious species and dioecious species represented 7% and 4% of the total, respectively; this is probably an overestimate. A recent compilation by Renner and Ricklefs (1995) based on 240,000 species indicated that 6% were dioecious, with 7% of the 13,479 genera (Brummitt, 1992) and 43% of the 365 angiosperm families containing one or more dioecious species (Cronquist, 1988). Families with the highest concentrations of dioecious genera are the Menispermaceae, Myristicaceae, Monimiaceae, Euphorbiaceae, Moraceae, Cucurbitaceae, Anacardiaceae, and Urticaceae (Renner and Ricklefs, 1995). Dioecy appears to be slightly more common among dicot genera than among monocot genera (Renner and Ricklefs, 1995). Variables that were significantly and strongly associated with dioecy were monoecy, climbing growth form, abiotic pollination, shrub habit, and tropical distribution. Some geographic regions are particularly rich in dioecious species; in Hawaii, for example, more than 27% of species are dioecious (Bawa, 1980).

III. Evolution and Developmental Biology of Plant Species with Unisexual Flowers

A. Primitive Plants and Sex

Analysis of floral biology and breeding mechanisms in plants leads us to believe that, in almost all cases, unisexuality (monoecy and dioecy) in angiosperms has arisen from hermaphroditism. In order to set the scene for a discussion of the evolution of unisexuality in higher plants, it is worth considering the sexual systems in more primitive plants.

In the simplest plants, such as the green alga *Chlamydomonas*, the unicellular flagellated organism is haploid throughout most of its life cycle, with the vegetative cells reproducing asexually by mitotic division. Sexual reproduction is by fusion of two cells (gametes) of opposite mating type (derived from the vegetative cells), followed immediately by a meiotic division. In more evolved plants, the bryophytes (mosses and liverworts), pteridophytes (ferns), and spermatophytes or higher plants (both gymnosperms and angiosperms), the life cycle includes alternation of a diploid sporophytic and a haploid gametophytic generation, the latter being the sexual generation (i.e., gametes are formed by differentiation after mitotic division of haploid cells).

In bryophytes, the gametophyte is thalloid or leafy in liverworts or an erect structure with spirally arranged "leaves" on a "stem" in mosses; it is the dominant phase of the life cycle. In pteridophytes, the gametophyte is generally short lived and often forms a small, heart-shaped thallus. When mature, the gametophyte carries the male (antheridia) and female (archegonia) sex organs, which are usually spatially separated on the thallus or may be unisexual. The flagellate sperm produced by the antheridia swim into the archegonial neck and fertilize the single egg cell. Although terrestrial, these plants require water for reproduction. The existence of the male and female organs on the same thallus in ferns and liverworts may argue for their classification as hermaphrodites, but because the male and female sex organs are physically separate, these plants, and certainly mosses, are considered to be most comparable to monoecious flowering plants (although the terms monoecious and dioecious should, strictly speaking, be applied only to the flowers of flowering plants). A complication in many homosporous ferns such as *Ceratopteris richardii*, is that the gametophyte may be so-called hermaphrodite or male (Näf *et al.*, 1975). This choice is under the control of sex-determining genes (see later discussion). Many bryophytes have unisexual (male or female) gametophytes, and sex chromosomes have been identified in these species (Chattopadhyay and Sharma, 1991).

Moving in evolutionary terms to the gymnosperms, vascular plants with naked seeds, we find increased complexity; the sporophyte generation (the plant itself) is the dominant generation, and the gametophyte generation is reduced and is contained within the male and female cones. Gymnosperms are either monoecious, as in *Pinus sylvestris*, or dioecious, as in *Cycas* spp. or *Taxus baccata* (yew). The male microsporophylls (cones) produce pollen grains (microspores) from microsporangia. Pollen grains are wind dispersed and fertilize the single egg cell on the ovuliferous scale of the female cone. After fertilization, development of the sporophyte is characterized by polyembryony, in which two embryos begin development but one usually aborts.

Therefore, in the algae, bryophytes, pteridophytes, and gymnosperms, self-fertilization is usually prevented by temporal or spatial separation of the male and female gametophytes. In the angiosperms, the gametophyte generation is reduced to the pollen grains of the male and the embryo sac of the female. Hermaphroditism is the rule rather than the exception, and a wide variety of mechanisms have evolved to prevent inbreeding. A naive view of plant evolution would envision a direct line of increasing complexity through the algae, bryophytes, pteridophytes, gymnosperms, and angiosperms. Under this scenario, it would be difficult to explain the marked change from unisexual to hermaphrodite flowers between the gymnosperms and angiosperms unless hermaphroditism was derived from unisexuality. The finding of Taylor and Hickey (1990) that a fossil angiosperm plant carried pistillate flowers provides some evidence that the first angiosperms may have been unisexual and that hermaphroditism evolved from unisexuality. Although it is likely that the gymnosperms and angiosperms are derived from a common ancestor, there is much debate about the actual line of evolution of angiosperms from the fossil plants (Doyle, 1994).

B. Evolution of Unisexuality in Flowering Plants

Examination of the flower structure of dioecious flowering plants compared with taxonomically closely related species indicates that the likely evolutionary route to dioecy is from a hermaphrodite. Since monoecy and dioecy occur in many families and genera, and it is accepted that the angiosperms are of monophyletic origin (Doyle, 1994), then these unisexual conditions must have arisen independently many times. Examination of the floral organs present in the unisexual flowers of different species reveals that there are many different routes to monoecy and dioecy, in terms of the divergence from the perfect flower (see the review of Dellaporta and Calderon-Urrea, 1993). The differences relate to the timing during flower development of the action of sex determination genes which cause the suppression of one set of floral organs. Male and female flowers of a range of monoecious and dioecious plant species are shown in Fig. 4.

In most cases, both sets of sex organs are initiated and the inappropriate set of organs develops to some extent before being aborted. In white campion (*Silene latifolia*), a rudimentary gynoecium is formed in male flowers; in the female flower, rudimentary stamens are produced which abort and degenerate (Grant, Hunkirchen et al., 1994). In sorrel (*Rumex acetosa*), there is much less development of the inappropriate set of organs, but these organs are initiated (Ainsworth et al., 1995). Similarly, in Z. *mays* (maize), the inappropriate sex organs are initiated in both male and female inflorescences (reviewed by Veit et al., 1993). Such is also the case with pistachio (*Pistacia vera*), in which rudimentary organs of the opposite sex are initiated in both sexes but the vestiges are incorporated completely into the growing appropriate sex organs (Hormaza and Pollito, 1996).

In some dioecious systems, the arrest in development of the inappropriate sex organs occurs very late, so that male and female flowers are indistinguishable from each other and from perfect flowers. In *A. officinalis*, gynoecium size in males varies, but the gynoecium can develop to the same size as that found in the female (Galli *et al.*, 1993; Caporali *et al.*, 1994). In *A. deliciosa* (kiwifruit) the inappropriate organs are of normal size, the difference being that the pollen produced by anthers in the female is sterile, as is the gynoecium in the male (Schmid, 1978).

In *Mercurialis annua* (annual mercury; Durand and Durand, 1991), *C. sativa* (hemp; Heslop-Harrison, 1958; Mohan Ram and Nath, 1964), *Spinacia oleracea* (spinach; Sherry *et al.*, 1993), and *Humulus* spp. (hop; H. L. Shephard, P. Derby, J. S. Parker, and C. C. Ainsworth, unpublished data), the monomorphic or bisexual stage common to most dioecious species, in which both sets of sex organs develop in both males and females before the inappropriate set aborts, is not apparent in flowers of either sex. Male and female flowers are strikingly different in these plant species, and in all these cases the male flowers resemble perfect flowers but the female flowers are quite different. Interestingly, these species in which male and female flowers are developmentally very different from each other are in taxa that display mainly monoecy, a fact which argues for an evolutionary line from monoecy to dioecy (see later discussion).

The stage at which there is an arrest in organ development, therefore, varies considerably between the different dioecious and monoecious plants, indicating independent evolution in each species. The genetics and detailed developmental biology of sex determination in a number of plants are discussed later, but one question to address is how, in evolutionary terms, these species arose.

C. Evolutionary Models

1. Dioecy

The two main factors involved in the evolution of dioecy are considered to be inbreeding avoidance and selection on allocation of reproductive resources (Lewis, 1942; Ross, 1978; Mather, 1940; Charlesworth and Charlesworth, 1978a). Inbreeding is disadvantageous to a population because progeny are produced that suffer from inbreeding depression, and separation of the sexes may enable resources to be used more efficiently than in hermaphrodite plants, where, for example, pollen and ovule may limit each other's production. A number of models involving different breeding systems have been proposed for the evolution of dioecy (see Fig. 3).

Lewis (1942) proposed that dioecy could evolve in a hermaphrodite species by the occurrence of two separate mutations in genes that were tightly linked: one for male sterility and one for female sterility. Linkage would prevent the generation of hermaphrodites. Ross (1978) proposed a similar model but invoked partial male and female sterility, again in tightly linked genes or, alternatively, in a multilocus model. It is considered that the occurrence and establishment of two independent mutations must be extremely rare and that a more likely route for the evolution of dioecy is through gynodioecy or androdioecy with the initial establishment of a single mutant form (Charlesworth and Charlesworth, 1978a). Androdioecy (the coexistence of males and hermaphrodites) is rare, and its very existence was questioned by Charlesworth (1984), who proposed that androdioecy describes the morphologic features only and that androdioecious species are actually functionally dioecious. One well-documented example of an androdioecious species is *D. glomerata* (Liston *et al.*, 1990). Evidence from the analysis of chloroplast DNA in the family Datiscaceae, in which all species are dioecious except *D. glomerata*, clearly indicated that androdioecy in this species was derived from breakdown of dioecy and not the converse (Rieseberg *et al.*, 1992). The reason that androdioecy is rare may be that male sterility causes less loss of fitness than female sterility, given the large number of pollen grains relative to the number of ovules (Charlesworth and Charlesworth, 1978a). It has been calculated that androdioecy could be established in an outcrossing population only if the female steriles had more than twice the pollen fertility of the hermaphrodites (Charnov *et al.*, 1976; Charlesworth and Charlesworth, 1978a).

Gynodioecy in plants (the coexistence of females and hermaphrodites) is well documented, and some supposedly dioecious plants are in fact gynodioecious. Gynodioecy may represent a stable condition in itself (Lewis, 1942), or it may simply be an intermediate step to dioecy (Carlquist, 1974). Gynodioecy must be considered to be a more likely intermediate step to dioecy than is androdioecy, with the first mutation causing total or partial male sterility. It is not uncommon that dioecious species have gynodioecious relatives (Charlesworth and Charlesworth, 1978a), an example being the genus *Silene* L., which contains several hundred species that are mainly hermaphrodite but also display dioecy, gynodioecy, or trioecy (subdioecy). Gynodioecy is common, and it is suggested that dioecy has arisen several times independently in *Silene* (Desfeux *et al.*, 1996).

A mutation causing male sterility will spread through a population if mutant plants are fitter (and produce more seed) than the hermaphrodites (Lewis, 1941; Lloyd, 1974; Ross and Shaw, 1971), and it will also spread in a population suffering inbreeding depression even if the mutant is no fitter (Charlesworth and Charlesworth, 1978a; Lloyd, 1975a; Valdeyron *et al.*, 1973). A further mutation, causing female sterility in hermaphrodites, would convert the gynodioecious population to a dioecious one. These changes would probably be gradual, and the causal mutations would be more likely to be recessive than dominant (Charlesworth and Charlesworth, 1978a). An alternative possibility is that nuclear-cytoplasmic male sterility is involved as an intermediate state (Maurice *et al.*, 1993, 1994). There is strong evidence for this mode of inheritance of male sterility in the genus *Silene* (Desfeux *et al.*, 1996).

Another alternative route to dioecy is through monoecy, which has been implicated in the evolution of a number of dioecious species including *Mercurialis* (Westergaard, 1958; Charlesworth and Charlesworth, 1978b). As with angiosperms in general, dioecy in the British flora is more frequently associated with monoecy than with hermaphroditism (Lewis, 1942). The selective forces leading to the evolution of dioecy from monoecy may be similar to those outlined for the pathway from hermaphroditism. A monoecious population could evolve into a dioecious one through a series of mutations that alter the ratio of male to female flowers (Charlesworth and Charlesworth, 1978b). The salient difference from the gynodioecious route to dioecy is that, after the establishment of monoecy, only a single developmental switch is required to alter the sex ratio of flowers, rather than the two needed to suppress gynoecium and androecium development. In the genus *Cotula*, there is evidence that dioecy has arisen from monoecy (Lloyd, 1975b).

Although prevention of inbreeding may be a primary selective pressure in the evolution of dioecy, other factors may influence the frequency with which dioecy has evolved in different plant groups. Dioecy is correlated with perennial habit and with fleshy fruits (biotic dispersal), which may explain the high proportion of dioecious species in tropical floras (Bawa, 1980). It is argued by Renner and Ricklefs (1995) that the correlation between fleshy fruits and dioecy is itself a reflection of the high proportion of tree species with fleshy fruits. Dioecy is also strongly correlated with the climbing growth habit and abiotic (wind) pollination (Lloyd, 1975a). The evolutionary mechanisms underlying these correlations are not yet clearly understood. The association of climbing habit is proposed to occur because of differential selection for optimal resource allocation to sexual function; that is, it may be advantageous to have fruit-bearing flowers on plants with stronger stems and pollen-bearing flowers on thin stems (Renner and Ricklefs, 1995). To explain the association of dioecy with small flowers, it is suggested that the structures of flowers pollinated by insects and other animals have evolved to suit the specific pollinator, and modifications that suppress sex organ production are likely to be selected against (Renner and Ricklefs, 1995).

2. Monoecy

As compared with dioecy, the evolution of monoecy has been the subject of very little scientific attention and is not well understood. Darwin (1876) noted that the flowers of monoecious plant species often show traces of opposite sex function, and it is suggested that monoecy has evolved from hermaphroditism through the intermediate states of gynomonoecy or andromonoecy (female and hermaphrodite flowers on a plant and male and hermaphrodite flowers on a plant, respectively) (Charlesworth and Charlesworth, 1978b). Mutation of a hermaphrodite to the gynomonoecious state may promote outcrossing, but it is less easy to argue a case for andromonoecy. However, although gynomonoecious species do occur, it is only rarely that they are in genera that also contain monoecious species (Charlesworth and Charlesworth, 1978b). The initial mutation may cause only some flowers to be converted to females, which could promote outcrossing. How

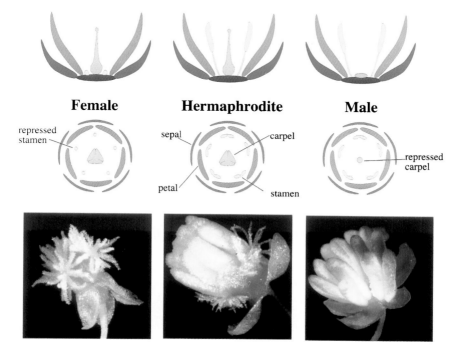

Fig. 1 Hermaphrodite and unisexual flower patterns. The typical pentamerous (organ numbers based on five) hermaphrodite flower pattern and the derived unisexual states for male and female flowers, where the stamens are suppressed in the female and the carpels are suppressed in the male, are shown. This is illustrated by a hermaphrodite *Rumex* species (*R. scutatus*) and male and female flowers of the related dioecious species, *R. acetosa*. These species have organ numbers based on three and have two similar whorls of perianth parts as opposed to distinct sepals and petals. The female flower of *R. acetosa* is reproduced from Ainsworth *et al.* (1995) with permission.

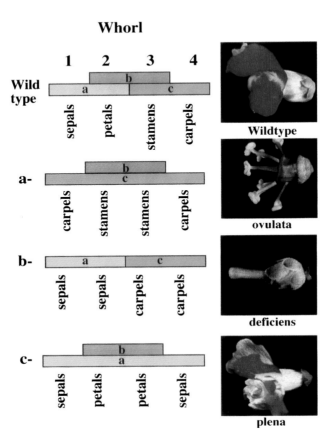

Fig. 2 The ABC model of Coen and Meyerowtz (1991). Organ identity is determined by the combinatorial action of the A, B, and C functions active in three overlapping domains of the floral meristem. Sepals are specified by A; petals are specified by A plus B; stamens are specified by B plus C; carpels are specified C. The wild-type *Antirrhinum* flower and three mutants, *ovulata*, *deficiens*, and *plena*, which represent losses of the A, B, and C functions, respectively, are shown. (The *ovulata* phenotype is actually the result of suppression of A expression in whorls 1 and 2 and is a *plena* mutant; Bradley *et al.* (1993). Adapted from Coen and Meyerowitz (1991) with permission.

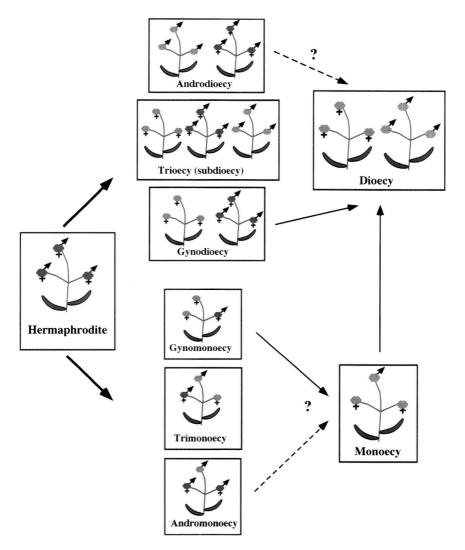

Fig. 3 Sex strategies in plants. The three main sex strategies in plants are hermaphroditism, monoecy, and dioecy. Various mixed states exist; they may be stable states, or they may be intermediates in the evolution of monoecy and dioecy. See text for description.

Fig. 4 Mature male and female flowers in a range of hermaphrodite, monoecious, and dioecious plants. (a) *Arabidopsis thaliana* (hermaphrodite), (b) Cucumber (male and female flowers), (c) Maize tassel, (d) Maize ear, (e) Annual mercury (male), (f) Annual mercury (female), (g) *Humulus lupulus* (male), (h) *H. lupulus* (female), (i) Sorrel (male), (j) Sorrel (female), (k) White campion (male), (l) White campion (female), (m) Kiwifruit (male), (n) Kiwifruit (female). [(i) and (j) from Ainsworth *et al.* (1995), with permission. (k) and (l) from Grant *et al.* (1994a), with permission. Photographs of *Arabidopsis*, maize, and kiwifruit kindly provided by Hong Ma, Erin Irish, and Raffaele Testolin.]

Fig. 4 Continued.

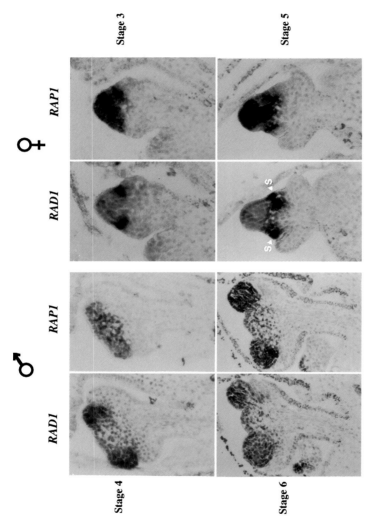

Fig. 9 Expression of B and C function genes in sorrel flowers. Developing inflorescences from male and female plants were sectioned and adjacent sections were hybridized with antisense RNA probes of *RAD1* (a B function gene) and *RAP1* (a C function gene). B function gene expression in male and female primordia is confined to the stamen whorl in all developmental stages. C function expression starts in stamen and carpel whorls in the male and female primordia but refines to a single whorl in older primordia, stamens in the male, and carpels in the female. S, Stamen primordia in the female. Developmental stages are as described by Ainsworth *et al.* (1995). Adapted from Ainsworth *et al.* (1995), with permission.

Fig. 10 Floral variation in hermaphrodite sorrel. Mature flowers exhibit increasing degrees of gynoecium development, from almost none (a) to full (e). The arrow in (a) indicates the rudimentary gynoecium.

hermaphrodite flowers are converted to males is a problem, and it is proposed that there must first be selection for reduced female fertility of the hermaphrodite flowers, by one or more mutations, before full monoecy is established (Charlesworth and Charlesworth, 1978b).

A problem with this model is that, in contrast to the evolution of dioecy, the main selective force in the evolution of monoecy does not appear to be the promotion of outcrossing. In a survey of ability to self-pollinate and type of breeding system in 588 species, it was found that the sexual system was independent of self-fertility, indicating that avoidance of inbreeding is not a major factor in the evolution of monoecy (Bertin, 1993). This notion is supported by the analysis of monoecious and dioecious populations of *Ecballium*; the monoecious populations were found to be more inbred than the dioecious ones (Costich, 1995). It may, therefore, take a considerable amount of further research before the evolution of monoecy is understood.

IV. Control of the Determination of Sex

There are clearly many different routes to unisexual flowers, as illustrated by plants that have floral development blocked at different developmental stages. There is good evidence that there can be a genetic basis for sex in dioecious plants and that males and females are genetically distinct (e.g., the existence of plants with heteromorphic sex chromosome systems). It is also clear that this cannot be the case in the determination of male and female flowers on monoecious plants, since each individual is genetically uniform.

Several fundamental questions arise from this: what are the genes involved in determination; how is sex determined in dioecious plants that do not carry differentiated sex chromosomes; and, how are the sexes of individual flowers in monoecious plants determined? In plants that do not have sex chromosomes, are males and females different in genetic terms, or is sex the result of the differential activation of genes in identical genotypes? The latter must clearly be the case in monoecious species. A distinction needs to be made between the primary sex determination genes of dioecious plants and those genes (i.e., the sex differentiation genes) that are activated as a consequence of the expression of the sex determination genes and that result in the observed morphologic differences between the sexes. There may be few or many steps between these "cause" and "effect" genes. It seems unlikely, however, that the primary determination gene would also be the one that resulted in organ differences in the flowers.

In monoecious species, there must be an internal developmental cue that sets in motion the cascade of events leading to the floral differences. The nature of this cue remains a mystery, although plant growth substances have been shown to play a role in sex determination in dioecious and monoecious species, as discussed later. Another set of factors that may affect the process of sex determination are those pertaining to the environment.

The mechanisms used by different plants exhibiting monoecy or dioecy are likely to be different from plant to plant given the range of floral differences between male and female flowers in the different species. It is likely, therefore, that the underlying molecular mechanisms are quite different and that there is no simple and general molecular model that can be proposed for sex determination in plants. Indeed, rather few molecular biologic studies have been undertaken in monoecious and dioecious plant species. Insight into the control of sex determination in plants can be gained by examination of the best-studied systems.

A. Sex Determination in Monoecious Plants

Two monoecious plants have been the subject of extensive research in the area of sex determination, cucumber (*C. sativus*) and maize (*Z. mays*), and this review will focus on these species. Sex determination in these two very different plants is similar in two respects: males and females pass through a bisexual stage (the inappropriate organs are initiated but do not develop very far before they are aborted), and plant growth substances seem to be implicated. Although there are many other monoecious plant species, there has been little research, other than that of a purely descriptive or theoretical nature, addressed at understanding the mechanisms that underlie sex differences. A possible exception is *Ecballium elaterium* (the squirting cucumber), a mainly monoecious species, in which a single locus with three alleles determines sex (Mather, 1949).

1. Maize: *Zea mays*

Maize, together with all other members of the Maydeae, is monoecious (Kellogg and Birchler, 1993), and the terminal (male) and lateral (female) inflorescences—tassels and ears (see Fig. 4), respectively—develop as a result of the conversion of vegetative shoot meristems (Bonnett, 1948). The male and female inflorescences each consist of numerous florets arranged on short branches called spikelets. The florets are not synchronous in their development, each tassel and ear carrying florets at a range of developmental stages. The maize spikelet consists of a pair of glumes (bracts) that subtend two florets, each of which is associated with additional bracts, the palea and lemma (Fig. 5A). Each individual floret develops three whorls of organ primordia; from the outside of the floret inwards these whorls contain the two lodicules (petal homologues), three stamens, and three carpels, which fuse to give the pistil. The first sex-related difference, the presence of the basal branches on tassels but not on ears, is apparent when the inflorescences are 2 mm in length. The sex of the basal, and developmentally most advanced, florets is detectable as size differences in car-

A

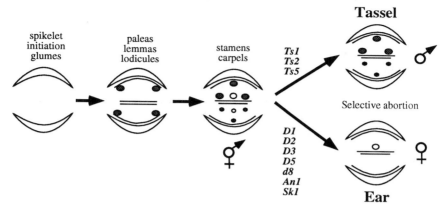

B

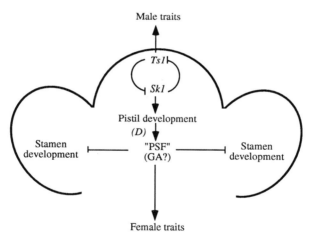

Fig. 5 Floral development in maize. (A) Floral diagrams of development of male and female spikelets. The maize spikelet consists of a pair of glumes that subtend two florets (upper and lower), each of which is associated with additional bracts, the palea and lemma. Each individual floret initially develops three whorls of organ primordia: the two lodicules, three stamens, and three carpels which fuse to give the pistil. The *Tasselseed* and *Dwarf* genes bring about the selective abortion of the pistil and stamens, respectively. Note the differences between upper and lower florets. Adapted from Irish *et al.* (1994) and Dellaporta and Calderon-Urrea (1994) with permission. (B) Model for sex determination in maize. *SK1* promotes development of pistils, which produce a pistil-specific factor (PSF) that inhibits stamen development and possibly prevents *TS2* action (*TS2* normally causes pistil abortion). When active, *TS2* suppresses *SK1*, blocking pistil development and PSF production. Redrawn from Dellaporta and Calderon-Urrea (1994) with permission.

pels and stamens when the inflorescences are 10 mm in length (Irish and Nelson, 1993), after which time the inappropriate organs arrest and later degenerate (Cheng *et al.*, 1982). The fundamental difference between ear and tassel is that ear florets suppress stamen primordia and tassel florets suppress pistils. Pistils are also usually suppressed in the lower florets on ears, such that these florets are sterile, but this is not uniform among all maize varieties. Both upper and lower tassel florets are staminate (see Irish, 1996).

Determination of sex in maize has been the subject of several reviews (Irish and Nelson, 1989; Irish, 1996; Irish *et al.*, 1994; Dellaporta and Calderon-Urrea, 1993, 1994; Lebel-Hardenack and Grant, 1997). From the many mutations that affect floral structure in maize, a number have been identified that disrupt the determination process (i.e., the suppression of one set of organs), rather than affecting floral organ identity or number; the genes affected can be classified as masculinizing or feminizing genes (see Fig. 5A).

A number of mutations are masculinizing in that they affect the abortion of the stamen primordia; these are the *dwarf* (*d*) mutants, *d1*, *d2*, *d3*, *d5*, *anther ear* (*an1*), and *D8*, which, in addition to being short due to reduced internode lengths, have broader leaves and are andromonoecious. They have hermaphrodite flowers in the primary ear florets and male flowers in the tassel and secondary ear florets (reviewed in Irish, 1996). The *silkless* (*sk*) mutation is masculinizing because of suppression of pistil development in the ear so that sterile florets result (Jones, 1925). Two of the *dwarf* mutations, *d1* and *d5*, result in phenotypes that are clearly caused by blocks in the biosynthesis of gibberellin, one of the major plant growth substances. This was demonstrated by measurement of endogenous gibberellin levels and by gibberellin application (Hansen *et al.*, 1976; Spray *et al.*, 1984; Fujioka *et al.*, 1988).

There are five classes of plant growth substances, and all have been implicated in sex determination in plants. The auxins and cytokinins represent two of the three major classes of plant growth substance, the third class being the gibberellins (reviewed in Davies, 1995). The main roles of these growth substances are in cell enlargement, cell division, root initiation, and tropistic responses (auxins); cell division, morphogenesis, and leaf expansion (cytokinins); and stem growth, fruit set, seed germination, and flowering (gibberellins). Two other plant growth substances, abscisic acid (ABA) and ethylene, have more specific roles in stress responses and in ripening and senescence of fruits and leaves, respectively.

The *an1* mutation, which results in a semidwarf phenotype in addition to its effect on the sex of the flower, has been shown to be a lesion in the gibberellin biosynthetic pathway. The *AN1* gene has been cloned, and its product is involved in the synthesis of *ent*-kaurene, the first tetracyclic intermediate in the pathway, from GGPP (Bensen *et al.*, 1995). This result is supported by earlier physiology experiments which showed that the *an1* phenotype is gibberellin-responsive and is reversible if *ent*-kaurene is supplied (Katsumi, 1964).

Gibberellin, therefore, seems to be essential for the abortion of stamens in the female flowers. The reduced gibberellin levels found in the *an1* mutant and in the other gibberellin-responsive maize mutants do not, however, affect whorl identity in the flowers, and, consequently, the effects on floral development must occur after the initiation of the floral organs (Bensen *et al.*, 1995).

The feminizing mutations are of the *TASSELSEED* (*TS*) loci; five have been mapped genetically. These are *ts1*, *ts2*, *ts4*, *ts5*, and *ts6*; in the most extreme cases (*ts1*, *ts2*), they cause the reversal of sex in tassel florets in that pistils develop and stamen development is suppressed (Emerson, 1920). The *TS1* and *TS2* genes are thought to function to cause the abortion of the gynoecium in male flowers and to determine the secondary sexual characters (Dellaporta and Calderon-Urrea, 1994). The *TS2* gene has been cloned; it encodes an alcohol de-hydrogenase–like protein that has similarity to steroid dehydrogenases (DeLong *et al.*, 1993). The *TASSELSEED* genes may act antagonistically to the feminizing effects of gibberellic acid, either by direct inactivation or indirectly by inducing pistil abortion and blocking gibberellin production as a consequence (Dellaporta and Calderon-Urrea, 1994). An interesting difference between the feminizing *D* genes and masculinizing *TS* genes is that the mutations in the former lead to hermaphrodite flowers through failure to suppress the stamens, while mutations in some of the latter result in suppression of a different organ type and in sex reversal, a fact that argues for the monoecious phenotype and independent deter-mination of the sex of a flower (Stevens *et al.*, 1986). Analysis of double mu-tants of *TS* and *D* genes suggests that the loci act independently and additively; in *ts2/d1* double mutants, florets in the ears and tassels have well-developed stamens (Irish *et al.*, 1994). This, in addition to the observations that gibberellin-deficient ears have well-developed stamens and that environmentally feminized tassels have increased levels of gibberellin-like activity (Rood *et al.*, 1980), has led to the suggestion that a stamen-suppressing substance, possibly gibberellin, is produced by the developing pistil (Dellaporta and Calderon-Urrea, 1994; Irish, 1996).

Dellaporta and Calderon-Urrea (1994) proposed the following model for sex determination in maize (see Fig. 5B). The *SILKLESS1* (*SK1*) gene promotes pistil development; the pistil produces a pistil-specific factor (PSF), which is possibly a gibberellin, that inhibits stamen development and feminizes floral tissues. *SK1* may also prevent *TS2* action in the primary ear florets. *TS2* normally induces abortion of the gynoecium and suppresses *SK1*, thereby blocking pistil develop-ment, PSF production, and stamen arrest. There is considerable supporting evi-dence for this model and, given the wide range of well characterized mutants, it is likely that the details will become clear in the not so distant future. A key, more difficult question, however, involves the identity of the primary sex-determining signal: is there a gene that controls whether an ear or a tassel develops, and how is it controlled?

2. Cucumber: *Cucumis sativus*

Cucumber (*C. sativus*) generally displays monoecy, although a wide range of flower types based on male, female, and hermaphrodite forms can be found in the various cucumber genotypes (Malepszy and Niemirowicz-Szczytt, 1991). In male and female buds both sets of sex organs are initiated and stamen and pistil primordia are present (Atsmon and Galun, 1960). The inappropriate organs arrest during development, while the stamens in the male and pistils in the female continue growth. The vestigial organs are present in the mature cucumber flowers (see Fig. 4).

Sex expression in cucumber occurs in three phases; initially, male flowers are produced; there is then a phase in which male and female flowers are produced; and finally there is a phase of female flower production. Clearly, there must be genes that regulate the sex of a flower and also determine its position on the plant. Seven genes have been identified that affect the sexual phenotype; combinations of the alleles at three unlinked loci account for the majority of the cucumber sex phenotypes (Malepszy and Niemirowicz-Szczytt, 1991). The dominant allele at the *M* locus (originally designated *G*) specifies unisexual flowers; the partially dominant allele at the *F* locus (originally designated *Acr*) controls the production of female flowers, and the recessive allele of the *A* gene intensifies maleness in homozygotes for the recessive allele at the *F* locus (Robinson *et al.*, 1976; Galun, 1959; Malepszy and Niemirowicz-Szczytt, 1991).

Levels of ethylene are elevated in female cucumber lines (relative to monoecious ones), whereas gibberellin levels are lower (Rudich *et al.*, 1972; Atzmon *et al.*, 1968). Levels of enzymes in the plant growth substance biosynthetic pathways may be important, as is suggested by metabolite feeding experiments in which female lines were shown to have higher capacities to convert ACC to ethylene (Trebitsh *et al.*, 1987). The sex of cucumber plants can be shifted in either direction by the application of exogenous plant growth substances. Femaleness is promoted by ethylene (and ethylene-releasing compounds) and auxins, whereas maleness is promoted by gibberellins and compounds that are ethylene antagonists (Rudich *et al.*, 1969; Shannon and de la Guardia, 1969; Malepszy and Niemirowicz-Szczytt, 1991; Galun, 1961; Atsmon and Tabbak, 1979). Gibberellin not only induces male flowers on female plants but also induces abortion of female flowers and the formation of adventitious male flower buds (Fuchs *et al.*, 1977). Ethylene has also been implicated in the determination of sex in the dioecious teasle gourd, *Momordica dioica*, because the application of silver nitrate, an inhibitor of ethylene biosynthesis, causes production of stamens with fertile pollen on flowers of female plants (Hossain, *et al.*, 1996).

There are two hypotheses for the determination of sex in plants with unisexual flowers by plant growth substances. In the two-growth-substance system, each would promote one sex (Chailakhan, 1979). In the single-growth-substance system, a growth substance would have male and female receptors to inhibit one sex and induce the other independently (Yin and Quinn, 1992). The single-growth-substance model has been tested in cucumber and in the dioecious or monoecious grass, *Buchloe dactyloides* (Yin and Quinn, 1995a, 1995b). These experiments supported the notion that one plant growth substance only is involved in sex determination in these species. In cucumber, it is ethylene and not gibberellin that is important, with ethylene having overriding effects on gibberellin. The *F* gene controls the endogenous concentration of ethylene, and the *M* gene controls the male sensitivity to ethylene. Combinations of dominant and recessive alleles at these loci lead to different levels of ethylene and different sensitivities, accounting for the sex phenotypes (Yin and Quinn, 1995b). This model must await isolation of the relevant genes for validation.

There is no evidence, at the molecular level, as to the mechanisms that operate to determine sex in cucumber. Several members of the MADS box gene family have been cloned (Rosenmann *et al.*, 1996). Although analysis of three *AGAMOUS*-like genes suggested that there were differences in transcript levels between male and female flowers, expression patterns in male and female flowers have not been described in sufficient detail to ascribe a role for these genes in sex determination (Rosenmann *et al.*, 1996).

B. Sex Determination in Dioecious Plants

Of the many dioecious plant species, only a small number have been studied in any detail with respect to the mechanism of sex determination. These are annual mercury (*M. annua*), hop (*Humulus* species) and its relative cannabis (*C. sativa*), sorrel (*R. acetosa*), white campion (*S. latifolia*), asparagus (*A. officinalis*), and kiwifruit (*A. deliciosa*), species that belong to six different families (see Table I). We will examine these dioecious systems in this order, which reflects the stage at which the male and female flowers diverge during development, from very early (annual mercury, hop) to very late (kiwifruit). In doing so, a number of issues are raised, for example, whether sex chromosomes are present, the involvement of plant growth substances, and the stability of dioecy in response to environmental variations. Consideration of these topics may allow us to see whether there are general mechanisms that operate in this diverse group of plants.

1. Chromosomal Basis of Dioecy

Sex chromosomes in plants were first detected by Allen (1919) in the liverwort *Sphaerocarpus donnelii*. A large X chromosome was found in female gametophytes and a much smaller Y in the male gametophytes. Sporophytes, therefore, are all XY, and sex differentiation is determined by the segregation of the X-Y bivalent at meiosis. Heteromorphic sex chromosomes were reported in flowering plants a few years later with the discovery of chromosome heteromorphism in males and homomorphism in females of *R. acetosa* and *Melandrium rubrum* (*Silene dioica*) by Kihara and Ono (1923) and Blackburn (1923). This discovery precipitated a search for sex chromosomes in dioecious flowering plants and their subsequent reporting in 80 or 90 species. However, most of these reports were based on flimsy evidence in cytologically intractable species, and the current list of species with authenticated sex chromosomes is extraordinarily small. Differentiated sex chromosomes have been established in only five families—Cannabinaceae, Caryophyllaceae, Cucurbitaceae, Loranthaceae, and Polygonaceae—and in a total of only about seven species and two major species groups. Thus, the *acetosa* section of the genus *Rumex* contains about 10 species characterized by the same sex chromosome system (Wilby and Parker, 1988), while the family Loranthaceae contains many species in the genus *Viscum* that carry sex-specific chromosome rearrangements (Barlow and Martin, 1984). It is remarkable, however, that very few sex-chromosome systems have been reported in the last 40 years, despite the intense cytologic activity over this period. This suggests that sex chromosome systems in flowering plants are rare phenomena. Their rarity, however, makes them intriguing, and more recently they have come under increasing scrutiny by modern molecular cytology.

Characteristically, sex chromosomes in flowering plants are found in only one or two species of a family. For example, *S. dioica* and *S. latifolia* form fully fertile hybrids and are clearly closely related. They are the only two species known to have sex chromosomes in a genus of about 500 species and a family (Caryophyllaceae) of more than 2000 species. Cannabinaceae is the only family in which all species have sex chromosomes; this very small family comprises the two hop species, *H. lupulus* and *H. japonicus*, and *C. sativa*. Even in this family, however, the sex chromosome systems are diverse: *H. japonicus* has an XX/XY_1Y_2 system, while *H. lupulus* is polytypic, with XX/XY found most commonly and a translocation derivative, $X_1X_1X_2X_2/X_1X_2Y_1Y_2$, found in the Japanese subspecies *H. lupulus* subsp. *cordifolius*. Sex chromosomes in flowering plants, then, have evolved independently, and comparative studies should enable us to uncover common elements in the evolutionary pathway leading to chromosome differentiation.

A number of features differentiate plant and animal sex chromosome systems. In animals the Y (or W) chromosomes are often among the smallest members of the complement and are always smaller than the X (or Z). In flowering plants, by contrast, the Y chromosomes or summed Y-multiples are much larger than the X in all species except *H. lupulus* and *Viscum*. In addition, where comparisons are possible, the X and Y chromosomes are the largest in the entire genus (*Silene*, *Rumex*, *Coccinia*). The evolution of sex chromosomes in flowering plants therefore seems to have been associated with an increase in the DNA amount in those chromosomes. The degeneracy of Y chromosomes in animals is not a feature of these plant systems, which may reflect their sporadic and presumably recent evolution. It is not clear, however, why the Y chromosomes should grow to exceed the X in DNA content, unless this is a random event associated with rarity.

Despite the rarity of sex chromosome systems in flowering plants, the main categories characterizing sex determination in animals are represented. For example, the *Silene* species have an active-Y system similar to that in mammals, while the *R. acetosa* group has a *Drosophila*-type dosage system (a recessive-X system in the terminology of Bull, 1983). Sex determination in the American species *Rumex hastatulus*, which evolved independently of the *R. acetosa* system (Smith, 1967), is distinct in that both X and Y are involved in an intermediate system. The ready manipulation of plants by polyploidy and their resilience to chromosome imbalance have been supported by our own studies (Parker, 1990).

Since their first discovery in 1923, the sex chromosome systems of *Silene* (*Melandrium*) and the *R. acetosa* group have intrigued cytologists. Their study offers an opportunity to investigate the mechanisms of chromosomal evolution and the genetic control of sex determination in plants using a combination of cytologic and DNA technologies. These studies enable a fine analysis of these sex-determining systems. Yamamoto (1938), for example, showed in *R. acetosa* that three autosomes have male-promoting properties while two are female promoting. This has been supported by our studies (Parker, 1990). A more detailed analysis of these two systems is given later.

2. Annual Mercury: *Mercurialis annua*

Male and female flowers of annual mercury (*M. annua*, Euphorbiaceae) show no rudiments of the sex organs of the opposite sex (Durand and Durand, 1991). Both European species of *Mercurialis*, annual mercury and dog's mercury (*Mercurialis perennis*), exhibit dioecy and sometimes monoecy or intermediate states. Although *M. annua* and *M. perennis* have the same floral morphology, it is only *M. annua* that has been the subject of investigation into the mechanism of sex determination.

Male and female inflorescences develop in the leaf axils adjacent to a vegetative bud. The male flower consists of a whorl of three perianth segments and a whorl of 8–15 stamens, arranged in three groups. The female flower consists of a whorl of three perianth segments and two nectar glands alternating with a pistil formed by two carpels. There is no evidence of the rudiments of the inappropriate sets of organs (Durand and Durand, 1991). Male and female flowers of *M. perennis* are shown in Fig. 4.

Combinations of two alternative alleles at each of three unlinked loci, A, B_1, and B_2, are implicated in sex determination (Louis, 1989; Durand and Durand, 1991). These loci affect the levels of the plant growth substances, auxins and cytokinins, and it is probable that sex in annual mercury is brought about by modification of the plant growth substance biosynthetic pathways. Maleness is induced by the presence of the dominant allele at A together with one or more of the dominant alleles at the B_1 and B_2 loci. The strength of maleness, in terms of resistance to the feminizing effects of cytokinins, is a reflection of the number of B dominant alleles. Femaleness is induced by the presence of the dominant A allele plus recessive alleles at both B loci, or by the presence of the recessive allele at the A locus plus a dominant allele at one of the B loci (Durand and Durand, 1991).

Sex expression in annual mercury is sensitive to exogenously applied plant growth substances. Auxins have a masculinizing effect and cytokinins have a feminizing effect (Durand, 1969). *Trans*-zeatin, the specific cytokinin responsible for femaleness, which accumulates to high levels in female shoot apices, was found to be undetectable as a free base in male shoot apices, *trans*-zeatin riboside accumulating instead (Louis *et al.*, 1990). Auxin levels (IAA) were three to six times higher in male flowers than female flowers (Louis *et al.*, 1990). Whatever the precise mechanism of sex determination in annual mercury, it is clear that it is the auxin–cytokinin balance that is important.

As well as the obvious morphologic differences between male and female flowers of annual mercury, they exhibit different patterns of peroxidase isozymes; feminization of genetically male plants by cytokinin application causes reestablishment of the female pattern (Boissay *et al.*, 1996). The male pattern is one of predominantly anionic peroxidases. As female flowers develop, the anionic peroxidases decrease in abundance (Boissay *et al.*, 1996). Whether the anionic peroxidase expression is controlled directly or indirectly by cytokinin is not known.

3. Cultivated Hop and Its Relatives: *Humulus lupulus*

All three species in the family Cannabidaceae are dioecious and carry sex chromosomes. This fact alone might suggest that dioecy is of ancient origin in the Cannabidaceae.

As in annual mercury, male and female hop flowers are strikingly different; the male flower closely resembles a typical perfect flower, but the female flower is almost unrecognizable as a flower (see Fig. 4). Scanning electron microscopy and light microscopy have revealed distinctly different patterns of flower development in the sexes of *H. lupulus* (H. L. Shephard, unpublished data). In the vegetative phase, the meristems of both sexes are indistinguishable from each other. On transition to the reproductive phase, however, there are major developmental differences between the male and female floral primordia (Fig. 6). In the male, the emergence of five perianth segment primordia is followed by the appearance of five stamen primordia in the same developmental pattern. Developing female flowers arise with two separate carpel primordia differentiating on the floral buttress (see Fig. 6). Each female flower, comprising a pair of carpels and an ovary enclosed within the perianth at the base of the flower, is

♂   ♀

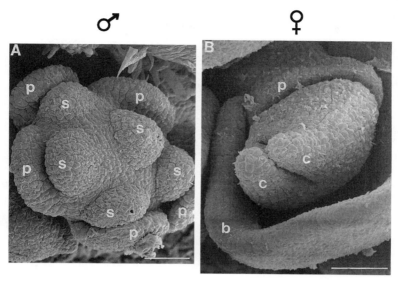

Fig. 6 Male and female hop floral development. Scanning electron micrographs of single male and female flower primordia from *Humulus lupulus*. (A) The male primordium shows five perianth segment primordia (p) and five stamen primordia (s). Note the lack of development in the center of the flower. (B) The female primordium is enclosed within a bracteole (b) and comprises a pair of carpels (c) and an ovary enclosed within the perianth (p) at the base of the flower. Bars, 50 μm. Photographs kindly provided by Helen Shephard.

enclosed within a bracteole and pairs of flowers, arranged alternately, and subtended by a bract (see Fig. 6). A bisexual stage, common to most dioecious species, in which both sets of sex organs develop in both males and females before the inappropriate set aborts, is not apparent in developing hop flowers of either sex.

H. lupulus, the cultivated hop, has 2n=20 in both sexes. It is the female hop plants that are economically valuable; the essential oils and resins produced by the abundant lupulin glands on the female cones are used to flavor beer (Neve, 1991). There are at least five different sex chromosome systems in *H. lupulus*. In plants of European origin, the females have sex chromosomes of type XX and the males have XY (Fig. 7A), with each sex possessing nine autosomal bivalents (see Fig. 7B). There are differences in the proportions of X and Y between races, and some are completely homomorphic. In *H. lupulus* subsp. *cordifolius* from Japan, an X–autosome translocation has led to the development of a multiple system with $X_1X_1X_2X_2$ in females and $X_1X_2Y_1Y_2$ in males (Fig. 7C). Remarkably, the disjunction of the quadrivalent in the male is inefficient, with about 30% irregular segregation at anaphase I of meiosis (J. S. Parker, unpublished data). With the

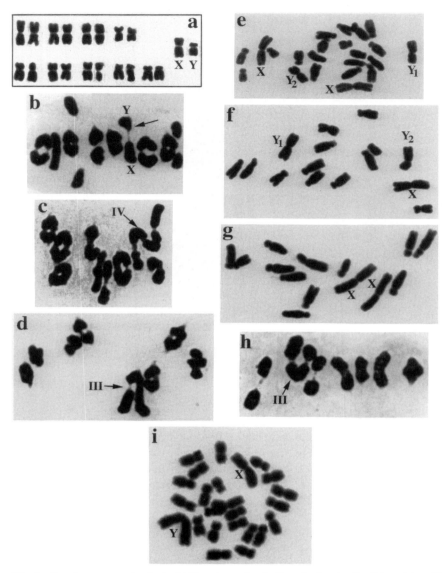

Fig. 7 Sex chromosomes in plants. (a) Karyotype of male *Humulus lupulus*, English variety, with 2n=18+XY. Note that the Y is smaller than the X. (b) Metaphase I in pollen mother cell (PMC) of male *Humulus lupulus*, English variety, 2n=18+XY with heteromorphic XY bivalent (arrow). (c) Metaphase I in PMC of the Japanese subspecies *H. lupulus* subsp. *cordifolius* male, 2n=16+X$_1$X$_2$Y$_1$Y$_2$. Note the sex-quadrivalent in alternate orientation (IV). (d) Metaphase I in PMC of *Humulus japonicus* male, 2n=14+XY$_1$Y$_2$ with sex-trivalent (III). (e) Mitotic metaphase chromosomes of *Rumex acetosa* hermaphrodite, 2n=18+XXY$_1$Y$_2$. (f) Mitotic metaphase chromosomes of *R. acetosa* male, 2n=12+XY$_1$Y$_2$. (g) Mitotic metaphase chromosomes of *R. acetosa* female, 2n=12+XX. (h) Metaphase I in PMC of *R. acetosa* male, 2n=12+XY$_1$Y$_2$, showing orientation of the sex-trivalent (III). (i) Karyotype of *Silene latifolia* male, 2n=2x=22+XY.

exception of the homomorphic races, the Y chromosomes of hop are unique in flowering plants in that they are smaller than the X chromosome.

The annual hop, *H. japonicus*, has a sex chromosome system similar to that of *R. acetosa*. The chromosome constitution of the female is $2n=14+XX$ and that of the male is $2n=14+XY_1Y_2$ (Fig. 7D).

A dosage sex chromosome system operates in both *H. lupulus* and *H. japonicus*. An X:autosome (X:A) ratio of 0.5 or less is associated with a male phenotype, while females have a ratio of 1.0. Intermediate ratios give rise to intersexual, monoecious phenotypes. As in sorrel, the production of the male phenotype is independent of the Y chromosome, although male plants lacking this chromosome fail to produce viable pollen. Sex can also be manipulated in certain cultivars of *H. lupulus* (e.g., Fuggle) by application of the weak synthetic auxin, alpha(2-chlorophenylthio) propionic acid, which induced some male flowers to form on genetically female plants (Weston, 1960). The timing of spraying was critical for its effectiveness in sex reversal. A reasonably common phenomenon in hop is the development of terminal female flowers on inflorescences of male plants of *H. japonicus*.

In the related species, cannabis, plant growth substances also have effects on sex. Auxins and ethylene have feminizing effects (Heslop-Harrison, 1956; Mohan Ram and Jaiswal, 1970), whereas cytokinins and gibberellins have masculinizing effects (Chailakhan, 1979; Atal, 1959).

Analysis of a number of monoecious plants of *H. lupulus* revealed a range of different patterns of male and female flower production, including tetraploid plants that carried approximately equal numbers of male and female flowers (H. L. Shephard, P. Darby, and C. C. Ainsworth, unpublished data). Other diploid monoecious plants were either of predominantly male or predominantly female phenotype. The mainly male monoecious plants carried a small number of laterals that terminated in a female inflorescence, while predominantly female plants bore a large number of hermaphrodite flowers that were clustered in panicles near the top of the plant. In hermaphrodite flowers, organs were sometimes found where the features of carpel and stamen were combined, a condition described more than a century ago by Wehrli (1892) in other plant species. Mixed stamen/carpel organs have also been found in cannabis in environmentally and chemically modified sex phenotypes (Heslop-Harrison, 1963) and in annual mercury (Durand and Durand, 1991).

To date, there are no reports of experiments that address the molecular mechanisms of sex determination in hop, with the exception of some genetic mapping (Pillay and Kenny, 1996). We are currently investigating the roles of the floral MADS box genes in hop and have cloned putative homologues of *DEFICIENS/APETALA3* and a number of other MADS box genes, with a view to assessing their expression patterns by in situ hybridization (H. L. Shephard, P. Darby, and C. C. Ainsworth, unpublished data). The expectation is that the expression patterns of the B and C function MADS box genes will be different in male and female floral primordia since the inappropriate organ sets are not initiated.

The rather leaky nature of gender in hop, and the fact that gender can be affected by synthetic auxins, argues for there being a plant growth substance component to sex determination. This has been more convincingly demonstrated in cannabis, which shows a similar pattern of floral development (Mohan Ram and Nath, 1964), although auxins, cytokinins, and ethylene all have feminizing effects only (Heslop-Harrison, 1956; Chailakhan, 1979; Mohan Ram and Jaiswal, 1970).

4. Sorrel: *Rumex acetosa*

Sorrel, *R. acetosa*, is a strictly dioecious perennial member of the Polygonaceae and is cultivated in some countries for its acidic leaves. The subgenus *acetosa* of *Rumex* contains 10 dioecious species characterized by a distinctive sex chromosome system but also has a number of hermaphrodite species such as *R. scutatus* and *R. vesicarius*. In contrast to the previously described plants, annual mercury and hop, males and females of sorrel pass through a bisexual stage that contains all the organ primordia expected in a hermaphrodite or perfect flower.

The basic Polygonaceae flower type consists of four whorls of organs. In hermaphrodite *Rumex* species, the two outer whorls belong to the perianth; then follows one stamen whorl, with the gynoecium occupying the central whorl. The organs in the perianth whorls are very similar, small and sepaloid, there being no clear distinction between sepals and petals as there is with most plant species. *R. scutatus* is shown, as an example, in Fig. 1. The small, nonshowy flowers are associated with wind pollination. In *R. acetosa*, both male and female flowers have only three whorls of developed organs: the male flower consists of a whorl of stamens and two whorls of perianth segments; the female flower consists of three fused carpels with a single ovule and two whorls of perianth segments, one of which encloses the ovary (see Figs. 1 and 4). The mature flowers differ greatly in size; the male flower bud before opening is 2 mm in diameter and the female bud is 1 mm. In the vegetative phase, plants of the two different sexes of sorrel are indistinguishable from one another.

The first differences between male and female are manifested very early during the development of the flowers, at a stage soon after initiation of the organ primordia (Ainsworth *et al.*, 1995). There is very little development of the inappropriate organs. In the developing male flower, there is no significant proliferation of cells in the center of the flower, in the position normally occupied by the carpels of an hermaphrodite plant (Fig. 8). In the female flower, groups of cells in the third whorl proliferate to form small stamen primordia, which develop no further and are soon overgrown by the developing carpels (see Fig. 8). In flowers of both sexes, the arrest in development does not appear to be accompanied by cell death and tissue degeneration.

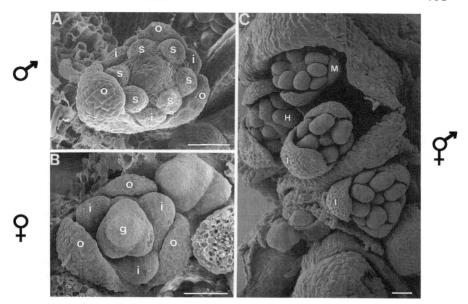

Fig. 8 Development of male, female, and hermaphrodite sorrel flowers. Scanning electron micrographs of single (A) male and (B) female flower primordia, and (C) part of an inflorescence from a hermaphrodite plant carrying flower primordia with differing degrees of gynoecium development, from male (M, with no gynoecium development) to intermediate (I) to full hermaphrodite (H). i, outer perianth segments; o, inner perianth segments; s, stamens; g, gynoecium. Note the lack of development in the center of the male flower. Bars, 50 μm. Stamen primordia in the female are hidden by the gynoecium. [(A) and (B) from Ainsworth *et al.* (1995), with permission.]

Sorrel has a multiple sex system, with $2n=12+XX$ in females and $2n=12+XY_1Y_2$ in males (see Figs. 7G,F). In the male, the two Y chromosomes and the X chromosome form a trivalent at metaphase I of meiosis (Fig. 7H). The system is one of X:A dosage. When the X:A ratio is 1.0 or higher the individuals are female, whereas X:A ratios of 0.5 and lower result in males. Ratios between 0.5 and 1.0 can give an intermediate, hermaphrodite phenotype. The primary sex determination, therefore, is independent of the presence or absence of the Y chromosomes. The production of the male floral parts is also independent of the Y chromosomes, because near-tetraploids, in which $2n=24+XXX$ (with a ratio of 0.75), are hermaphrodite. The two Y chromosomes, however, are required for the successful progress of meiosis in pollen mother cells (Parker and Clark, 1991). Y deletions cause arrest of meiosis at zygotene and death of pollen mother cells. Premeiotic control of cell size is also disturbed.

Sex determination in sorrel is influenced not only by the X:A ratio but also by the number and types of autosomes. Analysis of near-triploids, which are tetrasomic for particular autosomes, has enabled classification of the autosomes as male-promoting, female-promoting or sex-neutral (Yamamoto, 1938). For example, triploids in which $2n = 18 + XXY_1Y_2$ (see Fig. 7E) are hermaphrodite, whereas triploids in which $2n = 19 + XXY_1Y_2$, which include an extra copy of chromosome 2, flower as females. Parts of chromosomes also affect sex determination. In plants that carry a translocation resulting in disomy for a quarter of the X chromosome in addition to XY_1Y_2, hermaphrodite flowers are formed that are female-fertile but have shrivelled anthers (Parker, 1990). Plants lacking this X chromosome segment are male. The extra X chromatin is able to give a functionally female phenotype, although it is insufficient to suppress the development of an androecium. These observations point to the fact that a number of genes, carried on autosomes and sex chromosomes, interact in a complex way to determine sex. Genes on the X chromosome and autosomes determine the onset of the cascade of events that results in development of maleness and femaleness. It is only after the establishment of maleness that genes on the Y chromosome are involved in pollen development.

Ainsworth *et al.* (1995) asked the question, are the differences between male and female flowers the result of differences in expression of the organ identity genes? The putative B and C function homeotic genes were isolated from sorrel, and their patterns of gene expression as revealed by *in situ* hybridizations were shown to be strikingly different from those seen with homologous genes in other plant systems (Fig. 9).

A surprising finding, unrelated to sex determination, was that the expression of two putative B function genes, *RAD1* (*RUMEX ACETOSA DEFICIENS 1*) and *RAD2*, which are *DEFICIENS* homologues was confined to a single whorl—whorl 3, the stamen whorl—in both male and female flowers, rather than in the stamen and petal whorls, as would be expected for a hermaphrodite plant such as *Antirrhinum* or *Arabidopsis* (see Fig. 2; Bradley *et al.*, 1993; Jack *et al.*, 1992; Goto and Meyerowitz, 1994; Sommer *et al.*, 1990; Schwarz-Sommer *et al.*, 1992; Tröbner *et al.*, 1992). In view of this observation, it was proposed that the lack of expression of the B function gene causes the second whorl organs to become sepaloid and that sorrel displays a "mutant" phenotype (relative to plants whose flowers contain typical sepals and petals). It is possible that all the species in the genus have two whorls of sepals (and no petals) and the expression pattern of the *DEFICIENS* homologues is restricted to one whorl in all of them. Furthermore, it is possible that this phenomenon may extend to the entire Polygonaceae family, because its members all have two similar whorls of perianth segments. This situation is not without parallel in other plants; in petunia the whorl 2 organs of the homeotic *green petals* (*gp*) mutant have the same shape and color as the sepals in whorl 1, and petaloid cells occur on the stamen filaments (van der Krol and Chua, 1993). Analysis of the expression of the petunia MADS box genes shows that in the wild type the *DEFICIENS* homologue, *pMADS1*, is expressed in the petal and stamen whorls (Angenent *et al.*, 1995) whereas, in the *gp* mutant, expression of *pMADS1* mRNA is not detectable in any whorl (van der Krol and Chua, 1993; Angenent *et al.*, 1995).

The expression of the putative B function genes in sorrel is coincident with the initiation and development of the stamens in the male and with the formation of the stamen primordia, which abort after a short period of development in the female (see Fig. 9). B function gene expression in male and female flowers, therefore, differs only in the area of the domain of expression.

By contrast, *RAP1* (*RUMEX ACETOSA PLENA 1*), the putative C function gene, shows a sex-specific expression pattern (see Fig. 9). The gene is expressed in the young male flower primordia in whorls 3 and 4, the stamen and carpel whorls, respectively. The expression in the carpel whorl is, however, transient, and as soon as the stamen primordia begin to enlarge significantly, expression in the center of the flower becomes undetectable. Because in normal flowers the ABC model predicts that the C function alone is needed for carpel development (Coen and Meyerowitz, 1991), the lack of continued carpel development could be caused by the inactivation of the C function gene. Alternatively, repression of C function gene expression may be a consequence of the fact that carpel development is prevented by the activity of other genes. A gene similar to the sex-determining *TASSELSEED2* gene of maize, which results in the abortion of the androecium tissue in the tassel (DeLong *et al.*, 1993) may operate to cause the arrest of carpel development in the male, which itself (or by cell death) would result in the inactivation of *RAP1* expression. However, obvious signs of cell death were not observed during the development of male or female flowers.

RAP1 is expressed in the young female flower primordia in whorls 3 and 4, the stamen and carpel whorls. Later than this, when the stamen primordia are visible as hemispherical structures, the expression is retained in the carpel whorl but is lost from the stamen primordia. The inactivation of *RAP1* expression is coincident with the cessation in further development of the stamen primordia. The ABC model predicts that both the B and C functions are needed for stamen development (Coen and Meyerowitz, 1991). As is the case with carpel development in the male, the lack of expression of this gene in the female may be a consequence of cessation of stamen development or a cause of it. However, the latter seems likely, because the B function genes *RAD1* and *RAD2* are still expressed in the stamen primordia after *RAP1* transcripts become undetectable. Therefore, in the female at least, the arrest in development is not caused by cell death. Further experiments are needed to investigate this phenomenon in both males and females: is the mechanism of arrest the same in flowers from both sexes, and is arrest simply a result of the switching off of the C function gene, or is there a more extensive inactivation of gene activity in these tissues?

Analysis of a genetically hermaphrodite line of sorrel, a triploid with the chromosome constitution $2n = 3x = 18 + XXY_1Y_2$ (see Fig. 7E), has provided interesting insights into the timing of sex determination (C. C. Ainsworth, V.

Buchanan-Wollaston, and J. S. Parker, unpublished data). All flowers produced on inflorescences of these hermaphrodite plants are staminate, although there is variation between flowers in the degree of anther development. Some anthers produce pollen, only a proportion of which is fertile (this is normal for a triploid), while others develop much less, are shriveled, and produce no pollen. The flowers, however, exhibit quite astonishing variation in the degree of development of the gynoecium (Fig. 10). Development of the gynoecium ranges from a simple rod-like structure or filament (varying in length from a few cells to the height of the normal gynoecium), to a filament with stigmas, to a gynoecium consisting of one, two or three separate carpel elements, each developing its own stigma. The full hermaphrodite flower has six stamens and a tricarpellate, mono-ovulate gynoecium with three stigmas. This hermaphrodite plant can really be considered to be andromonoecious. The range of female development can be seen in adjacent flowers at the scanning electron microscopic level (see Fig. 8). Interestingly, mixed organs are sometimes found in these hermaphrodite flowers, such that a carpel-like structure makes up the upper part and a stamen the lower part. This may indicate that whatever gene products are required for determination of sex (i.e., which organ sets are promoted and which suppressed), others may also be needed to maintain the correct development of the organs in the set. However, the key implication is that it is the sex of the individual flowers that is determined, rather than that of the whole inflorescence.

It is clear that sex determination in monoecious species is at the level of the flower and that determination is a property of buds rather than of whole plants or inflorescences. This has been discussed previously by Huala and Sussex (1993). Additional evidence for determination of the phenotype of the individual flower is provided by mutants such as *centroradialis* in *Antirrhinum*, in which the normal indeterminate inflorescence terminates with a flower that shows radial symmetry, as opposed to the usual bilateral symmetry shown by all flowers lower down on the inflorescence (Bradley *et al.*, 1996).

Why is there variation among the sorrel flowers formed on hermaphrodite plants? These plants have an X:A ratio of 0.66, between the ratio of 0.5 required for maleness and the ratio of 1.0 required for femaleness. We have no clear picture of the number of genes, or even the number of autosomes, which are involved in the X:A system of determination. There must be an interplay between genes on the X chromosome and those on the autosomes, and the levels of transcripts or their products must be important. In the hermaphrodite plants, intermediate dosages of genes lead to molecular confusion. The fact that all flowers produce stamens and only some produce carpels may reflect the fact that the X:A ratio of 0.66 is closer to that of normal males than that of normal females. These hermaphrodite individuals should prove invaluable in the identification of the key autosomes that are involved in the X:A sex determination, because crosses with a normal male should yield all possible trisomics and sex chromosome combinations (XX, XXX, XXY, XXYY, XYY, XYYY, XYYYY) in the progeny.

5. White Campion: *Silene latifolia*

The white campion, (*S. latifolia*; formerly known as *Melandrium album* and *Silene alba*), is a member of the family Caryophyllaceae and is one of a small group of dioecious plants that have an active-Y system of sex determination. Of the members of this group, white campion is undoubtedly the best characterized and has been the subject of considerable research effort for many years, since the first genetic studies in the 1930s and 1940s (Winge, 1931; Warmke and Blakeslee, 1939; Westergaard, 1940, 1946). White campion has been the subject of a number of reviews (for example, Ye *et al.*, 1991; Grant, Hunkirchen *et al.*, 1994, 1996).

The floral differences between male and female white campion flowers were described in detail by Grant, Hunkirchen *et al.* (1994). Floral organ development is similar in primordia from both sexes up to the stage at which petal and stamen primordia arise, when the first differences become apparent; the size of the undifferentiated fourth carpel whorl in the female is much larger than in the male. During further development, differences between the sexes become more pronounced: development of the stamens in the female is arrested abruptly (at the time of tapetum initiation in the male stamen), whereas in the male gynoecium development arrests soon after its initiation and results in an undifferentiated rod 2–3 mm in length (Fig. 11). Mature male flowers have five sepals, five petals, 10 stamens, and the rudimentary gynoecium (see Figs. 4 and 11). Mature female flowers have five sepals, five petals, five carpels that fuse to form a free central ovary, and little sign of the 10 rudimentary stamens, which degenerate during the later stages of floral development (see Figs. 4 and 11). The gynoecium in the female is 5–10 times larger than in the male and attains a length of 10–30 mm. When compared to sorrel, therefore, the inappropriate organs in campion develop considerably further. Differences in timing of the arrest of the inappropriate sex organs suggest that the developmental program leading to suppression of gynoecium development in the male is independent of that leading to suppression of stamen development in the female (Grant, Hunkirchen *et al.*, 1994). The analysis of mutants suggests that the default developmental pathway is that leading to femaleness and that maleness is substituted for this pathway.

Male white campion plants have a chromosome constitution of $2n = 2x = 22 + XY$; females have $2n = 2x = 22 + XX$ (Warmke and Blakeslee, 1939; Westergaard, 1940) (see Fig. 7I). In contrast to the system in sorrel, where the Y chromosome is irrelevant to sex determination, in campion it is the presence of the Y chromosome that has the major effect on determination of sex, maleness being dominant to femaleness.

200   Charles Ainsworth *et al.*

Fig. 11 Development of male and female white campion flowers. (a) Scanning electron micrograph (SEM) of young male flower (stage 6). (b) SEM of almost mature male flower (stage 9). (c) SEM of mature male flower (stage 12). (d) SEM of young female flowers (upper flower at stage 7, lower flower at stage 6). Sepals have been removed from the stage 7 flower. (e) SEM of almost mature female flower (stage 10). (f) SEM of almost mature female flower (stage 10). Developmental stages are as described by Grant *et al.* (1994a). Bars, 200 μm. c, carpel; g, gynoecium; st, stamens. [(b), (d), and (e) from Grant *et al.* (1994a), with permission. (a), (c), and (f) kindly provided by P. M. Gilmartin.]

Genetic analysis of Y chromosome deletion mutants has shown that the Y chromosome carries dominant genes in three regions which suppress carpel development, promote stamen development, and allow development of the stamens once initiated (Westergaard, 1946, 1958; Ye *et al.*, 1991) (Fig. 12). The loss of the distal part of one arm of the Y chromosome results in hermaphroditism; that is, the gynoecium is no longer suppressed. Deletion of part of the same arm as was involved in gynoecium suppression resulted in plants with no stamens or carpels (asexual mutants); therefore, a region for stamen initiation had been lost.

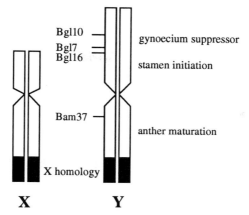

Fig. 12 Sex-determining functions on the white campion Y chromosome. The campion Y chromo-some carries loci for stamen initiation and gynoecium suppression on one arm and for anther maturation on the other arm. The black areas indicate the pseudoautosomal pairing regions of X/Y homology. The RDA markers of Donnison *et al.* (1997) are shown. Redrawn from Donnison *et al.* (1996) with permission.

Deletion of part of the other arm, however, resulted in plants that were male but carried degenerated sterile anthers; that is, a region for stamen maturation had been lost. Although three regions were identified, it is not clear how many genes are involved in each region or how large the regions are.

Recently, the technique of representational difference analysis (RDA; Lisitsyn *et al.*, 1993), a method of genomic subtraction, has allowed male-specific DNA sequences to be isolated (Donnison *et al.*, 1996). The positions of the RDA markers on the campion Y chromosome were assessed in Y chromosome–linked her-maphrodite and asexual mutations (deletions of the Y chromosome) which were generated by X-ray mutagenesis. Using this technique, four RDA Y-specific sequences, Bg17, Bg110, Bg116, and Bam37, enabled the regions of the Y chromosome that carry loci involved in carpel suppression, stamen initiation, and stamen maturation to be identified. From this analysis it was clear that the carpel suppression and stamen initiation regions were linked on the same arm (see Fig. 12).

There is evidence that the effects of the Y chromosome are not confined to the floral whorls containing the sex organs, and an interesting difference between the flowers of male and female campion plants, unrelated to sex organ development, was noted by Scutt *et al.* (in press). *Men-6* (for **M**ale **en**hanced), one of a series of cDNA clones isolated from a subtracted male flower library, represented a gene expressed in anthers and petals of male buds that was not expressed in petals from female buds.

The X and Y chromosomes are homologous over part of their lengths and pair over this pseudoautosomal region to form a bivalent at meiosis (Grant, Hunkirchen *et al.*, 1994). The role of the campion X chromosome and the autosomes in sex determination is less well understood. The Y chromosome has been shown to be central in sex determination and, if diploid plants carry an additional copy of the X chromosome (XXY), they predominantly flower as males. Further excesses of X chromosome copies reduce the frequency of male plants (Westergaard, 1948, 1958). A ratio of 1X to 1Y gave 100% males in the progeny, 2X:1Y gave 89% males, 4X:1Y gave 35% males, and with 8X:1Y all plants were female (Westergaard, 1958). The X chromosome, therefore, plays some part in the process. This is supported by earlier work of Westergaard (1946), who suggested that the differential region of the X chromosome contains male suppressor functions, and also by the work of Veuskens *et al.* (1992), who identified female-promoting genes in this region of the X chromosome.

The autosomes are also implicated in sex determination. Van Nigtevecht (1966) selected a line of true-breeding XY hermaphrodites which, on outcrossing to normal females, segregated for normal males and females in the progeny. In polyploids with 2n=42+XXXY or 2n=44+XXXY, the difference of two autosomes was shown to give essentially male plants and essentially female plants, respectively. The specific autosomes have not yet been identified.

The active-Y system is, therefore, not simple. Ye *et al.* (1991) proposed that sex in campion depends on interaction between five classes of genes: (1) male-promoting genes on the proximal part of the differential or nonpairing arm of the Y chromosome (epistatic to the male supressor genes on the X chromosome), (2) male-promoting genes on the autosomes, hypostatic to the male supressor genes on the X chromosome, (3) female supressor genes on the distal part of the differential arm of the Y chromosome (partly epistatic to the female-promoting genes), (4) female-promoting genes on the autosomes and probably the X chromosome, and (5) a male suppressor in the differential region of the X chromosome.

As in sorrel, the genetic basis of sex determination in campion is strong, and there is little evidence for environmental effects of the standard kind (Grant, Hunkirchen *et al.*, 1994). Plant growth substances are ineffective (Ye *et al.*, 1991), although testosterones and estrogens have been suggested to have an effect on sex (Lyndon, 1985). In the related but hermaphrodite species, *Silene pendula*, auxin treatment was shown to suppress stamen development and promote gynoecium development, but similar effects were not observed in *S. latifolia* unless "substantial doses" were applied (Heslop-Harrison, 1963).

The striking exception to the stability of the campion sex system to external stimuli is that infection with the anther smut fungus *Ustilago violacea* (also known as *Microbotryum violaceum*) can modify sex expression. Stamen development in chromosomally female plants can be induced such that the anthers of infected plants contain spores of *U. violacea* rather than pollen (Audran and Batcho, 1981). Infection of male plants can lead to development of the gynoecium further than is normal in the male (Batcho and Audran, 1981). Despite considerable research efforts, it has not been possible to identify the causative agent provided by *U. violacea* infection. However, an interesting and possibly relevant observation is that the hypomethylating drug 5-azacytidine can suppress female differentiation in males (by, it was proposed, methylation of specific DNA sequences) and furthermore was able to induce male plants to develop as hermaphrodites (Janousek et al., 1996). This androhermaphroditism may be caused by a metastable epigenetic process that abolishes the function of Y chromosome female-suppressing genes or activates autosomal female-determining or -promoting genes (Janousek et al., 1996).

The fact that infection with U. violacea causes stamens to develop on chromosomally female plants suggests that the fungus is able to substitute for one or more of the Y chromosome functions. Differential screening of subtracted cDNA libraries has allowed the isolation of eight cDNA clones, Men-1 to Men-8 which show male-specific patterns of gene expression (Scutt et al., 1997). The genes represented by these clones, which encode a rather diverse range of proteins of largely unknown function, all appeared to be expressed in male plants in response to signals derived from the Y chromosome but were not thought to be Y-located. Interestingly, three of these genes showed induced expression in the stamens of smut-infected female plants. The genes that did not show smut-induced expression are probably involved in the development of the tapetum or male germ line cells (Scutt et al., in press).

Although cytologic and molecular studies on deletion mutants affecting sex in campion have allowed regions of the Y chromosome that contain loci involved in sex determination (gynoecium suppression, stamen initiation) or sex differentiation (stamen maturation) to be identified, there is as yet no information on the nature of the genes. Studies using 5-azacytidine, which promotes female differentiation in males, indicated that female sex suppression in males depends on methylation of specific DNA sequences (Janousek et al., 1996).

One of the first molecular targets was the MADS box organ identity genes. Campion homologues of the B and C function genes have been isolated and their expression patterns during male and female flower development assessed (Hardenack et al., 1994). It was clear from this study that the MADS box genes did not play a key role in sex determination; expression patterns were much as predicted for model hermaphrodite plants, and there were not sufficient differences between male and female flowers. The C function gene, SLM1, showed no difference in expression pattern until after meiosis in the female flowers, when expression in the stamens became undetectable. This observation was considered to be attributable to cell death in the degenerating stamens. This contrasts with the situation in sorrel, where sex-related differences in C function gene expression occur and there is no evidence for cell death (Ainsworth et al., 1995). A difference in the spatial expression patterns of the B function genes between males and females was noticed; SLM2 and SLM3 were expressed closer to the center of the male floral primordium than of the female primordium, resulting in a smaller fourth whorl, correlated with the reduction in size of the gynoecium. The organ identity genes, therefore, do not appear to have a direct role in sex determination, and the key genes must act downstream of them.

6. Asparagus: *Asparagus officinalis*

Asparagus (*A. officinalis*, Liliaceae) can be considered to be a subdioecious species; females are consistent in their sexual phenotype, while staminate plants range from fully male to hermaphrodite with substantial berry production (Machon *et al.*, 1995). In terms of development, the bisexual stage persists in male and female floral primordia until quite late in development. Male and female flowers are indistinguishable until close to male meiosis (Lazarte and Palser, 1979; Bracale *et al.*, 1990). Mature male flowers have a rudimentary pistil containing nonfunctional ovules, and female flowers have stamens with collapsed anthers that lack fertile pollen (Machon *et al.*, 1995; Caporali *et al.*, 1994). Gynoecium size in males varies, but it can develop to the same size as that found in females. In the female, the stamens attain a length of 0.8 mm, in contrast to the 1.3 mm found in males (Galli *et al.*, 1993; Caporali *et al.*, 1994).

In asparagus, an XX/XY sex chromosome system has been claimed by some authors, but most studies have shown chromosomal homomorphism (e.g., Loptien, 1979). The genetic evidence, however, indicates a system of male heterogamety and the presence of YY individuals (Marks, 1973). Sexes occur in a 1:1 ratio, and a bipartite gene of the male activator/female suppresser type found in white campion has been proposed to operate (Marks, 1973). In a molecular biology study, a number of MADS box genes were cloned, including the putative C function homologue (Miller *et al.*, 1995). However, no experiments have been reported that attempt to correlate the expression of homeotic genes with floral phenotype in asparagus.

In asparagus, there is not such a clear distinction in endogenous levels of plant growth substance as in annual mercury; levels of cytokinins, IAA, and ABA were found to be generally higher in male than in female plants (Bracale *et al.*, 1991). It is probably the balance of plant growth substances, rather than their precise levels, that is important in sex determination. Gibberellins have been shown to have a masculinizing effect (Lazarte and Garrison, 1980). It was suggested that ABA has a role in determining the balance of plant growth substances in asparagus (Bracale *et al.*, 1991).

The observations of great pistil variability in asparagus males supports the hypothesis that dioecy in asparagus has arisen from the hermaphrodite condition by the establishment of females and then males (Galli *et al.*, 1993). The lack of heteromorphic sex chromosomes, the rather advanced development of the inappropriate organs, the variability in sex expression, and the fact that this is the only dioecious species in the Liliaceae led Machon *et al.* (1995) to suggest that dioecy in asparagus has evolved relatively recently.

7. Kiwifruit: *Actinidia deliciosa*

The kiwifruit (*A. deliciosa*, Actinidiaceae) is remarkable in that mature male and female flowers are superficially identical (see Fig. 4). Male plants carry staminate flowers that produce viable pollen grains and a small ovary that has vestigial styles and does not contain ovules. Female plants carry pistillate flowers, the ovaries of which contain many ovules, but also stamens that produce empty pollen grains (McNeilage, 1991a). For detailed accounts of floral development in *Actinidia* the reader is directed to Harvey *et al.* (1987), Harvey and Fraser (1988) and McNeilage (1991a). The structural differences between male and female flowers occur very late in development, coincident with microsporogenesis (Messina, 1993). An analysis of microsporogenesis in females showed that formation of the exine wall of the pollen grain was normal but intine wall synthesis was impaired, leading to pollen inviability (Messina, 1993).

This condition, termed functional dioecy or cryptic dioecy (Mayer and Charlesworth, 1991) is not peculiar to kiwifruit and is found in all 60 or so species in the genus *Actinidia* (Testolin *et al.*, 1995). Cryptically dioecious plant species are present in a number of unrelated plant families, many families being represented by just one genus or species, which argues for independent evolution of this type of behavior (Mayer and Charlesworth, 1991). The production of sterile pollen appears wasteful but may be explained by the fact that even sterile pollen is a reward for foraging insect pollinators which might not visit a flower without stamens (Cane, 1993). Mutations that abolish organ production in insect-pollinated plant species are likely to be selected against.

Although mainly dioecious, some male plants show variable ovary and style development and can be considered hermaphrodite because they bear fruit, albeit with less seed than in the females (Harvey and Fraser, 1988). The frequency of production of hermaphrodite flowers appears to be sensitive to environmental conditions (McNeilage, 1991b). In diploid and polyploid *Actinidia*, controlled crosses gave a 1:1 male–female ratio in the progeny, and the male sex appeared to be the heterogametic one (Testolin *et al.*, 1995). It was therefore proposed that a monofactorial system based on one or more linked genes or an X/Y chromosome pair must be controlling sex expression. However, no evidence of sex chromosomes in kiwifruit has been found. Minor modifying genes seemed to be responsible for the feminization of males, and the expression of these appeared to be enhanced by environmental conditions. Masculinizing genes are not thought to be present in females (Testolin *et al.*, 1995). Little information on endogenous levels of plant growth substances or the responses of male and female plants to exogenously applied plant growth substances is available, but the general instability of the system seems to suggest their involvement in the determination of sex in kiwifruit.

8. Environmental Effects and Sex Determination

Analysis of sex determination in maize and other plant species shows that, in addition to a genetic component to the determination events, plant growth substances may be implicated, and that the genes that determine sex may be involved in the biosynthesis of plant growth substances. There also appears to be an environmental component. In some monoecious and dioecious plant species (and variants of both), the sex of an individual or of a single flower can be determined environmentally or as a result of genotype by environment interactions (Charnov, 1982; Lloyd and Bawa, 1984; Meagher, 1988). Although plant growth substances have been implicated in sex determination in several other plant species, some, such as the dioecious species sorrel and white campion, appear to be insensitive to supplied plant growth substances (C. C. Ainsworth and J. S. Parker, unpublished data; Grant, Houben *et al.*, 1994).

Examples of monoecious species that are sensitive to environmental stimuli are the oil palm *E. guineensis* (Corley *et al.*, 1995), cucumber (Frankel and Galun, 1997; Malepszy and Niemirowicz-Szczytt, 1991), and maize (Heslop-Harrison, 1961). In maize, feminization of tassels can be induced by short days and by cool nights (Heslop-Harrison, 1961); such feminized tassels have 100 times the normal level of gibberellin (Rood *et al.*, 1980). Among the dioecious plants, sex in kiwifruit and Canada yew, *Taxus canadensis*, is sensitive to environment (McNeilage, 1991b; Allison, 1991). In an androdioecious population of annual mercury, male plants appear to switch their development to the hermaphrodite condition, the switch being more frequent at lower population densities (Pannell, 1997). Other dioecious species, such as *R. acetosa* (C. C. Ainsworth and J. S. Parker, unpublished data) and *S. latifolia* (Heslop-Harrison, 1963; Ye *et al.*, 1991; Grant, Houben *et al.*, 1994), are particularly resistant to the effects of environment. A remarkable exception is the induction of stamen development in chromosomally female plants by infection with *U. violacea*.

An interesting observation in the usually hermaphrodite plant, *Salsola komarovii* (saltwort), is that photoperiod has a profound effect on flower development. *Salsola* is a facultative intermediate-day-length plant, but under long days all flowers formed are hermaphrodite and under short days all flowers are females (Takeno *et al.*, 1995).

V. Plant Growth Substance–Mediated Sex Determination in Ferns

A parallel exists in the determination of sex in ferns in that growth substances are also implicated. In many homosporous ferns, the choice between the alternative developmental fates, male or hermaphrodite, is determined by the pheromone antheridiogen, which is secreted into the surroundings by hermaphrodites (Näf *et al.*, 1975). Male gametophytes result from spore germination in the presence of antheridiogen, whereas hermaphrodite gametophytes result if antheridiogen is absent. The antheridiogens of several ferns are gibberellins (Banks, 1994). Although the structure of the antheridiogen from *Ceratopteris richardii*, (A_{CE}), is not known, it also is likely to be a gibberellin (Banks, 1994). The window of competence to respond to A_{CE} is narrow, and once established, the male program is irreversible even if A_{CE} is supplied (Banks *et al.*, 1993).

A simple model for sex determination in *Ceratopteris* has been proposed, based on evidence from the mutants *her1* (*hermaphroditic*), *tra1* (*transformer*) and *fem1* (*feminization*), and their responses to antheridiogen (Banks, 1994). Three genes *HER1*, *TRA1*, and *FEM1*, are involved. *FEM1* is constitutively expressed and causes development of antheridia (the male organs) in the presence or absence of antheridiogen. In the presence of antheridiogen, the *HER1* gene is activated and represses the expression of the *TRA1*, gene which is required for the formation of archegonia (the female organs). In the absence of antheridiogen, both *FEM1* and *TRA1* are expressed, leading to the development of archegonia and antheridia. However, the situation is more complex, because linkage analysis and tests of epistasis among the different mutants have indicated that sex determination in *Ceratopteris* involves at least seven interacting genes in addition to antheridiogen, the primary sex-determining signal (Eberle and Banks, 1996).

VI. Are There Analogies with Sex Determination in Animals?

In animals, the sum of individual cellular decisions in the soma, referred to as somatic sex determination, is seen in the whole organism as the morphologic differences between male and female. The overall strategy for mammalian sex determination is relatively simple: the Y chromosome locus, *SRY* (*TDY*), acts as a signal for testis development and subsequent male sexual differentiation (Sinclair *et al.*, 1990; reviewed in Capel, 1996); In an apparently independent event, the organism recognizes the number of X chromosomes and inactivates all but one to achieve dosage compensation (Grant and Chapman, 1988; reviewed in McElreavey, 1996). The system in plants such as white campion, which have an active-Y system of sex determination, may be similar.

The systems in *Drosophila* and *Caenorhabditis* are more complex (reviewed in Mittwoch, 1996). The Y chromosome is not an important signal in *Drosophila* and is absent in *Caenorhabditis*. The chromosomal signal in these organisms is the X:A ratio; in females of both organisms, the X:A ratio is 1.0, in males it is 0.5.

In *Drosophila*, the primary binary switching gene, *Sxl*, is transcribed in both sexes throughout development but is functionally active only in females and directs female development through its effect on at least two separate cascades of genes that regulate sexual differentiation and dosage compensation (Keyes *et al.*, 1992). *Sxl* is activated by a system that assesses the X:A ratio and includes X-located numerator genes and autosomal denominator genes. Several numerator genes, *sis-a*, *sis-b*, *sis-c*, and *runt*, have been identified, all of which encode transcription factors and which must increase the probability of activating the *Sxl* promoter. This system is paralleled by the X:A dosage systems of sorrel, hop, and cannabis. In sorrel, dosage differences of both the X chromosome and the autosomes affect plant sex, suggesting that the X chromosome and the autosomes carry similar numerator and denominator genes, respectively. The equivalent gene to *Sxl* may be autosomal or X-located.

One question raised by the existence of systems of sex determination in plants that use sex chromosomes is whether systems of dosage compensation operate. In many eukaryote species of both animals and plants, sex is determined by differences in the number of copies of a single chromosome. In animals, dosage compensation regulates the activity of genes carried on the X chromosomes; several different systems have evolved to allow the different number of copies of genes in the two sexes to produce the same amount of functional product (reviewed in Lyon, 1996). These include completely inactivating one of the two X chromosomes in the female (as in mammals; Lyon, 1961), decreasing the rate of transcription of genes on both the X chromosomes in the female (as in *Caenorhabditis*; Villeneuve and Meyer, 1990), and increasing the rate of transcription of genes on the X chromosome in the male (as in *Drosophila*, reviewed by Jaffe and Laird, 1986; Lucchesi and Manning, 1987; Kelley and Kuroda, 1995). In some animals, such as birds and butterflies, there is no obvious mechanism of dosage compensation (Chandra, 1991).

Little information is available as to whether dioecious plant species have analogous systems, but it would be surprising if some species did not. Only from white campion is there any experimental data on this subject. One of the two campion X chromosomes carried by females was shown to be late replicating and was hypermethylated and inactive, suggesting that a system similar to that used in mammals may be employed (Choudhuri, 1969; Vyskot *et al.*, 1993; Siroky *et al.*, 1994). In female mammals, and possibly in insects, one of the X chromosomes is late replicating, hypermethylated, and transcriptionally silent (Grant and Chapman, 1988; Rao and Padmaja, 1992). In sorrel there is no evidence for late replication of one of the X chromosomes (J. S. Parker, unpublished data), suggesting that, if a system of dosage compensation is used, it is different from the one in campion. Zuk (1969) showed by autoradiography in *Rumex thyrsiflorus*, a member of the *R. acetosa* group, that the DNA replication pattern of X chromosomes and autosomes was similar and that the two X chromosomes were identical in pattern. In the X chromosomes of mammals there is a marked difference in DNA replication pattern between the active and inactive X chromosome. This may indicate a lack of dosage compensation in *R. thyrsiflorus*.

VII. Sex-Related Differences and Molecular Markers for Sex

In dioecious species there may be morphologic differences associated with one sex or the other that are not directly related to sexual organ differences and reproductive effort. For example, in asparagus, the male plants yield more spears, yield earlier, and live longer (Prickett and Walker, 1989; Franken, 1970). In yam (*Dioscorea*) species, the tuber yield from females is greater than from males (Akorodo *et al.*, 1984). In the dioecious tropical annual, *Telfairia occidentalis*, the females are the more vigorous sex (Emebiri and Nwufo, 1996). In other plants, such as sorrel, the male and female individuals are indistinguishable before flowering. The flowering stems of sorrel females are taller and more robust than those of males. After anthesis, male flower stems collapse and decay much more rapidly than the female stems that bear the seeds.

Of the economically important dioecious plant species, it is usually the female plant that is cultivated for its fruit or seed. Examples are kiwifruit, hop, date palm, and pistachio. Asparagus is unusual in that it is the higher-yielding male plants that are preferred. Therefore, considerable effort is being put into the development of systems for early gender diagnosis in fruiting dioecious plants. In cannabis, a male-specific DNA sequence has been identified and cloned (Sakamoto *et al.*, 1995). In pistachio and asparagus, RAPD and RFLP markers have been used successfully to distinguish between the sexes (Hormaza *et al.*, 1994; Biffi *et al.*, 1995). In *Gleditsia triacanthos* there is a nonrandom association between sex and 6-phosphogluconate dehydrogenase isozyme genotypes in five natural populations (Schnabel and Hemrick, 1990). For those dioecious species carrying heteromorphic sex chromosomes, cytogenetic methods provide a convenient means of gender identification. In date palm, the sexes can readily be distinguished by cytologic examination of interphase nuclei in root tip cells. Cells from male plants carry two fluorescent blocks of unequal intensity, whereas female cells carry two equal blocks (Siljak-Yakovlev *et al.*, 1996).

VIII. Sequence Organization of Plant Sex Chromosomes

We know little about the nature of the DNA sequences carried on plant sex chromosomes. The Y chromosomes of sorrel appear as chromocenters in interphase nuclei and are positively heteropycnotic during mitotic prophase. The Y chromosomes, consist of facultative heterochromatin and are transcriptionally active during a few cell cycles immediately before meiosis in pollen mother cells. Attempts by Clark *et al.* (1993) to isolate Y-specific repetitive sequences proved unsuccessful, although seven nonhomologous repeats were isolated. These sequences were all dispersed with no areas of localization with the genome. However, a repetitive sequence of 180 bp, isolated by Ruiz Rejon *et al.* (1994) from sorrel, mapped by *in situ* hybridization to the sex chromosomes alone. Both Y_1 and Y_2 were heavily labeled, as were both arms of the X. There was no evidence of the sequence in the autosomes at this level of discrimination. The distribution of this sequence suggests a common origin of both Y chromosomes from the X chromosome or its evolutionary precursor.

A direct approach to isolation of chromosome-specific sequences can be made following flow cytometry or by microdissection. Flow cytometry was applied to *S. latifolia* by Veuskens *et al.* (1995) in an effort to isolate Y chromosomes. A wild-type Y and a deleted Y from an asexual mutant were both sorted at a purity of about 90%, which is a 21-fold enrichment of the Y. This now provides the opportunity to prepare enriched libraries and to isolate DNA sequences that are Y specific. Chromosome microdissection has also been applied to the X chromosome of *S. latifolia* (Buzek *et al.*, 1997). A library was constructed by DOP-PCR amplification, and six clones were further characterized by Southern, Northern, and fluorescence *in situ* hybridization analyses. All six clones were repetitive and derived from the telomeric region of the X. They mapped by *in situ* hybridization to subterminal regions on the majority of chromosome arms but were absent from the nonpairing (differential) arm of the Y. The Y chromosome, therefore, has diverged from the rest of the genome. None of the clones was informative when sequenced, although one contained short direct repeats of the conserved plant telomere sequence (Buzek *et al.*, 1997). The sequences tested were not informative in checking the DNA methylation status of the X chromosomes, as proposed by the dosage studies of Vyskot *et al.* (1993) and Siroky *et al.* (1994).

IX. Lessons for Hermaphrodites from Analysis of Unisexual Plants

Almost all of our understanding of the control of the processes involved in flower development in hermaphrodite plants is derived from the analysis of mutants that deviate from the wild-type pattern and comparison of these with the wild type. The minority of plants, which have unisexual flowers, are clearly different from the majority, which carry hermaphrodite flowers. Analysis of monoecious and dioecious plants is likely to provide insight into molecular mechanisms in hermaphrodite plants that may not be apparent from studies of hermaphrodites themselves.

In sorrel, for example, expression of the C function gene *RAP1* appears to be required throughout the development of the carpel in the female and the stamen in the male (Ainsworth *et al.*, 1995). All A, B, and C function mutants examined so far in hermaphrodite species are complete loss-of-function mutations; there are no examples of mutants in which there is temporally reduced expression of a specific organ identity gene. In sorrel, transient expression of the C function gene is insufficient to allow growth of the stamens in the female or the carpel in the male; this observation supports the argument that the C function is required for the initial determination of an organ and its continued presence is required for continued organ development.

In addition to organ conversion in whorls 3 and 4, the complete loss of C function, as in the *plena* and *agamous* mutants, also leads to a lack of determinacy in the center of the flower and to the development of "flowers within flowers" (Coen and Meyerowitz, 1991). In sorrel, the suppression of floral meristem growth appears to have been separated from the organ identity function; transient expression of *RAP1* is sufficient to suppress further growth in the center of the flower but is not sufficient for full organ development. Studies of transgenic *Arabidopsis* plants expressing *AGAMOUS* antisense RNA have provided evidence that organ identity and determinacy are separate processes that require different levels of *AGAMOUS* expression (Mizukami and Ma, 1995).

X. Future Directions

It is clear that the identification of the key genes that determine sex in monoecious and dioecious plants will not be easy. The identification of downstream genes, the sex differentiation genes, represents an easier but less significant target. Attempts at cloning the sex-determining genes have focused on genes that are involved in the control of flower development in hermaphrodites. This approach is likely, in the main, to be unsuccessful since we can not predict easily which genes will be important simply from expression patterns and their roles in hermaphrodites. An exception may be the isolation of genes involved in the biosynthesis of plant growth substances in plants where there is evidence of a strong growth substance component in sex determination (e.g., maize, cucumber, annual mercury). In addition, although the primary sex-determining genes may be present in hermaphrodite plants, they may have completely different roles.

A problem in the isolation of genes from strictly dioecious plants such as sorrel and white campion is that standard genetic mapping and chromosome walking is not possible. It is difficult to generate useful mapping populations in species that cannot be self-pollinated, because mutations affecting sex that are not caused by polyploidy or large deletions will be rare. Standard molecular methods involving library screening are not likely to succeed because sex-determining genes are probably expressed at low levels. While the nature of sex-determining genes may be elusive, genes involved in sex differentiation are easier to identify and characterize.

Several different approaches are being taken to isolate sex-determining genes, including chromosome microdissection methods, preparative *in situ* hybridization, differential display, and RDA.

Differential display and subtracted library techniques are being used in several laboratories to identify genes that are differentially expressed during the development of males and females in campion (Gilmartin, 1997; Scutt *et al.*, in press; Robertson *et al.*, in press; Veuskens, 1996) and in asparagus (Caporali *et al.*, 1996). In white campion, differential screening has mainly identified several cDNA clones that represent genes involved in stamen development (Matsunaga *et al.*, 1996; Gilmartin, 1997; Scutt *et al.*, in press). Flow sorting of sex chromosomes from dioecious species (Veuskens *et al.*, 1995) and standard or laser-mediated microdissection techniques (Gilmartin, 1997; Ruiz Rejon *et al.*, 1994) are being used to generate chromosome-specific libraries in order to characterize the sequences carried on these chromosomes. Preparative *in situ* hybridization, which combines *in situ* hybridization with chromosome microdissection, has been used to clone genes from specific regions of chromosome 2 from mouse and human (Hozier *et al.*, 1994). This method is currently being attempted in sorrel to isolate genes from the X chromosome that are expressed during floral development (J. Lu, C. C. Ainsworth, J. S. Parker, unpublished data). RDA of Y-linked sex mutants, in which the deletions are small, may enable finer mapping of the sex-determining loci and allow the genes to be isolated (Donnison *et al.*, 1996).

At present, neither white campion nor sorrel has been transformed. The development of transformation systems for these plant species is critical if the roles of putative sex-determining genes are to be confirmed once they have been isolated.

XI. Conclusions

Sex determination systems in plants, leading to unisexuality as monoecy and dioecy, have clearly evolved independently many times and are just one of the strategies found in plants to promote outcrossing and avoid inbreeding. In these unisexual systems, the standard hermaphrodite flower has been modified to give separate male and female flowers. In the case of monoecy in maize and cucumber, which represent the only well-studied plants, there seems to be a strong plant growth substance basis underlying the determination of sex, although the systems are very different. The manipulation of sex by alteration of the level of a plant growth substance or a balance of more than one growth substance would seem an obvious way of positioning two types of flower on the same plant. How differences in the levels of plant growth substances along a stem are controlled is, however, a key question. The transduction pathway is completely unknown.

In dioecious plant species, the point of divergence from the hermaphrodite pattern shows wide variation among species, implying that the underlying genetic bases are very different. We can draw some general conclusions about the mechanisms and can put dioecious plants into two groups. First, there are those species such as mercury, spinach, cannabis, and hop, which come from taxa in which the relatives are mainly monoecious rather than hermaphrodite (or all dioecious, in the case of cannabis and hop) and in which the differences between male and female flowers are programmed early in floral development, probably by regulation of the levels of plant growth substances. Second, there are species such as white campion and sorrel, in which the relatives are mainly hermaphrodite and the male and female flowers resemble hermaphrodites in that they possess rudimentary organs of the opposite sex. There is little evidence for involvement of plant growth substances in these species. Dioecy in these plants is very stable with little environmental influence. We propose, therefore, that dioecy arises either from monoecy as an environmentally unstable system controlled by plant growth substances or from hermaphroditism where the underlying mechanisms are highly stable and control does not involve plant growth substances. The fact that the same plant growth substance may have quite different effects in different plants illustrates that the mechanisms involved in the former class are fundamentally different.

There does not seem to be any correlation between the evolution of heteromorphic sex chromosomes and the mechanism of sex determination. In evolutionary terms, the oldest extant systems may be those with highly evolved heteromorphic sex chromosomes. However, the willows and poplars of the family Salicaceae, in which all the species are dioecious, carry single-sex flowers that are highly evolved and far removed from perfect flowers. The flowers basically comprise either an ovary or a group of stamens, yet these species show no sex chromosome heteromorphy. Perhaps modern cytologic methods will reveal a difference between the sexes in this family.

Sex determination is an ideal system in which to study cellular commitment because the developmental decision is made between only two states, male or female. However, the isolation of the key genes involved in determining this decision will not be trivial.

Acknowledgments

We are very grateful to Phil Gilmartin, Erin Irish, Hong Ma, Helen Shephard, and Sarah Grant for providing unpublished information and photographs, for critically reading the manuscript, and for helpful comments and suggestions. We are very grateful to Deborah Charlesworth and Blanche Capel for critically reading the manuscript and for helpful comments. We are also very grateful to Raffaele Testolin for providing unpublished photographs of kiwifruit and to Mark Bennett for help with figures.

References

Ainsworth, C., Crossley, S., Buchanan-Wollaston, V., Thangavelu, M., and Parker, J. (1995). Male and female flowers of the dioecious plant sorrel show different patterns of MADS box gene expression. *Plant Cell* **7,** 1583–1598.

Akorodo, M. O., Wilson, J. E., and Chheda, H. R. (1984). The association of sexuality with plant traits and tuber yield in white yam. *Euphytica* **33,** 435–442.

Allen, C. E. (1919). The basis of sex inheritance in *Sphaerocarpos. Proc. Am. Phil. Soc.* **58,** 298–316.

Allison, T. D. (1991). Variation in sex expression in Canada yew (*Taxus canadensis*). *Am. J. Bot.* **78,** 569–578.

Angenent, G. C., Busscher, M., Franken, J., Dons, H. J. M., and van Tunen, A. J. (1995). Functional interaction between the homeotic genes fbp1 and pMADS1 during Petunia floral organogenesis. *Plant Cell* **7,** 507–516.

Anton, A. M., and Connor, H. E. (1995). Floral biology and reproduction in *Poa* (Poeae, Gramineae). *Aust. J. Bot.* **43,** 577–599.

Atal, C. K. (1959). Sex reversal in hemp by application of gibberellin. *Curr. Sci.* **28**, 408–409.

Atsmon, D., and Galun, E. (1960). A morphogenetic study of staminate, pistillate and hermaphrodite flowers in *Cucumis sativus* L. *Phytomorphology* **10**, 110–115.

Atsmon, D., Lang, A., and Light, E. N. (1968). Contents and recovery of gibberellins in monoecious and gynoecious cucumber plants. *Plant Physiol.* **43**, 806–810.

Atsmon, D., and Tabbak, C. (1979). Comparative effects of gibberellin, silver nitrate and aminoethoxyvinyl glycine on sexual tendency and ethylene evolution in the cucumber plant (*Cucumis sativus* L.). *Plant Cell Physiol.* **20**, 1547–1555.

Audran, J. C., and Batcho, M. (1981). Microsporogenesis and pollen grains in *Silene dioica* (L.) Cl. and alterations in its anthers parasitised by *Ustilago violacea* (Pers.) Rouss. (Ustilaginales). *Acta Soc. Bot. Pol.* **50**, 29–32.

Banks, J. A. (1994). Sex-determining genes in the homosporous fern *Ceratopteris*. *Development* **120**, 1949–1958.

Banks, J. A., Webb, M., and Hickok, L. (1993). Programming of sexual phenotype in the homosporous fern *Ceratopteris richardii*. *Int. J. Plant Sci.* **154**, 522–534.

Barlow, B. A., and Martin, N. (1984). Chromosome evolution and adaptation in mistletoes. *In* "Plant Biosystematics" (W. F. Grant, Ed.), pp. 117–140. Academic Press, Toronto.

Batcho, M., and Audran, J. C. (1981). Sporulation de l'*Ustilago violacea* dans les ovaires de *Silene dioica*. *Phytopathology* **101**, 72–79.

Bawa, K. S. (1980). Evolution of dioecy in flowering plants. *Ann. Rev. Ecol. Syst.* **11**, 15–39.

Bensen, R. J., Johal, G. S., Crane, V. C., Tossberg, J. T., Schnable, P. S., Meeley, R. B., and Briggs, S. P. (1995). Cloning and characterisation of the maize *An1* gene. *Plant Cell* **7**, 75–84.

Bernier, G., Havelange, A., Houssa, C., Petitjean, A., and Lejeune, P. (1993). Physiological signals that induce flowering. *Plant Cell* **5**, 1147–1155.

Bertin, R. I. (1993). Incidence of monoecy and dichogamy in relation to self-fertilization in angiosperms. *Am. J. Bot.* **80**, 557–560.

Bertin, R. I., and Newman, C. M. (1993). Dichogamy in angiosperms. *Bot. Rev.* **59**, 112–152.

Biffi, R., Restivo, F. M., Tassi, F., Caporali, E., Carboni, A., Marziani, G. P., Spada, A., and Falavigna, A. (1995). A restriction fragment length polymorphism probe for early diagnosis of gender in *Asparagus officinalis*. *Hort. Sci.* **30**, 1463–1464.

Blackburn, K. B. (1923). Sex chromosomes in plants. *Nature* **112**, 687–688.

Boissay, E., Delaigue, M., Sallaud, C., Esnault, R., and Kahlem, G. (1996). Predominant expression of a peroxidase gene in staminate flowers of *Mercurialis annua*. *Physiol. Plantarum* **96**, 251–257.

Bonnett, O. T. (1948). Ear and tassel development in maize. *Ann. Mo. Bot. Gard.* **35**, 269–287.

Bowman, J. L., Sakai, H., Jack, T., Weigel, D., Mayer, U. and Meyerowitz, E. M. (1992). *SUPERMAN*, a regulator of floral homeotic genes in *Arabidopsis*. *Development* **114**, 599–615.

Bowman, J. L., Smyth, D. R., and Meyerowitz, E. M. (1991). Genetic interactions among floral homeotic genes of *Arabidopsis*. *Development* **112**, 1–20.

Bracale, M., Caporali, E., Galli, M. G., Longo, C., Marziani-Longo, G., Rossi, G., Spada, A., Soave, C., Falavigna, A. Raffaldi, F., Maestri, E., Restivo, F. M., and Tassi, F. (1991). Sex determination and differentiation in *Asparagus officinalis* L. *Plant Sci.* **80**, 67–77.

Bracale, M., Galli, M. G., Falavigna, A., and Soave, C. (1990). Sexual differentiation in *Asparagus officinalis* L. II. Total and newly synthesized proteins in male and female flowers. *Sex. Plant. Reprod.* **3**, 23–30.

Bradley, D., Carpenter, R., Copsey, L., Vincent, C., Rothstein, S., and Coen, E. (1996). Control of inflorescence architecture in *Antirrhinum*. *Nature* **379**, 791–797.

Bradley, D., Carpenter, R., Sommer, H., Hartley, N., and Coen, E. S. (1993). Complementary floral homeotic phenotypes result from opposite orientations of a transposon at the *PLENA* locus of *Antirrhinum*. *Cell* **72**, 85–95.

Brummitt, R. K. (1992). "Vascular Plant Families and Genera." Royal Botanic Gardens, Kew.

Bull, J. J. (1983). "Evolution of Sex Determining Mechanisms." Benjamin Cummings, Menlo Park, California.

Buzek, J., Koutnikova, H., Houben, A., Riha, K., Janousek, B., Siroky, J., Grant, S., and Vyskot, B. (1997). Isolation and characterisation of X chromosomes–derived DNA sequences from a dioecious plant *Melandrium album. Chromosome Res.* **5,** 57–65.

Cane, J. H. (1993). Reproductive role of sterile pollen in cryptically dioecious species of flowering plants. *Curr. Sci.* **65,** 223–225.

Capel, B. (1996). The role of Sry in cellular events underlying mammalian sex determination. *Curr. Topics. Dev. Biol.* **32,** 1–37.

Caporali, E., Carboni, A., Galli, M. G., Rossi, G., Spada, A., and Longo, G. P. M. (1994). Development of male and female flower in *Asparagus officinalis*: Search for point of transition from hermaphroditic to unisexual developmental pathway. *Sex. Plant Reprod.* **7,** 239–249.

Caporali, E., Carboni, A., Nicoloso, L., Portaluppi, P., Spada, A., and Marziani, G. (1996). Search for messenger specific male and female flowers of *Asparagus officinalis* L. through mRNA differential display. *In* "Molecular Mechanisms to the Plant: An Integrated Approach," p. 69. 10th FESPP Congress, special issue. Cauthier-Villars, Avenel, NJ.

Carlquist, S. (1974). "Island Biology." Columbia University Press, New York.

Chailakhan, M. K. (1979). Genetic and hormonal regulation of growth, flowering and sex expression in plants. *Am. J. Bot.* **66,** 717–736.

Chandra, H. S. (1991). How do heterogametic females survive without gene dosage compensation? *J. Genet.* **70,** 137–146.

Charlesworth, D. (1984). Androdioecy and the evolution of dioecy. *Biol. J. Linn. Soc.* **23,** 333–348.

Charlesworth, B., and Charlesworth, D. (1978a). A model for the evolution of dioecy and gynodioecy. *Am. Nat.* **112,** 975–997.

Charlesworth, B., and Charlesworth, D. (1978b). Population genetics of partial male sterility and the evolution of monoecy and dioecy. *Heredity* **41,** 137–153.

Charnov, E. L. (1982). "The Theory of Sex Allocation." Princeton University Press, Princeton, NJ.

Charnov, E. L., Maynard Smith, J., and Bull, J. J. (1976). Why be an hermaphrodite? *Nature* **263,** 125–126.

Chattopadhyay, D., and Sharma, A. K. (1991). Sex determination in dioecious species of plants. *Feddes. Rep.* **102,** 29–55.

Cheng, P. C., Greyson, R. I., and Walden, D. B. (1982). Organ initiation and the development of unisexual flowers in the tassel and ear of *Zea mays. Am. J. Bot.* **70,** 450–462.

Choudhuri, H. C. (1969). Late replication pattern in sex chromosomes of *Melandrium. Can. J. Genet. Cytol.* **11,** 192–198.

Clark, M. S., Parker, J. S., and Ainsworth, C. C. (1993). Repeated DNA and heterochromatin structure in *Rumex acetosa. Heredity* **70,** 527–536.

Coen, E. S. (1996). Floral symmetry. *EMBO J.* **15,** 6777–6788.

Coen, E. S., and Meyerowitz, E. M. (1991). The war of the whorls: Genetic interactions controlling flower development. *Nature* **353,** 31–37.

Coen, E. S., Romero, J. M., Doyle, S., Elliott, R., Murphy, G., and Carpenter, R. (1990). *Floricaula*: A homeotic gene required for flower development in *Antirrhinum majus. Cell* **63,** 1311–1322.

Corley, R. H. V., Ng, N., and Nonough, C. R. (1995). Effects of defoliation on sex differentiation in oil palm clones. *Exp. Agric.* **31,** 177–189.

Correns, C. (1906). Ein Vererbungsversuch mit *Dimorphotheca pluvialis. Ber. Deut. Bot. Ges.* **24,** 162–173.

Costich, D. E. (1995). Gender specialization across a climatic gradient: Experimental comparison of monoecious and dioecious *Ecballium. Ecology* **76,** 1036–1050.

Cronquist, A. (1988). "The Evolution and Classification of Flowering Plants." New York Botanical Gardens, Bronx, NY.

Darwin, C. (1876). "Effects of Cross and Self-Fertilisation in the Vegetable Kingdom." John Murray, London.

Davies, P. J. (1995). "Plant Hormones. Physiology, Biochemistry and Molecular Biology," 2nd ed. Kluwer Academic Publishers, Dordrecht, Netherlands.

Dellaporta, S. L., and Calderon-Urrea, A. (1993). Sex determination in flowering plants. *Plant Cell* **5,** 1241–1251.

Dellaporta, S. L., and Calderon-Urrea, A. (1994). The sex determination process in maize. *Science* **266,** 1501–1505.

DeLong, A., Claderon-Urrea, A., and Dellaporta, S. (1993). Sex determination gene TASSELSEED2 of maize encodes a short-chain alcohol dehydrogenase required for stage-specific floral organ abortion. *Cell* **74,** 757–768.

Desfeux, C., Maurice, S., Henry, J. P., Lejeune, B., and Gouyon, P. H. (1996). Evolution of reproductive systems in the genus *Silene. Proc. R. Soc. Lond. B Biol. Sci.* **263,** 409–414.

Donnison, I. S., Siroky, J., Vyskot, B., Saedler, H., and Grant, S. (1996). Isolation of Y chromosome-specific sequences from *Silene latifolia* and mapping of male sex determining genes using representational difference analysis. *Genetics* **144,** 1891–1899.

Doyle, J. A. (1994). Origin of the angiosperm flower: a phylogenetic perspective. *Plant. Syst. Evol.* **8,** 7–29.

Durand, B., and Durand, R. (1991). Sex determination and reproductive organ differentiation in *Mercurialis. Plant Sci.* **80,** 49–65.

Eberle, J. R., and Banks, J. A. (1996). Genetic interactions among sex-determining genes in the fern *Ceratopteris richardii. Genetics* **142,** 973–985.

Emebiri, L. C., and Nwufo, M. I. (1996). Occurrence and detection of early sex-related differences in *Telfairia occidentalis. Sex. Plant Reprod.* **9,** 140–144.

Emerson, R. A. (1920). Heritable characters in maize. II. Pistillate flowered maize plants. *J. Hered.* **11,** 65–76.

Fleming, T. H., Maurice, S., Buchmann, S. L., and Tuttle, M. D. (1994). Reproductive-biology and relative male and female fitness in a trioecious cactus, *Pachycereus pringlei* (Cactaceae). *Am. J. Bot.* **81,** 858–867.

Frankel, R., and Galun, E. (1977). "Pollination Mechanisms, Reproduction and Plant Breeding" (R. Frankel, G. A. E. Gall, M. Grossman, H. F. Linskens, and D. de Zeeuw, Eds.), pp. 141–157. Springer-Verlag, Heidelberg.

Franken, A. A. (1970). Sex differences and inheritance of sex in asparagus (*Asparagus officinalis* L.). *Euphytica* **19,** 277–287.

Fuchs, E., Atsmon, D., and Halevy, A. H. (1977). Adventitious staminate flower formation in gibberellin treated gynoecious cucumber plants. *Plant Cell Physiol.* **18,** 1193–1201.

Fujioka, S., Yamane, H., Spray, C. R., Gaskin, P., MacMillan, J., Phinney, B. O., and Takahashi, N. (1988). Qualitative and quantitative analyses of gibberellins in vegetative shoots of normal, *dwarf-1, dwarf-2, dwarf-3,* and *dwarf-5* seedlings of *Zea mays* L. *Plant Physiol.* **88,** 1367–1372.

Galli, M. G., Bracale, M., Falavigna, A., Raffaldi, F., Savini, C., and Vigo, A. (1993). Different kinds of male flowers in the dioecious plant *Asparagus officinalis* L. *Sex. Plant Reprod.* **6,** 16–21.

Galun, E. (1959). Effect of gibberellic acid and napthaleneacetic acid in sex expression and some morphological characters in the cucumber plant. *Phyton* **13,** 1–8.

Galun, E. (1961). Study of the inheritance of sex expression in the cucumber: Interactions of major genes with modifying genetic and non-genetic factors. *Genetica* **32,** 134–163.

Gilmartin, P. M. (1997). University of Leeds, Centre for Plant Biochemistry and Biotechnology, WWW home page 〈http://www.leeds.ac.uk/biology/groups/pltctr.html〉, February.

Goodrich, J., Puangsomlee, P., Martin, M., Long, D., Meyerowitz, E. M., and Coupland, G. (1997). A Polycomb-group gene regulates homeotic gene expression in *Arabidopsis*. *Nature* **386,** 44–51.

Goto, K., and Meyerowitz, E. M. (1994). Function and Regulation of the *Arabidopsis* floral homeotic gene *pistillata*. *Genes Dev.* **8,** 1548–1560.

Grant, S. G., and Chapman, V. M. (1988). Mechanisms of X-chromosome regulation. *Annu. Rev. Genet.* **22,** 199.

Grant, S., Donnison, I. S., Hardenack, S., and Law, T. F. (1996). Studies of the genetics of sex determination in dioecious *Silene latifolia* by the front and the back doors. *Flowering Plant Newsletter* **21,** 21–26.

Grant, S., Houben, A., Vyskot, B., Siroky, J., Pan, W. H., Macas, J., and Saedler, H. (1994). Genetics of sex determination in flowering plants. *Dev. Genet.* **15,** 214–230.

Grant, S., Hunkirchen, B., and Saedler, H. (1994). Developmental differences between male and female flowers in the dioecious plant white campion. *Plant J.* **6,** 471–480.

Hansen, D. J., Bellman, S. K., and Sacher, R. M. (1976). Gibberellic acid–controlled sex expression of corn tassels. *Crop Sci.* **16,** 371–374.

Hardenack, S., Ye, D., Saedler, H., and Grant S. (1994). Comparison of MADS box gene expression in developing male and female flowers of the dioecious plant white campion. *Plant Cell* **6,** 1775–1787.

Harvey, C. F., and Fraser, L. G. (1988). Floral biology of two species of *Acinidia* (Acinidiaceae). II. Early embryology. *Bot. Gaz.* **149,** 37–44.

Harvey, C. F., Fraser, L. G., Pavis, S. E., and Considine, J. A. (1987). Floral biology of two species of *Acinidia* (Acinidiaceae). I. The stigma, pollination, and fertilization. *Bot. Gaz.* **148,** 426–432.

Herskowitz, I. (1989). A regulatory hierarchy for cell specialization in yeast. *Nature* **342,** 749–757.

Heslop-Harrison, J. (1956). Auxin and sexuality in *Cannabis sativa*. *Physiol. Plantarum* **4,** 588–597.

Heslop-Harrison, J. (1958). Unisexual flower: A reply to criticism. *Phytomorphology* **8,** 177.

Heslop-Harrison, J. (1961). The experimental control of sexuality and inflorescence structure in *Zea mays* L. *Proc. Linn. Soc. Lond.* **172,** 108–123.

Heslop-Harrison, J. (1963). Sex expression in flowering plants. *Brookhaven Symp. Quant. Biol.* **16,** 109–125.

Hickey, M., and King, C. (1988). "100 Families of Flowering Plants." Cambridge University Press, Cambridge.

Hormaza, J. I., Dollo, L., and Polito, V. S. (1994). Identification of a RAPD marker linked to sex determination in *Pistacia vera* using bulked segregant analysis. *Theor. Appl. Genet.* **89,** 9–13.

Hormaza, J. I., and Polito, V. S. (1996). Pistillate and staminate flower development in dioecious *Pistacia vera* (Anacardiaceae). *Am. J. Bot.* **83,** 759–766.

Hossain, M. A., Islam, M., and Ali, M. (1996). Sexual crossing between two genetically female plants and sex genetics of kakrol (*Momordica dioica* Roxb.). *Euphytica* **90,** 121–125.

Hozier, J., Graham, G., Westfall, T., Siebert, P., and Davis, L. (1994). Preparative *in situ* hybridisation of chromosome region-specific libraries on mitotic chromosomes. *Genomics* **19,** 441–447.

Huala, E., and Sussex, I. M. (1993). Determination and cell interactions in reproductive meristems. *Plant Cell* **5,** 1157–1165.

Huang, H., and Ma, H. (1997). *FON1*, an *Arabidopsis* gene that terminates floral meristem activity and controls flower organ number. *Plant Cell* **9,** 115–134.

Huang, H., Tudor, M., Su, T., Zhang, Y., Hu, Y., and Ma, H. (1996). DNA binding properties of two *Arabidopsis* MADS domain proteins: Binding consensus and dimer formation. *Plant Cell* **8,** 81–94.

Huijser, P., Klein, J., Lönnig, W-E., Meijer, H., Saedler, H., and Sommer, H. (1992). Brac-teomania, an inflorescence anomaly, is caused by the loss of function of the MADS-box gene *SQUAMOSA* in *Antirrhinum majus*. *EMBO J.* **4,** 1239–1249.

Ingram, G. C., Goodrich, J., Wilkinson, M. D., Simon, R., Haughn, G. W., and Coen, E. S. (1995). Parallels between *UNUSUAL FLORAL ORGANS* and *FIMBRIATA*, genes controlling flower development in *Arabidopsis* and *Antirrhinum*. *Plant Cell* **7,** 1501–1510.

Irish, E. E. (1996). Regulation of sex determination in maize. *Bioessays* **18,** 363–369.

Irish, E. E., Langdale, J. A., and Nelson, T. M. (1994). Interactions between *Tassel Seed* genes and other sex determining genes in maize. *Dev. Genet.* **15,** 155–171.

Irish, E. E., and Nelson, T. (1989). Sex determination in monoecious and dioecious plants. *Plant Cell* **1,** 737–744.

Irish, E. E., and Nelson, T. (1993). Developmental analysis of the inflorescence of the maize mu-tant *Tassel seed 2*. *Am. J. Bot.* **80,** 292–299.

Jack, T., Brockman, L. L., and Meyerowitz, E. M. (1992). The homeotic gene *APETALA3* of *Arabidopsis thaliana* encodes a MADS-box and is expressed in petals and stamens. *Cell* **68,** 683–687.

Jaffe, E., and Laird, C. (1986). Dosage compensation in *Drosophila*. *Trends Genet.* **2,** 316–321.

Janousek, B., Siroky, J., and Vyskot, B. (1996). Epigenetic control of sexual phenotype in a di-oecious plant, *Melandrium album*. *Mol. Gen. Genet.* **250,** 483–490.

Jofuku, K. D., den Boer, B. G. W., Van Montagu, M., and Okamuro, J. K. (1994). Control of *Arabidopsis* flower and seed development by the homeotic gene *APETALA2*. *Plant Cell* **6,** 1211–1225.

Jones, D. F. (1925). Heritable characters in maize. XXIII. Silkless. *J. Hered.* **16,** 339–341.

Katsumi, M. (1964). Gibberellin-like activities of certain auxins and diterpenes. Ph.D. dissertation, Los Angeles, University of California.

Kellogg, E. A., and Birchler, J. A. (1993). Linking phylogeny and genetics: *Zea mays* as a tool for phylogenetic studies. *Syst. Biol.* **42,** 415–439.

Kelley, R. L., and Kuroda, M. I. (1995). Equality for X-chromosomes. *Science* **270,** 1607–1610.

Keyes, L. N., Cline, T. W., and Schedl, P. (1992). The primary sex determination signal of *Drosophila* acts at the level of transcription. *Cell* **68,** 933–943.

Kihara, H., and Ono, T. (1923). The sex chromosomes of *Rumex acetosa*. *Z. Ind. Abst. Vererb.* **39,** 1–7.

Koelewijn, H. P., and van Damme, J. M. M. (1996). Gender variation, partial male sterility and labile sex expression in gynodioecious *Plantago coronopus*. *New Phytol.* **132,** 67–76.

Lazarte, J. E., and Garrison, A. (1980). Sex modification in *Asparagus officinalis* L. *J. Am. Hort. Sci.* **105,** 691–694.

Lazarte, J. E., and Palser, B. F. (1979). Morphology, vascular anatomy and embryology of pistil-late and staminate flowers of *Asparagus officinalis* L. *J. Am. Hort. Sci.* **66,** 753–764.

Lebel-Hardenack, S., and Grant, S. R. (1997). Genetics of sex determination in flowering plants. *Trends Plant Sci.* **2,** 130–136.

Levin, J. Z., and Meyerowitz, E M. (1995). UFO: An Arabidopsis gene involved in both floral meristem and floral organ development. *Plant Cell* **7,** 524–548.

Lewis, D. (1941). Male sterility in natural populations of hermaphrodite plants. *New Phytol.* **40,** 56–63.

Lewis, D. (1942). The evolution of sex in flowering plants. *Biol. Rev.* **17,** 46–67.

Liu, Z., and Meyerowitz, E. M. (1995). *LEUNIG* regulates *AGAMOUS* expression in *Arabidopsis* flowers. *Development* **121,** 975–991.

Lisitsyn, N., Lisitsyn, N., and Wigler, M. (1993). Cloning the difference between two complex genomes. *Science* **259,** 946–951.

Liston, A. H., Rieseberg, L. H., and Elias, T. S. (1990). Functional androdioecy in the flowering plant *Datisca glomerata*. *Nature* **343,** 641–642.

Lloyd, D. G. (1974). Theoretical sex ratios of dioecious and gynodioecious angiosperms. *Heredity* **32,** 11–34.

Lloyd, D. G. (1975a). The maintenance of gynodioecy and androdioecy in angiosperms. *Genetica* **45,** 325–339.

Lloyd, D. G. (1975b). Breeding systems in *Cotula*. IV. Dioecious populations. *New Phytol.* **74,** 109–123.

Lloyd, D. G., and Bawa, K. S. (1984). Modifications of the gender of seed plants in varying conditions. *Evol. Biol.* **17,** 255–336.

Loptien, H. (1979). Identification of the sex chromosome pair in asparagus (*Asparagus officinalis* L.). *Z. Pflanz* **82,** 162–173.

Louis, J-P. (1989). Genes for the regulation of sex differentiation and male fertility in *Mercurialis annua* L. *J. Hered.* **89,** 104–111.

Louis, J-P., Augur, C., and Teller, G. (1990). Cytokinins and differentiation processes in *Mercurialis annua*. *Plant Physiol.* **94,** 1535–1541.

Lucchesi, J. C., and Manning, J. E. (1987). Gene dosage compensation in *Drosophila melanogaster*. *Adv. Genet.* **24,** 371–429.

Lyndon, R. F. (1985). Silene. *In* "Handbook of Flowering." vol. 4 (H. A. Halevy, Ed.), pp. 313–319. CRC Press, Boca Raton, Florida.

Lyon, M. F. (1961). Gene action in the X chromosome of the mouse. *Nature* **190,** 372–373.

Lyon, M. F. (1996). Molecular genetics of X-chromosome inactivation. *Adv. Genome Biol.* **4,** 119–151.

Ma, H., Yanofsky, M. F., and Meyerowitz, E. M. (1991). *AGL1-AGL6*, an *Arabidopsis* gene family with similarity to floral homeotic and transcription factor genes. *Genes Dev.* **5,** 484–495.

Machon, N., Deletrelebouloch, V., and Rameau, C. (1995). Quantitative-analysis of sexual dimorphism in *Asparagus*. *Can. J. Bot.* **73,** 1780–1786.

Malepszy, S., and Niemirowicz-Szczytt, K. (1991). Sex determination in cucumber (*Cucumis sativus*) as a model system for molecular biology. *Plant Sci.* **80,** 39–47.

Mandel, M. A., Gustafson-Brown, C., Savidge, B., and Yanofsky, M. F. (1992). Molecular characterization of the *Arabidopsis* floral homeotic gene *APETALA1*. *Nature* **360,** 273–277.

Marks, M. (1973). A reconsideration of the genetic mechanism for sex determination in *Asparagus officinalis*. *In* "Proceedings of the Eucarpia Meeting on Asparagus," pp. 122–130. Versailles.

Mather, K. (1940). Outbreeding and separation of the sexes. *Nature* **145,** 484–486.

Mather, K. (1949). Genetics of dioecy and monoecy in *Ecballium*. *Nature* **163,** 926.

Matsunaga, S., Kawano, S., Takano, H., Uchida, H., Sakai, A., and Kuroiwa, T. (1996). Isolation and developmental expression of male reproductive organ-specific genes in a dioecious campion, *Melandrium album* (*Silene latifolia*). *Plant J.* **10,** 679–689.

Maurice, S., Belhassen, E., Couvet, D., and Gouyon, P. H. (1994). Evolution of dioecy: Can nuclear-cytoplasmic interactions select for maleness? *Heredity* **73,** 346–354.

Maurice, S., Charlesworth, D., Desfeux, C., Couvet, D., and Gouyon, P-H. (1993). The evolution of gender in hermaphrodites of gynodioecious populations with nucleo-cytoplasmic male-sterility. *Proc. R. Soc. Lond. Biol. Sci.* **251,** 253–261.

Mayer, S. S., and Charlesworth, D. (1991). Cryptic dioecy in flowering plants. *Trends Ecol. Evol.* **6,** 320–325.

McElreavey, K. (1996). Mechanisms of sex determination in mammals. *Adv. Genet.* **4,** 305–354.

McNeilage, M. A. (1991a). Gender variation in *Actinidia deliciosa*, the kiwifruit. *Sex. Plant Reprod.* **4,** 267–273.

McNeilage, M. A. (1991b). Sex expression in fruiting male vines of kiwifruit. *Sex. Plant Reprod.* **4,** 274–278.

Meagher, T. R. (1988). Sex determination in plants. *In* "Plant Reproductive Ecology: Patterns and

Strategies" (J. Lovett Doust and L. Lovett Doust, Eds.), pp. 125–138. Oxford University Press, New York.

Mena, M., Mandel, M. A., Lerner, D. R., Yanofsky, M. F., and Schmidt, R. J. (1995). A characterisation of the MADS-box gene family in maize. *Plant J.* **8,** 845–854.

Messina, R. (1993). Microsporogenesis in male-fertile cv. Matua and male-sterile cv. Hayward of *Actinidia deliciosa* var. *deliciosa* (kiwifruit). *Adv. Hort. Sci.* **7,** 77–81.

Mezitt, L. A., and Lucas, W. L. (1996). Plasmadesmal cell-to-cell transport of proteins and nucleic acids. *Plant Mol. Biol.* **32,** 251–273.

Miller, H. G., Kocher, T. D., and Loy, B. (1995). New MADS box domains in *Asparagus officinalis* L. *Sex. Plant Reprod.* **8,** 318–320.

Mittwoch, U. (1996). Genetics of sex determination: An overview. *Adv. Genet.* **4,** 1–28.

Mizukami, Y., and Ma, H. (1995). Separation of AG function in floral meristem determinacy from that in reproductive organ identity by expressing antisense *AG* RNA. *Plant Mol. Biol.* **28,** 767–784.

Mohan Ram, H. Y., and Jaiswal, V. S. (1970). Induction of female flowers on male plants of *Cannabis sativa* by 2-chloroethane phosphonic acid. *Experientia* **26,** 214–216.

Mohan Ram, H. Y., and Nath, R. (1964). The morphology and embryology of *Cannabis sativa* Linn. *Phytomorphology* **14,** 414–429.

Näf, U., Nakanishi, K., and Endo, M. (1975). On the physiology and chemistry of fern antheriodiogens. *Bot. Rev.* **41,** 315–359.

Nasrullah, J. B., and Nasrullah, M. E. (1993). Pollen-stigma signalling in the sporophytic self-incompatibility response. *Plant Cell* **5,** 1325–1335.

Neve, R. A. (1991). "Hops." Chapman & Hall, London.

Newbegin, E., Anderson, M. A., and Clarke, A. E. (1993). Gametophytic self-incompatibility systems. *Plant Cell* **5,** 1315–1324.

Norman, C., Runswick, M., Pollack, R., and Treisman, R. (1988). Isolation and properties of cDNA clones encoding SRF, a transcription factor that binds to the c-fos serum response element. *Cell* **55,** 989–1003.

Pannell, J. (1997). Mixed genetic and environmental sex determination in an androdioecious population of *Mercurialis annua*. *Heredity* **78,** 50–56.

Parker, J. S. (1990). Sex chromosomes and sexual differentiation in flowering plants. *Chromosomes Today* **10,** 187–198.

Parker, J. S., and Clark, M. S. (1991). Dosage sex-chromosome systems in plants. *Plant Sci.* **80,** 79–92.

Passmore, S., Maine, G. T., Elble, R., Christ, C., and Tye, B. K. (1988). *Saccharomyces cerevisiae* protein involved in plasmid maintenance is necessary for mating of *MAT* cells. *J. Mol. Biol.* **204,** 593–606.

Pillay, M., and Kenny, S. T. (1996). Random amplified polymorphic DNA (RAPD) markers in hop, *Humulus lupulus*: Level of genetic variability and segregation in F_1 progeny. *Theor. Appl. Genet.* **92,** 334–339.

Prickett, T. C. R., and Walker, J. R. L. (1989). Flavone compounds in male and female asparagus plants. *J. Sci. Food Agric.* **47,** 53–60.

Rao, S. R. V., and Padmaja, M. (1992). Mammalian-type dosage compensation mechanism in an insect, *Gryllotalpa fossor* (Scudder), Orthoptera. *J. Biosci.* **17,** 253–273.

Renner, S. S., and Ricklefs, R. E. (1995). Dioecy and its correlates in the flowering plants. *Am. J. Bot.* **82,** 596–606.

Rieseberg, L. H., Hanson, M. A., and Philbrick, C. T. (1992). Androdioecy is derived from dioecy in Datiscaceae: Evidence from restriction site mapping of PCR-amplified chloroplast DNA fragments. *Syst. Bot.* **17,** 324–336.

Robertson, S. E., Li, Y., Scutt, C. P., Willis, M. E., and Gilmartin, P. M. (in press). Spatial expression dynamics of Men-9 delineate the third whorl in male and female flowers of dioecious *Silene latifolia*. *Plant J*.

Robinson, R. W., Munger, H. M., Whitaker, T. W., and Bohn, G. W. (1976). Genes of the Cucurbitaceae. *Hort. Sci.* **11,** 554–568.

Rood, S. B., Paris, R. P., and Major, D. J. (1980). Changes of endogenous gibberellin-like substances with sex reversal of the apical inflorescence of corn. *Plant Physiol.* **66,** 793–796.

Rosa, J. T. (1928). The inheritance of flower types in *Cucumis* and *Citrullus. Hilgardia* **3,** 421–452.

Rosenmann, N., Kahana, A., Xiang, Y., and Perl-Treves, R. (1996). Expression of agamous-like genes in male and female flowers of cucumber. *In* "Proceedings of the VIth Eucarpia Meeting on Cucurbit Genetics and Breeding" (Gomez-Guillamon, M. L., C. Soria, J. Cuartero, J. A. Tores, and R. Fernandez-Munoz, Eds.). European Association for Research on Plant Breeding. Estacion Experimental La Mayora C.S.I.C., Malaga, Spain.

Ross, M. D. (1978). The evolution of gynodioecy and subdioecy. *Evolution* **32,** 174–188.

Ross, M. D., and Shaw, R. F. (1971). Maintenance of male sterility in plant populations. *Heredity* **26,** 1–8.

Rounsley, S. D., Ditta, G. S., and Yanofsky, M. F. (1995). Diverse roles for MADS box genes in *Arabidopsis* development. *Plant Cell* **7,** 1259–1269.

Rudich, J., Halevy, A. H., and Kedar, N. (1969). Increase of femaleness of three cucurbits by treatment with Ethrel, an ethylene-releasing compound. *Planta* **86,** 69–76.

Rudich, J., Halevy, A. H., and Kedar, N. (1972). The level of phytohormones in monoecious and gynoecious cucumbers as affected by photoperiod and ethephon. *Plant Physiol.* **50,** 585–590.

Ruiz Rejon, C., Jamilena, M., Garrido Ramos, M., Parker, J. S., and Ruiz Rejon, M. (1994). Cytogenetic and molecular analysis of the multiple sex chromosome system of *Rumex acetosa. Heredity* **72,** 209–215.

Sakamoto, K., Shimomura, K., Komeda, Y., Kamada, H., Satoh, S. (1995). A male-associated DNA sequence in a dioecious plant, *Cannabis sativa* L. *Plant Cell Physiol.* **36,** 1544–1554.

Schmid, R. (1978). Reproductive anatomy of *Actinidia chinensis* (Actinidiaceae). *Bot. Jahrb. Syst. Pflanzengesch. Pflanzengeog.* **100,** 149–195.

Schmidt, R. J., Veit, B., Mandel, M. A., Mena, M., Hake, S., and Yanofsky, M. F. (1993). Identification and molecular characterisation of *ZAG1,* the maize homolog of the *Arabidopsis* floral homeotic gene *AGAMOUS. Plant Cell* **5,** 729–737.

Schnabel, A., and Hemrick, J. L. (1990). Non-random associations between sex and 6-phosphogluconate dehydrogenase isozyme genotypes in *Gleditsia triacanthos. J. Hered.* **81,** 230–233.

Schwarz-Sommer, Z., Hue, I., Huijser, P., Flor, P. J., Hansen, R., Tetens, F., Lonnig, W-E., Saedler, H., and Sommer, H. (1992). Characterization of the *Antirrhinum* floral homeotic MADS-box gene *DEFICIENS*: Evidence for DNA binding and autoregulation of its persistent expression throughout flower development. *EMBO J.* **11,** 251–263.

Schwarz-Sommer, Z., Huijser, P., Nacken, W., Saedler, H., and Sommer, H. (1990). Genetic control of flower development by homeotic genes in *Antirrhinum majus. Science* **250,** 931–936.

Scutt, C. P., Robertson, S. E., Willis, M. E., and Gilmartin, P. M. (in press). Sex determination in dioecious *Silene latifolia*: Effects of the Y chromosome and the parasitic smut fungus, *Ustilago violacea,* on gene expression during flower development. *Plant Physiol.*

Shannon, S., and de la Guardia, M. D. (1969). Sex expression and the production of ethylene induced by auxin in the cucumber (*Cucumis sativus* L.). *Nature* **223,** 186.

Sherry, R. A., Eckard, K. J., and Lord, E. M. (1993). Flower development in dioecious *Spinacia oleracea* (Chenopodiaceae). *Am. J. Bot.* **80,** 283–291.

Shore, P., and Sharrocks, A. D. (1995). The MADS box family of transcription factors. *Eur. J. Biochem.* **229,** 1–13.

Siljak-Yakovlev, S., Benmalek, S., Cerbah, M., Coba de la Pena, T., Bounaga, N., Brown, S. C., and Sarr, A. (1996). Chromosomal sex determination and heterochromatin structure in data palm. *Sex. Plant Reprod.* **9,** 127–132.

Simon, R., Carpenter, R., Doyle, S., and Coen, E. (1994). *Fimbriata* controls flower development by mediating between meristem and organ identity genes. *Cell* **78,** 99–107.

Sinclair, A. H., Berta, P., Palmer, M. S., Hawkins, J. R., Griffiths, B. L., Smith, M. J., Foster, J. W., Frischauf, A-M., Lovell-Badge, R., and Goodfellow, P. N. (1990). A gene from the human sex-determining region encodes a protein with homology to a conserved DNA-binding motif. *Nature* **346,** 240–244.

Siroky, J., Janousek, B., Mouras, A., and Vyskot, B. (1994). Replication patterns of sex chromosomes in *Melandrium album* female cells. *Hereditas* **120,** 175–181.

Smith, B. W. (1967). The evolving karyotype of *Rumex hastatulus. Evolution* **18,** 93–104.

Sommer, H., Beltran, J-P., Huijser, P., Pape, H., Lonnig, W-E., Saedler, H. and Schwarz-Sommer, Z. (1990). *Deficiens*, a homeotic gene involved in the control of flower morphogenesis in *Antirrhinum majus*: The protein shows homology to transcription factors. *EMBO J.* **9,** 605–613.

Spray, C. R., Phinney, B. O., Gaskin, P., Gilmour, S. J., and MacMillan, J. (1984). Internode length in *Zea mays* L. The *dwarf-1* mutation controls the 3β-hydroxylation of gibberellin A_{20} to gibberellin A_{20}. *Planta* **160,** 464–468.

Stevens, S. J., Lee, K. W., Stevens, E. J., Flowerday, A. D., and Gardener, C. O. (1986). Organogenesis of the staminate and pistillate inflorescences of corn: A comparison between dent and popcorn. *Crop Sci.* **26,** 712–718.

Takeno, K., Watanabe, K., and Suyama, T. (1995). Sex determination of flowers of *Salsola komarovii* Iljin by photoperiod. *J. Plant Physiol.* **146,** 672–676.

Tandre, K., Albert, V. A., Sundås, A., and Engström, P. (1995). Conifer homologues to genes that control floral development in angiosperms. *Plant Mol. Biol.* **27,** 69–78.

Taylor, D. W., and Hickey, L. J. (1990). An aptian plant with attached leaves and flowers: Implications for angiosperm origin. *Science* **247,** 712–714.

Testolin, R., Cipriani, G., and Costa, G. (1995). Sex segregation ratio and gender expression in the genus *Actinidia. Sex. Plant Reprod.* **3,** 129–132.

Thiessen, G., and Seidler, H. (1995). MADS-box genes in plant ontogeny and phylogeny: Haeckel's 'biogenetic law' revisited. *Curr. Opin. Genet. Dev.* **5,** 628–639.

Trebitsh, T., Rudich, J., and Riov, J. (1987). Auxin, biosynthesis of ethylene and sex expression in cucumber (*Cucumis sativus*). *Plant Growth Reg.* **5,** 105–113.

Tröbner, W., Ramirez, L., Motte, P., Hue, I., Huijser, P., Lönnig, W.-E., Saedler, H., Sommer, H., and Schwarz-Sommer, Z. (1992). *Globosa*: A homeotic gene which interacts with *Deficiens* in the control of *Antirrhinum* floral organogenesis. *EMBO J.* **11,** 4693–4704.

Valdeyron, G., Dommee, B., and Valdeyron, A. (1973). Gynodioecy: Another computer simulation model. *Am. Nat.* **107,** 454–459.

van der Krol, A. R., and Chua, N.-H. (1993). Flower development in petunia. *Plant Cell* **5,** 1195–1203.

van Nigtevecht, G. (1966). Genetic studies in dioecious *Melandrium*. II. Sex determination in *M. album* and *M. dioicum. Genetica* **37,** 307–344.

Veit, B., Schmidt, R. J., Hake, S., and Yanofsky, M. (1993). Maize floral development: New genes and old mutants. *Plant Cell* **5,** 1205–1215.

Veuskens, J. (1996). 1993–1994 Report of the Plant Genetics Unit, Department of Biology, Free University of Brussels. WWW home page ⟨http://imol.vub.ac.be/PLAN/Report1993_1994.html⟩, November.

Veuskens, J., Marie, D., Brown, S. C., Jacobs, M., and Negrutiu, I. (1995). Flow sorting of the Y sex chromosome in the dioecious plant *Melandrium album. Cytometry* **21,** 363–373.

Veuskens, J., Ye, D., Oliveira, M., Ciupercescu, D. D., Installe, P., Verhoeven, H. A., and Negrutiu, I. (1992). Sex determination in the dioecious *Melandrium album*: Androgenic embryogenesis requires the presence of the X chromosome. *Genome* **35,** 8–16.

Villeneuve, A. M., and Meyer, B. J. (1990). The regulatory hierarchy controlling sex determination and dosage compensation in *Caenorhabditis elegans. Adv. Genet.* **27,** 117–188.

Vyskot, B., Araya, A., Veuskens, J., Negrutiu, I., and Mouras, A. (1993). DNA methylation of sex chromosomes in a dioecious plant, *Melandrium album. Mol. Gen. Genet.* **239,** 219–224.

Warmke, H. E., and Blakeslee, A. F. (1939). Sex mechanism in polyploids of *Melandrium album. Science* **89,** 391–392.

Wehrli, L. (1892). Ueber einen Fall von "vollständiger Verweiblichung" der männlichen Kätzchen von *Corylus avellana* L. *Flora* **76,** 245–264.

Weigel, D. (1995). The genetics of flower development: From floral induction to ovule morphogenesis. *Annu. Rev. Genet.* **29,** 19–39.

Weigel, D., Alvarez, J., Smyth, D. R., Yanofsky, M. F., and Meyerowitz, E. M. (1992). *LEAFY* controls floral meristem identity in *Arabidopsis. Cell* **69,** 843–859.

Weigel, D., and Clark, S. E. (1996). Sizing up the floral meristem. *Plant Physiol.* **112,** 5–10.

Weigel, D., and Meyerowitz, E. M. (1994). The ABCs of floral homeotic genes. *Cell* **78,** 203–209.

Westergaard, M. (1940). Studies on cytology and sex determination in polyploid forms of *Melandrium album. Dansk. Bot. Arkiv.* **5,** 1–131.

Westergaard, M. (1946). Aberrant Y chromosomes and sex expression in *Melandrium album. Hereditas* **32,** 419–443.

Westergaard, M. (1948). The relationship between chromosome constitution and sex in the offspring of triploid *Melandrium album. Hereditas* **34,** 257–279.

Westergaard, M. (1958). The mechanism of sex determination in dioecious flowering plants. *Adv. Genet.* **9,** 217–281.

Weston, E. W. (1960). Changes in sex in the hop caused by plant growth substances. *Nature* **188,** 81–82.

Wilby, A. S., and Parker, J. S. (1988). Recurrent patterns of chromosome variation in a species group. *Heredity* **61,** 55–62.

Wilkinson, M. D., and Haughn, G. W. (1995). *UNUSUAL FLORAL ORGANS* controls meristem identity and organ primordia fate in *Arabidopsis. Plant Cell* **7,** 1485–1499.

Winge, O. (1931). X- and Y-linked inheritance in *Melandrium. Hereditas* **15,** 127–165.

Yamamoto, Y. (1938). Karyogenetische Untersuchungen bei der Gattung Rumex VI. Geschlechtsbestimmung bei Eu- und Aneuploiden Pflanzen von *Rumex acetosa* L. *Kyoto Imp. Univ. Mem. Coll. Agr.* **43,** 1–59.

Yampolsky, C., and Yampolsky, H. (1922). Distribution of the sex forms in the phanerogamic flora. *Bibl. Genet.* **3,** 1–62.

Yanofsky, M. F. (1995). Floral meristems to floral organs: Genes controlling early events in *Arabidopsis* flower development. *Ann. Rev. Plant Physiol. Plant Mol. Biol.* **46,** 167–188.

Yanofsky, M. F., Ma, H., Bownan, J. L., Drews, G. N., Feldman, K. A., and Meyerowitz, E. M. (1990). The protein encoded by the *Arabidopsis* homeotic gene *agamous* resembles transcription factors. *Nature* **346,** 35–39.

Ye, D., Oliveira, M., Veuskens, J., Wu, Y., Installe, P., Hinnisdaels, S., Truong, A. T., Brown, S., Mouras, A., and Negrutiu, I. (1991). Sex determination in the dioecious *Melandrium*: The X/Y chromosome system allows complementary cloning strategies. *Plant Sci.* **80,** 93–106.

Yin, T. J., and Quinn, J. A. (1992). A mechanistic model of one hormone regulating both sexes in flowering plants. *Bulletin of the Torrey Botanical Club* **119,** 431–441.

Yin, T. J., and Quinn, J. A. (1995a). Tests of a mechanistic model of one hormone regulating both sexes in *Buchloe dactyloides* (Poaceae). *Am. J. Bot.* **82,** 745–751.

Yin, T. J., and Quinn, J. A. (1995b). Tests of a mechanistic model of one hormone regulating both sexes in *Cucumis sativus* (Cucurbitaceae). *Am. J. Bot.* **82,** 1537–1546.

Zuk, J. (1969). Autoradiographic studies in *Rumex* with special reference to sex chromosomes. *Chromosomes Today* **2,** 183–188.

6

Somitogenesis

Achim Gossler[1] and Martin Hrabě de Angelis[2]
The Jackson Laboratory[1]
Bar Harbor, Maine 04609

Institute for Mammalian Genetics[2]
85758 Neuherberg, Germany

I. Introduction

Somitogenesis is a fundamental pattern-forming process that generates in all vertebrate (and similarly in cephalochordate) embryos in the mesoderm a periodic pattern of homologous blocks of cells, the somites, which form bilaterally on either side of the neural tube. The resulting segmental (or metameric) arrangement of the somites along the anterior-posterior body axis underlies the metamer-

Current Topics in Developmental Biology, Vol. 38

ism of the somite-derived structures and in addition determines the segmented arrangement of parts of the peripheral nervous system. Somites have been studied in embryos from all vertebrate classes, with a variety of experimental means, and from a number of different perspectives. The majority of experimental studies were performed with avian and amphibian species, since their embryos are more readily amenable to experimental manipulations than mammalian embryos. For quite some time the analysis of somitogenesis in mammalian embryos lagged behind, although in mice mutations affecting somitogenesis have been described and analyzed for decades (reviewed in Johnson, 1986). It has become clear that many aspects of somitogenesis are comparable in all vertebrate classes, and they are particularly similar in avian and mammalian (amniote) embryos. The development of *in vitro* culture conditions for postimplantation mammalian embryos (New *et al.*, 1973; Beddington, 1987; Cockroft, 1990) has allowed embryologic studies concerning somitogenesis. Recently, the possibility of specifically manipulating the mouse genome opens the way for elucidating the underlying molecular mechanisms in mammals. We write this review from the perspective of our work with mice, aware that our current knowledge about somitogenesis would be scarce without the seminal contributions made by the experimental work in other species, and we will frequently refer to these studies. We will emphasize the principles and mechanisms that appear to be common to all vertebrates. This is not meant to neglect the importance of species-specific aspects of somitogenesis and the necessity to study their significance. It rather reflects our endeavor toward drawing a coherent picture of somitogenesis and the processes involved.

II. General Description of Somitogenesis

A. Overview

The term "somitogenesis," as we use it here, includes the generation and patterning of paraxial mesoderm, the formation of the genuine somites, and the early steps of their subsequent differentiation. These processes represent a series of distinct, morphologically discernible events that progress in a strict temporal and (since vertebrate embryos develop in a craniocaudal succession) spatial order. These events are (1) the generation of the paraxial mesoderm; (2) the appearance of cellular condensations in the mesenchymal paraxial mesoderm, the so-called somitomeres; (3) the formation of the mature epithelial somites; (4) the differentiation of somites into dermomyotome and sclerotome; and (5) morphogenesis of the somite-derived structures according to their position along the anterior-posterior body axis (Fig. 1). The first somites form at the anterior end of the paraxial mesoderm in the future occipital region shortly after the germ layers are

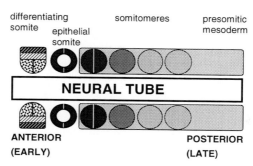

Fig. 1 Stages of somitogenesis. Somites develop from the paraxial mesoderm, which is arranged in two thick bands of mesenchymal tissue laterally on either side of the neural tube. Somites at different stages of somitogenesis are simultaneously present on developing embryos, anterior somites being more advanced than posterior ones. The paraxial mesoderm posterior to the most recently formed somite ("epithelial somite") is called presomitic mesoderm and contains discrete swirls of cells, called somitomeres, visible only by scanning electron microscopy. Anterior somites have already undergone differentiation into dermomyotome, sclerotome, and myotome.

formed (Fig. 2A). Thereafter, new somites are added at a relatively constant rate (depending on the species) in a strict anterior posterior sequence, while concomitantly new mesoderm is being generated and added at the posterior end. The paraxial mesoderm caudal to the last somite is called "presomitic mesoderm" in the mouse and "segmental plate" in the chick. During somite formation the loose mesenchyme of the presomitic mesoderm is transformed into a three-dimensional epithelium—with the exception of certain amphibia such as *Xenopus*, whose somites consist entirely of myotomal cells and form by segregation and rotation of groups of cells (Hamilton, 1969). The mature somite is a spherical ball of cells that form a simple three-dimensional columnar epithelium surrounding a central cavity, the somitocoel; this occurs also in the mouse (Fig. 2B), although it has been reported that mouse somites do not form a closed epithelial vesicle (Ostrovsky *et al.*, 1988). The somitocoel contains mesenchymal cells that during subsequent differentiation contribute to the sclerotome. The cells of the outer epithelial layer are connected to each other by junctional complexes close to the apical cell surface which face the somitocoel. The outside of the somites (made up of the basal cell surfaces) is covered by a basement membrane which connects each somite to, and separates it from, the neighboring tissues and adjacent somites. Shortly after their formation, epithelial somites differentiate further. During differentiation, the first cells at the ventral portion de-epithelialize and form the sclerotome (Fig. 2C,D). The sclerotomes of the anterior-most five somites contribute to the occipital bone, while the sclerotomes of the remaining somites are the precursors of the vertebral bodies, vertebral arches, and ribs (Theiler, 1988; Couly *et al.*, 1992; Christ and Ordahl, 1995). Dorsolateral to the

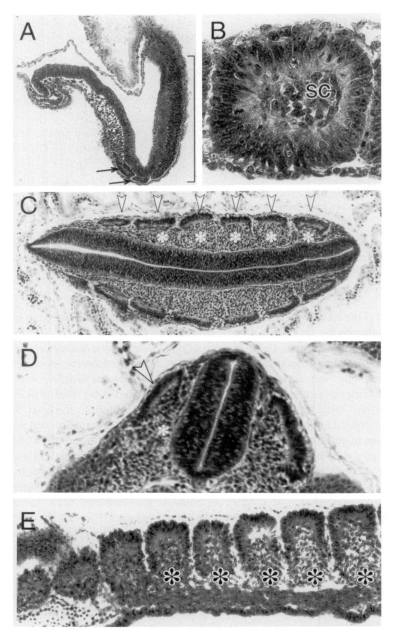

Fig. 2 Structure of somites. A, Day 8 mouse embryo with two somites (*arrows*). The bracket indicates the region of the presomitic mesoderm. B, Mature epithelial somite. Cells of the somitocoel (sc) form a loose mesenchyme. The nuclei of the surrounding epithelial cells are localized close to the outer (basal) surface. C, Frontal section of a day 10 mouse embryo. The dermomyotomes (*arrowheads*) form epithelial caps overlying the mesenchymal sclerotomes (*asterisks*). D, Transverse section of a day 10 mouse embryo. Dermomyotome (*arrowhead*) and sclerotome (*asterisk*) are indicated. E, Sagittal section through the posterior region of a day 9.5 mouse embryo. Progressive sclerotome formation is evident in the more anterior somites (posterior is to the left).

sclerotome, an epithelial cap of cells, the dermomyotome, remains (Fig. 2C,D). Cells from the dermomyotome give rise to the myotome, which underlies the dermomyotome medially, and subsequently cells of the dermomyotome contribute to the dermis of the dorsal skin. The myotome cells give rise to the epaxial and hypaxial muscles of the trunk and skeletal muscle of the limbs, and cells from all regions of the somite contribute to vascular endothelium. Mature anterior somites differentiate while new somites are still being formed posteriorly; therefore, early organogenesis-stage embryos simultaneously contain somites at different stages of their development (Fig. 2E). Although the basic cellular differentiation pattern of different somites is very similar, somites give rise to unique anatomic structures along the anterior-posterior axis, because they acquire early during their development specific identities according to their axial position.

This brief overview indicates that somitogenesis subsumes a number of complex and diverse events. These require ordered successions of changes in gene expression and cellular behavior, driving morphogenesis and pattern formation. These patterning processes include (1) the specification of paraxial mesoderm, (2) the regulated progression of mesenchymal–epithelial and epithelial–mesenchymal transitions, (3) patterning of the paraxial mesoderm (i.e., establishment and maintenance of segment borders) and anterior-posterior segment compartments, (4) positional specification, and (5) subdivision of somites in dorsoventral and mediolateral regions with particular cell fates and differentiation patterns (Fig. 3). How are these processes controlled, which tissues interact, how do they interact, and what are the molecular mechanisms underlying these interactions, are the questions of interest. These problems have been addressed on different levels, and by a variety of experimental means, including histologic and morphologic analyses, study of the fates and developmental potentials of cells, tissue transplantations and ablations *in vivo*, tissue reconstitutions *in vitro*, analyses of gene expression, and manipulation of gene expression and function in embryos. The results of such analyses and the current picture of "somitogenesis" are discussed below.

B. Staging of Somites

To allow for easy classification and comparison of somites of different developmental stages, a numbering system was proposed that takes into account that different somites are at the same developmental stage in embryos of different ages (Christ and Ordahl, 1995). According to this system, somites are counted beginning with the anterior-most somite in Arabic numbers. In addition, somites are numbered according to their developmental age in Roman numbers, the most recently formed somite being number I. Thus, for example, the 20th somite of a

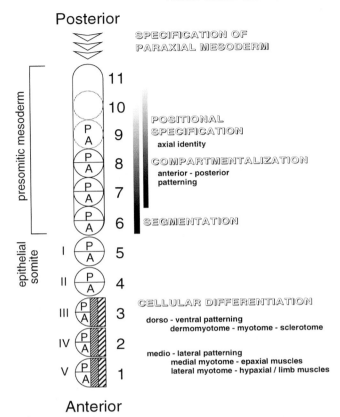

Fig. 3 Patterning processes during somitogenesis. Paraxial mesoderm cells become specified while they emerge from the primitive streak and are recruited to the posterior end of the presomitic mesoderm. In the presomitic mesoderm, before overt segmentation, the prospective somites acquire their axial identity and anterior-posterior (A–P) polarity. The exact timing and order of positional specification and compartmentalization are not known. Shortly after somites form, cells in different regions of the somites begin to differentiate according to their dorsoventral and mediolateral position. The Arabic numbers to the right of somites indicate their absolute number; the Roman numbers refer to the developmental stage of the somites, according to the staging system described in the text.

25-somite embryo would be somite VI/25, and the most recently formed somite in the same embryo would be somite I/25. In a 30-somite embryo, the 20th somite is XI/30, the 25th somite, is VI/30, and the most recent somite is I/30 (which would be equivalent in its development stage to I/25). Roman numbers alone indicate the developmental stage of a somite irrespective of its axial position. Whenever we refer to specific stages of somite development in this review, this numbering system is used.

III. Origin and Specification of Paraxial Mesoderm

A. Origin of Somitic Cells

Mesoderm formation is initiated during gastrulation, which in mouse embryos commences at about day 6.5 of development with the formation of the primitive streak (Poelmann, 1981; Snell and Stevens, 1991). Between days 6.5 and 9.5, cells invaginating through the primitive streak generate mesoderm. Thereafter, new mesoderm is generated from the tail bud until axis elongation ceases at about day 12.5 of development. Cells ingressing through different regions of the primitive streak (or the marginal zone or germ ring, its equivalent in amphibians and fish, respectively) along the anterior-posterior axis adopt different mesodermal fates. At the mid primitive streak stage (embryonic day 7.5 in the mouse), the first wave of cells has invaginated through the primitive streak. These cells give rise to extraembryonic, cardiac, and head mesoderm. At this time, the progenitors of the paraxial mesoderm are located in the primitive ectoderm (the epiblast in chick embryos), lateral to the primitive node, and in the anterior region of the primitive streak. During subsequent development cells ingressing at the anterior end of the streak (dorsal marginal zone) become paraxial mesoderm (dorsal mesoderm); cells ingressing successively more posteriorly adopt successively more ventral mesodermal fates (intermediate and lateral plate mesoderm); and, in amniotes, the posterior-most cells contribute (at least early during gastrulation) to extraembryonic mesoderm (Keller, 1975, 1976; Tam and Beddington, 1987; Lawson et al., 1991; Lawson and Pedersen, 1992; Smith, Gesteland, et al., 1994; Parameswaran and Tam, 1995). The paraxial mesoderm, which gives rise to the somites, lies on either side of the neural tube and is flanked laterally by the intermediate mesoderm, which gives rise to the embryonic (mesonephric) kidney (Butcher, 1929; Snell and Stevens, 1991).

Progenitor cells for the paraxial mesoderm are continuously found in the primitive streak, and then in the tail bud (Tam and Beddington, 1987; Lawson and Pedersen, 1992; Tam and Tan, 1992). Labeling of individual cells in transgenic embryos by intragenic homologous recombination that restored the enzymatic activity of a modified lacZ transgene demonstrated that clonal descendants of cells marked before gastrulation can contribute to paraxial mesoderm on both sides of the embryo at the same axial level (Bonnerot and Nicolas, 1993). This indicates that the lineages forming the right and left paraxial mesoderm separate only during gastrulation and that, despite the extensive cell mixing and lack of coherent clonal growth in the mouse egg cylinder–stage embryo (Lawson et al., 1991), clonal descendants of paraxial mesoderm precursors are not dispersed along the anterior-posterior axis. In mouse late primitive streak–stage embryos (embryonic day 8.5), the progenitors for the paraxial mesoderm constitute a much greater portion of the streak and are also located in its posterior region (Wilson and Beddington, 1996). Cells from the anterior streak colonize predominantly the medial portion of the paraxial mesoderm, whereas the majority of cells from the posterior streak region contribute to the lateral portion (Wilson and Beddington, 1996). This does not necessarily mean that the anterior and posterior late primitive streak contain different types of progenitor cells, but it may rather reflect different movements of cells emerging from different regions of the primitive streak (see later discussion). Progenitor cells that ingress through the primitive streak early contribute to more anterior somites than cells that ingress at later stages of development or are derived from the tail bud. Thus, new cells are continuously recruited to the paraxial mesoderm throughout gastrulation and early organogenesis. In addition, descendants of paraxial mesoderm progenitor cells remain in the primitive streak during its regression for an extended period (Tam and Beddington, 1987, 1992; Tam and Tan, 1992; Wilson and Beddington, 1996), suggesting that a resident population of cells with self-renewal capacity resides in the primitive streak and provides the stem cells for paraxial mesoderm (Stern et al., 1992).

Very similar to the situation described for mouse, in chick embryos the progenitor cells for the paraxial mesoderm are located at the mid and definitive primitive streak stages in the epiblast lateral to Hensen's node and in the anterior primitive streak, and clonal descendants or progenitor cells remain in the anterior primitive streak during its regression (Spratt, 1955; Ooi et al., 1986; Selleck and Stern, 1991; Hatada and Stern, 1994; Psychoyos and Stern, 1996). Furthermore, it was shown that progenitors in the lateral region of Hensen's node give rise to cells that colonize only the medial portion of the paraxial mesoderm; cells from the anterior-most region of the primitive streak colonize the medial and lateral portion of the paraxial mesoderm, and cells more posteriorly in the anterior streak region contribute only to the lateral portion (Selleck and Stern, 1991; Psychoyos and Stern, 1996). The contribution of cells from the node to the somites is somewhat debated, because grafted quail nodes did not contribute to the somites in chick embryos (Schoenwolf et al., 1992). The separation in medial and lateral cell populations is maintained during somite formation, and there appears to be very little mixing between medial and lateral cells. The differential contributions to the medial and lateral portions of the paraxial mesoderm by cells residing in the node or along the primitive streak were suggested to be caused by the different migration routes of these cell populations: cells from the lateral node and anterior-most part of the streak migrate through the node and stay close to the midline, whereas cells from further posterior migrate more laterally and never pass through the regressing node (Schoenwolf et al., 1992; Psychoyos and Stern, 1996).

At about day 9.5 in the mouse, the remnant of the primitive streak occupies a small zone dorsally at the caudal end of the embryo, and it is subsequently replaced by the tail bud (Schoenwolf, 1984; Griffith et al., 1992). The tail bud contributes to paraxial mesoderm posterior to the hind limb buds (Tam and Tan,

1992). It has been shown that avian and amphibian tail buds contain regions with distinct developmental fates and that paraxial mesoderm is derived from cells in the caudal mediodorsal part of the tail bud that "invaginate" and migrate laterally, similar to the progenitor cells in the anterior primitive streak (Gont *et al.*, 1993; Catala *et al.*, 1995). The overall fundamental similarity in topology among the fate maps of different vertebrate species and the gene expression patterns in gastrulating vertebrate embryos (Lawson and Pedersen, 1992; Beddington and Smith, 1993) suggests that also in mouse embryos distinct cell populations in the tail bud give rise to the paraxial mesoderm.

In summary, paraxial mesoderm cells, the precursors of somites, are generated during gastrulation and early organogenesis. There appears to be a resident population of "stem cells" in the primitive streak and tail bud that continuously produces paraxial mesoderm, and progenitor cells, which reside in distinct regions of the node, primitive streak, and tail bud, contribute to different portions of the paraxial mesoderm along the mediolateral and anterior-posterior axes.

B. Commitment and Specification of Somitic Cells

Paraxial mesoderm cells that originate from distinct regions of the gastrulating embryo still show considerable developmental plasticity and are not stably committed to the paraxial mesodermal (somitic) fate. Primitive streak cells that are transplanted to heterotopic sites develop according to their new position and differentiate into tissues normally derived from this site (Beddington, 1982). Also, cells in the presomitic mesoderm are not yet irreversibly committed to the paraxial mesoderm (somitic) fate, since descendants of orthotopic grafts in various regions of the presomitic mesoderm could contribute to lateral plate mesoderm (Tam, 1988). Similarly, in chick embryos, cells from the lateral node and anterior streak that normally contribute to the medial and lateral portions of the paraxial mesoderm (Selleck and Stern, 1991; Psychoyos and Stern, 1996) did not give rise to somites but altered their fate when transplanted to ectopic sites (Selleck and Stern, 1992). The developmental plasticity of these paraxial mesoderm precursors contrasts with cells of the anterior node, which are already committed to form the notochord (Selleck and Stern, 1992). Even cells in mature epithelial somites appear not to be irreversibly committed to the somitic fate. When quail somites were transplanted beneath the ectoderm of younger chick embryos, graft cells contributed not only to host somites but also to endoderm, lateral plate mesoderm, and endothelium (Veini and Bellairs, 1991). Therefore, the available evidence suggests that paraxial mesoderm cells may become determined to the somitic fate only after formation of distinct somites.

Whereas the fates of cells invaginating at different levels of the primitive streak are well established in mouse and chick embryos, the molecular mechanisms that specify these fates, and in particular those that commit cells to form paraxial mesoderm, are poorly understood. Secreted signaling proteins of the fibroblast growth factor (FGF), transforming growth factor-β (TGFβ), and Wnt gene families play essential roles in the induction and specification (dorsoventral patterning) of mesoderm in amphibian embryos (reviewed in Kessler and Melton, 1994; Slack, 1994) and are critical for multiple patterning processes in insect and vertebrate development (Dickinson and McMahon, 1992; McMahon et al., 1992; Hrabě de Angelis and Kirchner, 1993; Parr and McMahon, 1994; Ferguson, 1996; Hammerschmidt, Serbedzija, et al., 1996; Hogan, 1996). Members of these gene families, as well as homologues of other genes implicated in, or known to have, patterning functions during embryogenesis in diverse species, are expressed in the node and primitive streak of amniote embryos in restricted or graded patterns (reviewed in Tam and Trainor, 1994). This has led to the proposal that qualitatively or quantitatively different signals emanating from the node and different regions of the primitive streak act on nascent mesoderm cells and thereby modify gene expression and specify mesodermal cell fates (Sasaki and Hogan, 1993). However, signals from the node do not appear to be absolutely required for the specification of paraxial (dorsal) mesoderm, since mouse embryos lacking a node and notochord formed somites (Ang and Rossant, 1994; Weinstein et al., 1994). Mutational analysis in mice has begun to elucidate the role of signaling proteins for specifying paraxial mesoderm cells in mammals. These analyses clearly show that FGF, TGFβ, and Wnt signals are involved in patterning of mesoderm in mammals. However, only two among those proteins analyzed thus far appear to be involved in the specification of paraxial mesoderm.

Although TGFβ-like signals play essential roles in the initiation of gastrulation and mesoderm formation in mammals (Conlon et al., 1994; Mishina et al., 1995) and in the formation of lateral (ventral) mesoderm (Winnier et al., 1995), no distinct role for the specification of paraxial mesoderm in mammals could be assigned to a TGFβ-like activity, thus far. In contrast, FGF and Wnt signals appear to be specifically involved in specifying paraxial mesoderm. Four Fgf genes (Fgf3, Fgf4, Fgf5, and Fgf8) and two of the FGF receptor genes (Fgfr1 and Fgfr2) are expressed in distinct patterns in mouse primitive streak–stage embryos (Wilkinson et al., 1988; Haub and Goldfarb, 1991; Hebert et al., 1991; Niswander and Martin, 1992; Yamaguchi et al., 1992; Orr Urtreger et al., 1993; Crossley and Martin, 1995; Yamaguchi and Rossant, 1995). Thus far, results of mutational analyses have been reported for Fgf3, Fgf4, and Fgf5, as well as for Fgfr1. The phenotypes of FGFR1- and FGF3-deficient mice indicate potential functions of FGF signaling in specifying paraxial mesoderm cell fates. In midstreak embryos, Fgfr1 is expressed in migrating embryonic mesoderm and then becomes restricted to the paraxial mesoderm. Subsequently, highest levels are transiently found in the rostral region of the presomitic mesoderm (Yamaguchi et al., 1992). FGFR-1-deficient embryos have an expansion of the axial mesoderm, have primitive streak defects, and lack somites. However, they are able to generate some paraxial mesoderm, which has been interpreted as respecification of the normal paraxial fates of anterior primitive streak cells to axial fates (Yamaguchi *et al.*, 1994). Developmental plasticity between axial and paraxial cell fates is also supported by the finding that axial mesoderm is re-specified to a paraxial fate in the zebrafish mutant *floating head* (Halpern *et al.*, 1995). Since several ligands interact with FGFR1, the severe phenotype in *Fgfr1*-deficient mice most likely reflects the loss of combinatorial FGF signals (Yamaguchi *et al.*, 1994). Consistent with this idea, mice lacking only FGF3 have a much milder phenotype: shortened tails, with variable numbers of frequently deformed vertebrae, and inner ear defects (Mansour *et al.*, 1988). Beginning at the early primitive streak stage, *Fgf3* is expressed in anterior regions of the streak and in embryonic mesoderm in cells emerging from the primitive streak. Until day 9.5, *Fgf3* expression in cells emerging from the primitive streak is maintained. Expression in the embryonic mesoderm is confined to the presomitic mesoderm and, at least until day 11.5, is found in the tail bud (Wilkinson *et al.*, 1988; Niswander and Martin, 1992). In mutant embryos, the somites and tissues are severely disorganized in the tail beginning at about day 11.5, suggesting that *Fgf3* function is essential only in the tail bud, potentially for the proliferation or intercalation of mesodermal cells (Mansour *et al.*, 1993).

Three of the presently known *Wnt* genes, *Wnt3a*, *Wnt5a*, and *Wnt5b*, are expressed in the primitive streak of mouse embryos (Takada *et al.*, 1994). *Wnt5a* and *Wnt5b* are expressed in posterior regions of the streak beginning at the early streak stage (day 6.5); *Wnt3a* expression starts only at the extended streak stage (day 7.5), and it is detected throughout much of the streak, with an anterior border just posterior to the mode. Subsequently, all three genes are expressed in the regressing streak and tail bud, *Wnt3a* expression being more restricted dorsally and caudally than that of *Wnt5a* or *Wnt5b* (Takada *et al.*, 1994). In embryos lacking functional Wnt3a, axial development was truncated posterior to the forelimb buds, somites were disrupted or missing, and no tail bud formed. No evidence for paraxial mesoderm was found posterior to the (morphologically normal) first seven to nine somites, whereas ventral mesoderm was still present at these axial levels (Takada *et al.*, 1994). These results clearly indicate a requirement of Wnt3a for the formation of embryonic mesoderm, and they suggest a specific role for Wnt3a in specifying paraxial mesoderm and later for the generation of the tail bud, which is essential for continuous axis elongation (Smith, 1964). These results also indicate that the specification of the anterior paraxial mesoderm (giving rise to somites 1–7) is independent of *Wnt3a* function and suggest that different mechanisms for mesoderm specification may operate during the earliest stages of gastrulation. Vestigial tail (*vt*), a spontaneous mutation affecting the posterior axial skeleton (Heston, 1951), was found to be a hypomorphic allele of *Wnt3a*, demonstrating the importance of *Wnt3a* expression levels during development (Greco *et al.*, 1996). Embryos carrying different combinations of the mutant *Wnt3a^{vt}* and *Wnt3a* null alleles generated progressively lower levels of Wnt3a. Decreasing levels of Wnt3a resulted in arrest of somitogenesis at increasingly more anterior positions, suggesting that somite formation requires increasing levels of Wnt3a along the anterior-posterior axis (Greco et al., 1996). The requirement for increasing doses of Wnt3a along the anterior-posterior axis is reminiscent of the increased requirement for brachyury, a transcriptional regulator expressed in mesodermal precursors in the primitive streak and tail bud (Herrmann et al., 1990; Wilkinson et al., 1990) along the anterior-posterior axis (MacMurray and Shin, 1988; Stott et al., 1993). Similarly, the mouse homeobox-containing putative transcription factor Evx1 shows a graded pattern of expression in the primitive streak, with higher expression levels posteriorly (Dush and Martin, 1992), which appears to be of functional significance. In embryos homozygous mutant for the eed (embryonic ectoderm development) gene, a global regulator for anterior-posterior patterning in mice, Evx1 expression is deregulated and high throughout the primitive streak (Faust et al., 1995; Schumacher et al., 1996). No paraxial mesoderm is formed in eed embryos, suggesting that the ectopic high levels of Evx1 in the anterior primitive streak of mutants respecify paraxial (and probably also lateral) mesoderm to an extraembryonic fate (Schumacher et al., 1996).

It is becoming increasingly clear that signals from the primitive streak are involved in the specification of paraxial mesoderm and that these signals are, at least in part, mediated by members of known families of secreted signaling proteins. Thus far, no gene has been found to be specifically required for the specification of paraxial mesoderm along the entire body axis, and it appears likely that there is no single gene that specifies the whole paraxial mesoderm. The presence of multiple members of various families of signaling proteins, the more severe phenotypes of mutations in receptors than in single ligands, and the restriction of phenotypes to regions of the paraxial mesoderm along the anterior-posterior axis suggest rather that distinct signals act successively and/or in combinatorial fashion at different times during gastrulation and axis elongation. In addition, different levels of other regulatory proteins (e.g., transcriptional regulators) may be required along the anterior-posterior axis, and as yet unidentified signals could be required for the specification of paraxial mesoderm along the entire anterior-posterior axis of the embryo.

IV. Cell–Cell and Cell–Matrix Interactions during Somitogenesis

Nascent paraxial mesoderm cells emerging from the primitive streak migrate anterolaterally and generate a bilateral mesenchymal tissue. During somite formation, groups of mesenchymal cells condense; these cells become polarized and form a three-dimensional epithelium, the somite. Subsequently, somitic cells deepitheliarlize in a strict temporal and spatial sequence and become mesenchymal again. Thus, cells fated to become somites repeatedly undergo profound changes in their adhesive and migratory properties until they reach their final destination and differentiate according to their development program. Cell adhesion and cell migration are essential for morphogenetic processes. They involve specific interactions between components on the surface of adjacent cells that establish contacts, or between cell surface receptors and the extracellular matrix (ECM) (reviewed in Thiery *et al.*, 1985; Vallés *et al.*, 1991; Hynes, 1996). ECM components with major adhesive and migratory functions, their principal receptors, the integrins, and various cell adhesion molecules (CAMs) are expressed in the paraxial mesoderm and in somites. The dynamic patterns of their expression reflect that changing requirements for cell–matrix and cell–cell interactions during somitogenesis, and the results of perturbing these interactions demonstrate their significance.

A. Cell–Matrix Interactions

Laminin and fibronectin (as well as other ECM components) are present in the ECM of the paraxial mesoderm. Laminin and fibronectin are two well-characterized ECM components with known functions in cell–substratum adhesion and cell migration. Laminin is present in the basal membranes underlying the ectoderm and endoderm but not in the interstitial space between the mesenchymal cells; it therefore contacts the paraxial mesoderm dorsally and ventrally (Duband *et al.*, 1987; Ostrovsky *et al.*, 1988; Peterson *et al.*, 1993). Fibronectin colocalizes with laminin in these basal laminae but is also found in the interstitial space between condensing mesenchymal cells at the anterior end of the presomitic mesoderm, (Ostrovsky *et al.*, 1983, 1988; Duband *et al.*, 1987). Mature somites are completely surrounded by basal laminae, which are progressively remodeled as cells undergo epithelial–mesenchymal transitions (Crossin *et al.*, 1986; Duband *et al.*, 1987; Duband and Thiery, 1987; Peterson *et al.*, 1993).

Fibronectin is essential for mesodermal cell migration during *Xenopus* gastrulation (Thiery *et al.*, 1995). Both peptides containing the fibronectin-binding sequence RGDS (Ruoslahti and Pierschbacher, 1986) and exogenous fibronectin increased cell–cell adhesion in cultured chick presomitic mesoderm cells. This led to a "trigger" hypothesis that the binding of fibronectin to its receptor (or receptors) on mesenchymal cells may be causally related to compaction in the anterior presomitic mesoderm that precedes somite formation (Lash *et al.*, 1984, 1987). Mouse embryos lacking fibronectin initiate gastrulation normally but have shortened anterior-posterior axes, are mesoderm deficient, and fail to form organized somites (George *et al.*, 1993; Georges Labouesse *et al.*, 1996). This indicates that fibronectin is essential for normal cell migration and adhesion during gastrulation, most likely also in the paraxial mesoderm. Components of the ECM interact with cells by binding to specific receptors on the cell surface. A family of major ECM receptors with members expressed in developing somites are the integrins (Pow and Hendrickx, 1995). They act as non-covalently-linked heterodimers with specific and overlapping ligand specificities, and binding of integrins to the ECM triggers intracellular signals that lead to changes in cell behavior (Hynes, 1992). Antibodies that block the function of β1-integrin caused lateral displacement of segmental plate tissue and somites after injection in to chick embryos and altered somite cell shape but did not prevent somite segmentation and differentiation (Drake and Little, 1991; Drake et al., 1992). Mouse embryos lacking functional focal adhesion kinase (FAK), a nonreceptor tyrosine kinase implicated in transducing signals generated by cell–ECM interactions (Parsons et al., 1994; Schaller and Parsons, 1994), display a phenotype similar to that of fibronectin-mutant embryos, and, although they form paraxial mesoderm, somite formation is severely impaired (Furuta et al., 1995). From these functional studies it is clear that interactions between paraxial mesoderm cells and the extracellular matrix are crucial for normal somitogenesis.

B. Cell Adhesion

Selective adhesion between cells is mediated by specific CAMs that are expressed on the cell surface and are classified depending on their requirement for calcium to function. CAMs with described expression in the paraxial mesoderm early during somitogenesis are NCAM, which is a Ca-independent CAM, and two Ca-dependent CAMs, cadherin 11 (*Cdh11*) and N-cadherin (*Cdh2*). Prior to ingression through the primitive streak, cells of the embryonic ectoderm express N-cadherin, and E-cadherin, another family member of the Ca-dependent CAMs. The epithelial–mesenchymal transition of cells ingressing through the streak and their loss of adhesiveness are associated with the down regulation of both cell adhesion molecules. Subsequently, expression of NCAM and N-cadherin is low in the caudal presomitic mesoderm, whereas E-cadherin is undetectable. NCAM and N-cadherin expression is up regulated in the rostral presomitic mesoderm, coinciding with the compaction of mesenchymal cells and the formation of the epithelial somite (Duband et al., 1987, 1988; Probstmeier et al., 1994). Similarly, cadherin 11, which is not expressed in the caudal presomitic mesoderm, is activated at the rostral end in the condensing mesenchyme (Kimura et al., 1995). NCAM and N-cadherin are expressed in the entire condensing somite, whereas cadherin 11 expression is initially restricted to the caudal half. During formation of the epithelial somite, NCAM and N-cadherin are at first evenly distributed on the cell surface of presomitic mesoderm cells. Later, in cells of the nascent epithelial somite, NCAM remains distributed over the cell surface but N-cadherin becomes restricted to the apical region of the cells and enriched in the adherens junctions (Duband et al., 1987, 1988). The protein distribution of cadherin 11 during somite formation has not yet been described. Consistent with the expression of CAMs, cells in the anterior presomitic mesoderm show increased adhesive properties compared with cells from the posterior region (Cheney and

Lash, 1984). During subsequent somite development, down regulation of N-cadherin in the ventromedial portion of the somite, which is fated to differentiate to sclerotome, precedes morphologic signs of de-epithelialization and cellular differentiation. In contrast, NCAM expression is maintained in the sclerotome, although at lower levels. Cadherin 11 expression appears to be complementary to N-cadherin; it is down regulated in the dermomyotome but maintained in the caudal and cranial edges of the sclerotome of each somite (Kimura *et al.*, 1995).

Therefore, the increase of cell contacts and acquisition of epithelial morphology during somite formation is preceded and accompanied by expression of CAMs, which suggests their causal participation in these processes. Subsequently, consistent with changing adhesive properties, expression of CAMs is differentially down regulated prior to de-epithelialization and cell migration. The significance of cell adhesion mechanisms for somite formation and differentiation is further supported by the phenotypes of embryos mutant for these molecules. A null mutation of N-cadherin led to (among other adhesion defects) small and irregular somites whose epithelial structure was partially disrupted (Radice *et al.*, 1997). Loss-of-function mutations in NCAM had no effect on somite formation (Tomasiewicz *et al.*, 1993). However, a targeted mutation creating a soluble NCAM molecule caused dominant embryonic lethality and led to reduced numbers of irregular, small somites. Since this phenotype was observed in the absence of membrane-bound NCAM, it appears that the soluble mutant protein causes these defects by heterophilic interactions, potentially by perturbing mesoderm migration through the extracellular matrix and disturbing signals required for normal somite differentiation (Rabinowitz *et al.*, 1996).

Interaction between CAMs on the surface of somitic cells most likely triggers the condensation of mesenchymal cells and the establishment of the epithelial somite; this requires precise spatial and temporal control of expression of CAMs and their interacting partners and/or modulation of their function or activity. Whether this regulation is intrinsic to the somitic mesoderm or requires inductive signals from the surrounding tissues is presently unknown. In turn, interaction of CAMs may regulate gene expression. Recently, it was demonstrated that cadherins can direct cell differentiation and tissue formation and regulate gene expression *in vitro* (Larue *et al.*, 1996). This suggests that during somitogenesis CAMs may function beyond mere mediation of cell adhesion. The up and down regulation of various genes that occurs in the nascent somite may be brought about in part by cadherin interactions.

V. Segments and Borders

One of the most striking aspects of somitogenesis is the sequential generation of segmental subunits on both sides of the embryonic midline in a precise synchronous and bilaterally symmetric manner. How this ordered segmentation is brought about and regulated is, after decades of experimental work, still one of the major unsolved problems of somitogenesis. Based on experimental data and/or theoretic considerations, over time a number of models evolved to explain the formation and maintenance of the periodic somite pattern and segment borders (Slack, 1991). However, objections have been made to all of them, and at present no model is consistent with all experimental data from different vertebrate species (Polezhaev, 1992).

A. Tissue Interactions

Tissue interactions, particularly between the paraxial mesoderm and the midline structures, were suggested for a long time to control somite formation. This idea was tested extensively by extirpation and grafting experiments, predominantly in chick embryos. Over time, a number of candidate tissues and/or tissue interactions were postulated to be causally related to segmentation. However, in subsequent studies, conflicting results were frequently obtained. One by one, all tested tissues were found to be dispensable for segmentation, and no specific tissue interactions that are essential for somite formation could ultimately be defined.

Somite "centers" located on both sides of the posterior half of the node were suggested by Spratt to be required for somite formation in avian embryos (Spratt, 1955, 1957a, 1957b), but somites were found to develop after removal of these regions (Bellairs, 1963). Similarly, experiments suggesting that the neural tube was implicated in somite induction (Fraser, 1960; Butros, 1967) conflicted with studies that clearly showed that somites developed after removal of the developing spinal cord (Christ, 1970; Christ et al., 1972; Menkes and Sandor, 1977; Packard, 1980b), albeit these somites were fused across the midline (Christ, 1970). The presumptive notochord appeared not to be essential for segmentation (Spratt, 1955; Grabowski, 1956; Bellairs, 1963; Christ et al., 1972) However, it induced somites when transplanted into the posterior primitive streak after extirpation of the anterior part of the blastoderm (Nicolet, 1970), presumably by means of diffusible factors (Nicolet, 1971). Based on transection experiments, a role of the regressing node in shearing the prospective somitic mesoderm into left and right halves, thereby releasing somite-forming capabilities already present in the presomitic mesoderm, was proposed and was suggested to resolve the conflicting results of earlier studies (Lipton and Jacobson, 1974a, 1974b). These findings were not confirmed in subsequent studies (Stern and Bellairs, 1984b).

It is now generally accepted that the somitic mesoderm in the segmental plate does not require specific signals from axial tissues for segmentation. Consistent with this view, it has been shown that somitic mesoderm can segment in isolation from the neural tube, the notochord, the ectoderm, the endoderm, or the lateral plate (Bellairs, 1963; Lanot, 1971; Christ et al., 1972, 1974a; Brustis and Gipoulou, 1973; Packard and Jacobson, 1976; Packard, 1980a, 1980b; Sandor and Fazakas Todea, 1980; Tam, 1986). In addition, the original craniocaudal sequence of somite formation was maintained when the orientation of the segmented plate was reversed (Christ et al., 1974a; Menkes and Sandor, 1977). The formation of somites from explants of presomitic mesoderm has been interpreted to imply segmentation is an intrinsic property of the paraxial mesoderm. This view was further supported by the finding that lateral mesoderm transplanted to the medial region normally occupied by paraxial mesoderm did not segment (Christ, 1970). However, in all of these experiments, somites never developed from truly isolated presomitic mesoderm without any contact to some supporting tissue. When presomitic mesoderm was cultured without contact to any tissue, the mesenchyme rapidly dispersed and no somites formed (Packard, 1980b). In addition, under conditions in which explanted or grafted presomitic mesoderm segmented, resulting somites were abnormal in size and shape (Packard and Jacobson, 1976; Packard, 1980b) or were not fully epithelialized (Christ et al., 1972). Whereas the formation of somites seemingly can occur without continuous tissue interactions, surrounding tissues can modulate somite formation. When quail segmental plates were transplanted into chick embryos, the number and size of somites formed by the graft was adjusted to a number and size more compatible with the host's somite pattern, and intersomitic borders in the graft developed mostly in alignment with intersomitic borders of the host on the opposite side of the midline (Packard et al., 1993). Barriers between neural tube and graft impaired somite formation in the graft, suggesting that signals from the midline tissues acted under these experimental conditions (Packard et al., 1993).

In summary, no specific, continuous influence of a defined embryonic tissue or region (e.g., notochord, neural tube) appears to be responsible for the metameric organization of the paraxial mesoderm. Explants of presomitic mesoderm are able to form somites, suggesting that continuous inductive interactions are not needed for mesoderm segmentation. However, some unknown "trophic" factor from an epithelial tissue appears to be required. This influence may be merely supportive, that is, stabilizing the mesenchyme of the presomitic mesoderm and preventing cell dispersal. The "prepattern" (see next section) of somites present in the presomitic mesoderm appears to be plastic and can be respecified by appropriate surrounding tissues, most likely from the midline tissues. Therefore, it cannot unambiguously be concluded that the paraxial mesoderm is a truly self-organizing tissue. In addition, the alignment of intersomitic borders between graft- and host-derived somites on both sides of the embryos suggests some coordinating signal.

B. Prepatterns in the Paraxial Mesoderm

On the light microscopic level, the presomitic mesoderm is a morphologically homogenous mesenchyme. However, it is clear now that this seemingly amorphous tissue is regionalized and is patterned. We refer to these regionalizations as "prepatterns," merely to indicate that these patterns precede the formation of obvious segments.

The idea of autonomous segmentation of the presomitic mesoderm was further fueled by the discovery that the presomitic mesoderm displays, prior to overt segmentation, a metameric arrangement of groups of cells, called "somitomeres." These transient structures are visible only by stereoscanning electron microscopy and were first described in chick embryos in the paraxial head mesoderm, which never forms somites (Meier, 1979). Somitomeres are swirls of loose mesenchymal cells that are organized around a central point and are formed in a strictly anterior-to-posterior order. Since the discovery of somitomeres in chick embryos, such a prepattern has been described in the paraxial mesoderm of all species examined (Meier, 1979, 1981, 1982; Meier and Jacobson, 1982; Meier and Tam, 1982; Tam and Meier, 1982; Tam et al., 1982; Jacobson and Meier, 1984, 1986; Packard and Meier, 1984; Bellairs and Sanders, 1986; Jacobson, 1988). Despite the quite different morphology of gastrulating embryos of different vertebrate classes, onset and formation of somitomeres is similar in all of them (Jacobson and Meier, 1986), reflecting the overall principle of similarity of gastrulation in vertebrate embryos (Beddington and Smith, 1993) and suggesting that similar principles govern somite formation in different vertebrate classes.

There is a substantial controversy about the existence of somitomeres in the paraxial mesoderm of the head. In mammals, birds, reptiles, and teleosts seven somitomeres (i.e., shallow, superficial furrows) have been described in the head as well as four somites in the lower hindbrain area (Meier and Tam, 1982; Meier and Packard, 1984; Packard and Meier, 1984; Martindale et al., 1987), whereas only four somitomeres in the head and two somites in the lower hindbrain region were found in amphibians (Jacobson and Meier, 1984). The number of somitomeres in the head of amphibians was considered to resemble the mesodermal segments in the head of shark embryos (Jacobson, 1988), whereas teleosts have the same number of somitomeres as amniotes. This led to the suggestion that the ancestral forms had seven somitomeres, which were lost in the shark line and amphibian line during evolution (Jacobson, 1980). Somitomeres were always found in spatial register with the neuromeres (Jacobson, 1988; Tam and Trainor, 1994), the segmental subdivisions of the developing brain (Fraser, 1993). Since somitomeres were described to underlie the neural plate before it becomes obviously segmented (Jacobson and Tam, 1982; Tam and Meier, 1982), it was suggested that the somitomeres may influence the segmentation of the neural plate (Jacobson and Meier, 1986). The discovery of somitomeres in the paraxial mesoderm of the head was considered to have resolved the long-standing question as to whether the paraxial head mesoderm, which never forms somites, is segmented (Jacobson, 1988). However, doubts were raised about the existence of a metameric prepattern in the paraxial mesoderm rostral to the first (occipital) somite, since attempts to reveal a cellular pattern resembling somitomeres by light microscopic studies or by mapping of cell densities and arrangement failed (Jacob *et al.*, 1986). Regionalized cell fates in the cranial paraxial mesoderm, and codistribution of neural crest and cranial mesoderm cells derived from similar rostrocaudal positions during craniofacial morphogenesis (Trainor *et al.*, 1994; Trainor and Tam, 1995) are consistent with a prepattern and some common segmental register. However, fates of paraxial mesoderm cells were respecified after heterotopic grafting (Trainor *et al.*, 1994), and positional specification and patterning activity resides in the neural crest cells (Noden, 1988), suggesting that the neural crest cells pattern the paraxial mesoderm of the head.

In contrast, the existence of somitomeres within the presomitic mesoderm caudal to the last-formed somite is not disputed. The number of somitomeres in the presomitic mesoderm varies between species, but for a given species it remains constant, although the length of the presomitic mesoderm changes during development (Packard and Meier, 1984; Tam, 1986). Mouse, rat, and snapping turtle have 6–7; newt and frog, 5–7; chicken and quail, 10–12; and medaka, 10 somitomeres (Meier, 1979; Packard, 1980a, 1980b; Meier, 1981; Meier and Tam, 1982; Tam *et al.*, 1982; Jacobson and Meier, 1984, 1986; Meier and Packard, 1984; Packard and Meier, 1984; Jacobson and Meier, 1986; Tam, 1986; Martindale *et al.*, 1987; Packard *et al.*, 1993). As discussed earlier, somites formed in explants of presomitic mesoderm. In such explants, the number of somites that developed correlated with the number of somitomeres present in the presomitic mesoderm before transplantation (Packard, 1980b; Tam and Beddington, 1986), which suggests that somitomeres represent the direct precursors of the somites. If somitomeres represented the true precursors of the segments, one would expect no mixing of cells between adjacent somitomeres. However, cells move in the presomitic mesoderm between somitomeres (Tam and Beddington, 1987; Stern *et al.*, 1988; Tam, 1988), and injections of rhodamine dye into cells of chick segmental plates showed no evidence for compartments correlating with somitomeres (Bagnall *et al.*, 1992). Furthermore, when the segmental plate was distorted by stirring the mesenchyme with a needle and mingling the fragments (Menkes and Sandor, 1977), and when segmental plates were cut into small pieces, scrambled, and grafted to a different embryo (Packard *et al.*, 1993), relatively normal somites developed, raising additional doubts that the somitomeres represent a true static prepattern of the definitive segmental units.

A different kind of prepattern in the presomitic mesoderm was suggested by Bellairs (Bellairs and Veini, 1984; Bellairs, 1985), who found that if a gastrulating chick embryo was cut longitudinally so that only one side retained the primitive streak, somites formed in both embryo fragments but the somites in the piece lacking the streak were frequently smaller. To explain these findings it was suggested that a population of somite precursor cells is established and laid down in the presumptive somite area (i.e., lateral to Hensen's node) before regression of the primitive streak begins. This population, referred to as presomite clusters, consists of small groups of cells, each with the capability of initiating the formation of a somite. During regression of the streak, this presumptive somite area also regresses and leaves behind a trail of presomite clusters. The clusters subsequently recruit paraxial mesoderm cells that ingress through the primitive streak, and cells from each cluster and newly recruited paraxial mesoderm cells form the somitomeres and eventually the somites. Because the side lacking the primitive streak would not receive new paraxial mesoderm cells, fewer cells would be recruited by the clusters; hence, the resulting somites would be smaller. Whereas this idea can explain the smaller somites on the streak-deprived side of operated embryos, it cannot be reconciled with the capacity of grafted segmental plate to adjust the somite number according to the host (Packard *et al.*, 1993), since the number of forming somites should be predetermined by the existing number of clusters in the graft.

The notion of a prepattern in the unsegmented paraxial mesoderm is strongly supported by the growing number of spatially restricted or periodic patterns of gene expression in the presomitic mesoderm (reviewed in Tam and Trainor, 1994). There is increasing experimental evidence that the expression of transcriptional regulators of the basic helix-loop-helix (bHLH) protein family and of genes encoding homologues of the *Drosophila Delta* and *Notch* genes is of functional significance for somite formation. Delta and Notch mediate direct cell-to-cell communication and regulate cell fate decisions in various organisms (Artavanis Tsakonas *et al.*, 1995; Chitnis *et al.*, 1995), and, in *Drosophila* they mediate epithelial–mesenchymal transitions in derivatives of all germ layers (Hartenstein *et al.*, 1992; Artavanis Tsakonas *et al.*, 1995; Tepass and Hartenstein, 1995) and wing margin development (de Celis *et al.*, 1996; Doherty *et al.*, 1996). *Delta* and *Notch* encode transmembrane proteins (Wharton *et al.*, 1985; Vässin *et al.*, 1987; Haenlin *et al.*, 1990) that bind to each other (Fehon *et al.*, 1990) in the manner of a receptor (Notch) and a ligand (Delta) (Technau and Campos-Ortega, 1987; Heitzler and Simpson, 1993). Binding of Delta leads to activation of Notch and transduction of a signal via a conserved transcription factor [suppressor of hairless (*Su[H]*), called *RBPjκ* in the mouse], which activates bHLH genes encoded by the Enhancer-of-split complex [*E(spl)C*], which in turn regulate gene expression (Artavanis Tsakonas *et al.*, 1995).

Mouse genes expressed in distinct regions of, or at varying levels throughout, the presomitic mesoderm include the bHLH protein–encoding genes *paraxis* (Burgess *et al.*, 1995), *scleraxis* (Cserjesi *et al.*, 1995), *Mesp1,* and *Mesp2* (Sage *et al.*, 1996, and personal communication); the *Delta* homologues *Dll1* (Bettenhausen *et al.*, 1995) and *Dll3* (Dunwoodie *et al.*, 1997): and *Notch1* (Del Amo *et al.*, 1992; Reaume *et al.*, 1992; Williams *et al.*, 1995). Genes expressed in periodic patterns in the presomitic mesoderm include the zebrafish and *Xenopus* homologues of the *Drosophila* pair-rule gene *hairy, Her1* (Müller *et al.*, 1996) and *Hairy2A* (Jen *et al.*, 1997); a *Xenopus* homologue of *Drosophila Delta,* *X-Delta-2* (Jen *et al.*, 1997); and the mouse *Hes5* gene (de la Pompa *et al.*, 1997), a homologue of the *Drosophila E(spl)C* genes.

It is becoming evident that at least a subset of these genes is required for somite formation and patterning. *Mesp2* is expressed in a single stripe in the rostral presomitic mesoderm. Mouse embryos lacking *Mesp2* have severe defects in segmentation and polarity of somites. *Notch1* expression is severely reduced in the presomitic mesoderm of *Mesp2* mutants, suggesting that at least part of the phenotype is caused by loss of Notch-mediated cell interactions and that *Mesp2* may act upstream of *Notch1* (Y. Saga, personal communication). Interestingly, mouse embryos mutant for *Notch1* and *RBPjκ*, a transcriptional regulator activated by Notch (Jarriault *et al.*, 1995), also show defects in somitogenesis, although these have been less well defined (Conlon *et al.*, 1995; Oka *et al.*, 1995). The importance of local cell–cell signaling for the formation of segment borders is further supported by the analysis of *Xenopus X-Delta 2*. *X-Delta-2* is expressed in a dynamic periodic pattern in the presomitic mesoderm. Expression in the caudal presomitic mesoderm appears to fill most of a prospective segment with higher expression in the anterior half than in the posterior half. Toward the cranial end of the presomitic mesoderm *X-Delta-2* becomes progressively down regulated in the posterior part of the prospective somite and is ultimately confined to a narrow stripe at the anterior end of a prospective segment (Jen *et al.*, 1997). Injection of a dominant negative form of *X-Delta-2* into *Xenopus* embryos, as well as ectopic expression of *X-Delta-2*, led invariably to perturbations of segmentation and in the most severe cases abolished the segmental somite pattern (Jen *et al.*, 1997), suggesting that local and localized cell interactions mediated by the Notch signaling pathway in the presomitic mesoderm are essential for forming segment borders. Local cell interactions between Delta and Notch appear also to be essential for epithelialization of somites. Mouse embryos homozygous mutant for the *Dll1* gene, expressed in the presomitic mesoderm and in posterior halves of somites (Bettenhausen *et al.*, 1995), did not form fully epitheliarized somites but formed segmental subdivisions of the paraxial mesoderm (Hrabě de Angelis *et al.*, 1997), suggesting that the formation of epithelial somites is not essential for establishing segment borders. This is strongly supported by the phenotype of mouse embryos lacking the function of the bHLH gene *paraxis*, which is expressed in the paraxial mesoderm immediately preceding somite formation and in newly formed somites (Burgess *et al.*, 1995). *Paraxis* mutant embryos did not form epithelial somites; rather, the paraxial mesoderm was segmented into loose mesenchymal units (Burgess *et al.*, 1996). These findings suggest that the formation of epithelial somites is a consequence of segmentation of the paraxial mesoderm, but the establishment of segments is independent from the formation of somitic epithelia.

In summary, there is increasing evidence that a true prepattern is present in the "unsegmented" paraxial mesoderm. This prepattern may be gradually established and eventually fixed toward the cranial end of the presomitic mesoderm. Gene expression in distinct regions of the presomitic mesoderm or in periodic patterns correlates with somitic primordia, and the function of (at least some of) these genes appears to be essential for segmentation and/or normal somite formation. How these expression patterns are established, how the genes and their products interact, and how they ultimately generate borders between segments are questions that must be addressed to obtain further insights into the molecular mechanisms that underlie segmentation in the mesoderm of vertebrates.

C. Half-Somites and Borders

A subdivision of somites into anterior and posterior halves was first postulated by von Ebner, who described, in the sclerotomes of snake embryos, subtle clefts (von Ebner's fissures) separating the cranial and caudal sclerotomal portions (von Ebner, 1888). Since then, the significance of von Ebner's fissures has been the object of numerous disputes; their existence has been clearly confirmed by some (Keynes and Stern, 1984, 1988; Stern and Keynes, 1987) and refuted by others as histologic artifact (Baur, 1969; Verbout, 1976, 1985). It is now firmly established that somites are subdivided into anterior and posterior half-segments which differ in gene expression and cellular properties, and this subdivision is of high functional significance.

Motor axons emerging from the ventral spinal cord grow exclusively through the anterior portion of the adjacent sclerotome (Keynes and Stern, 1984). Similarly, neural crest cell migration between the neural tube and sclerotome cells, and the development of the neural crest cell–derived spinal ganglia and outgrowth of their sensory axons, are restricted to the anterior half of each segment (Rickmann et al., 1985; Bronner Fraser, 1986; Teillet et al., 1987), resulting in a regular segmented pattern of spinal ganglia and nerves. When parts of the segmental plate were transplanted with reversed anterior-posterior orientation prior to axonal outgrowth, without disturbing the neural tube, axonal outgrowth was confined to the posterior (formerly the anterior) halves of the somites that had developed from the grafts (Keynes and Stern, 1984), demonstrating that the segmented outgrowth of axons is governed by the adjacent paraxial mesoderm. When several young adjacent somites were removed from chick embryos and replaced by multiple rostral somite halves, wider spinal nerves, wider areas of neural crest cells, and wider nonsegmented dorsal root ganglia were observed. In contrast, spinal nerves, neural crest cells, and spinal ganglia were virtually absent when multiple caudal somite halves were grafted (Stern and Keynes, 1987; Kalcheim and Teillet, 1989), suggesting that the segmented pattern of axonal growth, neural crest cell migration, and development of dorsal root ganglia is caused by inhibitory influences in posterior-segment halves rather than by attractive or inductive properties of anterior halves. Indeed, a growth cone–collapsing activity has been demonstrated in posterior somite halves (Davies et al., 1990), and the presence of somites is not essential to promote axonal outgrowth from the spinal cord or neural crest cell migration (Tosney, 1988). However, rostral somite halves can stimulate neural crest cell proliferation, suggesting that "trophic" interactions may be important *in vivo* (Goldstein *et al.*, 1990). The essential patterning properties of paraxial mesoderm appear to reside in the sclerotome, since removal of the dermomyotome had no effect on the axonal pattern (Tosney, 1987). However, the determination of rostrocaudal compartments in somites is not restricted to the prospective sclerotome (i.e., ventral region). When the dorso-ventral axis of somites I and II was reversed, normal rostrocaudal patterned sclerotomes developed from the former dorsal region of the graft (Aoyama and Asamoto, 1988). Thus, in avian and presumably in mammalian embryos, the anterior-posterior alternation of properties in the sclerotome patterns spinal gan-glia and nerves. Also, in zebrafish embryos motor axons and migrating neural crest cells colocalize with the anterior sclerotome; however, in zebrafish the ablation of sclerotome did not affect the pattern of motor axons and dorsal root ganglia (Morin-Kensicki and Eisen, 1997). Based on these results, it was con-cluded that in vertebrates whose sclerotome constitutes a much smaller propor-tion of the somite and is not in contact with the neural tube (e.g., fish and some amphibians) (Kielbowna, 1981), interaction between the myotome and neuronal cells is important for patterning. Since the whole somite appears to contain the patterning information (Aoyama and Asamoto, 1988) and short-range signals appear to be important for the segmentation of the peripheral nervous system (Tosney, 1988), proximity to the neural tube may be the critical factor for deter-mining which differentiated cell types in the paraxial mesoderm pattern the nervous system (Morin-Kensicki and Eisen, 1997). Somites also impose lineage restrictions on cells in the spinal cord. When single cells in the spinal cord of chick embryos were injected with a fluorescent dye, clones derived from these cells spread across a line opposite to the adjacent intersomitic border but never crossed an invisible border opposite the middle of the adjacent somite (i.e., von Ebner's fissure) (Stern *et al.*, 1991). When somites adjacent to the site of dye injection were removed, clones no longer respected the border, indicating that the presence of the somites is required to maintain the lineage restriction in the spinal cord (Stern *et al.*, 1991).

Besides demonstrating that neuronal segmentation is governed by the somites, the anterior-posterior reversal of the segmental plate also showed that the subdi-vision into anterior and posterior compartments must be determined before somite formation (i.e., in the segmental plate) and that the polarity of the seg-ments cannot be reversed by surrounding tissues. When rostral and caudal somite halves of chick and quail embryos were placed microsurgically adjacent to each other *in ovo*, it was found that sclerotome cells from like half-somites (i.e., rostral–rostral or caudal–caudal) mixed, whereas cells from unlike somite halves (e.g., rostral–caudal) did not mix, and these cells were separated by a boundary (Stern and Keynes, 1987). Thus, anterior and posterior somite halves behave like cellular compartments whose cells differ in their adhesive properties. Based on these observations it has been suggested that the differences in the anterior and posterior somite halves not only serve to pattern the nervous system but also provide a mechanism to maintain the borders between segments after the formation of the mesenchymal sclerotome (Stern and Keynes, 1987). Studies on the function of the mouse *Dll1* gene provide strong genetic support for this idea (Hrabě de Angelis *et al.*, 1997). *Dll1* is expressed in the unsegmented paraxial mesoderm, overlapping with expression of mouse *Notch1*, and in newly formed somites transiently in the posterior halves (Del Amo *et al.*, 1992; Reaume *et al.*, 1992; Bettenhausen *et al.*, 1995). In *Dll1*-deficient mouse embryos, a primary metameric pattern is established in mesoderm, but the segments have no craniocaudal polarity and the segment borders are not maintained (Hrabě de Angelis *et al.*, 1997). Together with the studies on *X-Delta-2* function (Jen *et al.*, 1997), these data suggest that cell–cell communication mediated by the Notch signaling pathway is involved in the formation as well as the maintenance of segment borders.

Since confrontation of anterior with posterior somite compartments is sufficient for border formation (Stern and Keynes, 1987) and loss of detectable anterior-posterior polarity in somites of *Dll1* mutant embryos is accompanied by loss of segment borders (Hrabě de Angelis *et al.*, 1997), it is highly likely that polarity of somites is required to maintain segment borders and to prevent the free mixing of mesenchymal cells, which would destroy the periodic pattern of the somites. It is not clear, however, whether different cell states in half-segments are also a mechanism for establishing segment borders, as has been suggested based on theoretic considerations (Meinhardt, 1986b). In this model, the interaction between the different cranial and caudal compartments was proposed as a potential mechanism to form segment (somite) borders. Cells in the presomitic mesoderm would acquire different cell states due to the graded distribution of a morphogen which could be generated by local autocatalysis and long-range inhibition. Cells would initially oscillate between different cell states (e.g., anterior and posterior) before they acquired their final stable cell state. The confrontation of anterior (A) and posterior (P) cell states would generate a border, similar to the formation of parasegment boundaries in *Drosophila* (Meinhardt, 1986a; Ingham and Martinez Arias, 1992). If, however, only two cell states, A and P, existed, the placing of borders would be ambiguous; that is, borders could occur at each A–P confrontation, and an additional mechanism would be required to ensure that borders formed only at alternate confrontations, resulting in an ...AP/AP/AP... pattern. As a solution, Meinhardt proposed that a third cell state, S, alternates with A and P (i.e., ...APSAPSAPS...). In this case, borders could form unambiguously between two cell states; for example, border formation between P and S would result in the succession of .../SAP/SAP/... segments. However, a third cell state has not yet been described in somitic cells. Alternatively, Meinhardt proposed that the primary metameric pattern could be established by a system with a two-segment periodicity or by a combination of both mechanisms, analogous to the action of pair-rule genes in *Drosophila*. In this model, somites could alternate between two states, O (odd) and E (even); the confrontation of O and E cell states would generate intersomitic borders, and each O and E state would be subdivided further in an A and P region (Meinhardt, 1986). Pair-rule-like genes may function during somitogenesis and may be involved in the establishment of primary segment borders, as suggested by the expression of *Her1* and *Hairy2A* in somitic primordia within the presomitic mesoderm with a two-segment periodicity (Müller *et al.*, 1996; Jen *et al.*, 1997), but no data about the functions of these genes are yet available.

At present, there are not sufficient experimental data to conclusively support either of these models. Although in *Dll1*-mutant embryos segment polarity was lost, it is not known whether segment polarity was initially established in the presomitic mesoderm of *Dll1* mutants and lost after formation of distinct segmental somites. Therefore, at present, the analysis of this mutant does not allow the conclusion that different anterior-posterior cell states are involved in the initial border formation in the paraxial mesoderm. Further analysis of *Dll1*-mutant embryos as well as mutational analysis of pair-rule-like genes may eventually provide experimental evidence to support or refute these models for segmentation in vertebrates.

D. Cellular Oscillators

Cellular oscillations in the paraxial mesoderm were proposed as a means to synchronize groups of cells and to enable them to collectively respond to some segmentation signal and form a somite. These models have been developed to explain experimental results that demonstrated a considerable regulatory capability during somitogenesis in amphibian embryos and multiple periodic defects in somites after a single developmental perturbation in avian embryos, respectively. However, none of the models is consistent with the experimental results obtained in various vertebrates. Given that the molecular mechanisms underlying mesoderm segmentation are very likely conserved between different vertebrate classes, it is questionable that cellular oscillations, as proposed in these models, underlie somite formation.

When the size of *Xenopus* embryos was reduced at the blastula stage, qualitatively normal embryos of about two thirds of the regular size developed, with normal numbers and relative positions of somites. The somites in the manipulated embryos contained fewer cells than those in controls, indicating that the cell number per somite was adjusted to match the reduced body size (Cooke, 1975). Under the term "clock and wavefront," a model with two principle components has been developed to account for the sequential formation and proportional regulation of somite numbers observed in size-reduced amphibian embryos (Cooke and Zeeman, 1976). The first component of this model is a gradient that specifies the rates of development of cells in the paraxial mesoderm along the anterior-posterior axis toward segmentation (i.e., anterior cells acquire the ability to segment earlier than more posterior cells). This timing gradient is reflected directly in the visible progression of morphogenic processes along the anterior-posterior axis, giving it a wave-like appearance. The relative speed of the "wave," the time it takes to move along the entire body axis, is the same in different-sized embryos, provided regulation has occurred after experimental manipulation. In other words, a cell at 50% embryo length in a small embryo will be committed to execute a particular developmental program at the same time as a cell at 50% embryo length in a normal-sized embryo. The second component is a biochemical oscillator which operates synchronously throughout the paraxial mesoderm and switches with a species- and temperature-specific periodicity between alternate phases of inhibition and permissiveness (for somite formation). During the time the oscillator holds the paraxial mesoderm in the inhibited state, a certain number of cells (i.e., a region along the anterior-posterior axis) become determined to execute the appropriate steps to form a somite (as the wave passes through this region) but are prevented from actually executing the program by the inhibitory state of the oscillator. When the oscillator switches to the permissive state, all cells that have become competent to form a segment during the inhibitory phase of the cycle synchronously and rapidly execute the program as a group and form a somite. Thus, the smooth gradient of development rates (the progression of determination of paraxial mesoderm along the anterior-posterior axis) is transformed into a sequence of discrete steps by the interaction of a periodic change (the oscillator, or cellular clock) and the wave, which passes through each cell once. Since the oscillator frequency and the relative speed of the wave are identical in normal and small embryos in which regulation occurs, the same number of steps will be executed and the same number of segments produced (Cooke, 1977; Cooke, 1981).

Single regions of discrete abnormalities of segment borders, obtained after heat shocks were applied to amphibian (Cooke, 1978; Elsdale and Pearson, 1979) and zebrafish (Kimmel et al., 1988) embryos, were interpreted as experimental evidence supporting the clock and wavefront model (Elsdale et al., 1976; Cooke, 1978; Elsdale and Pearson, 1979; Pearson and Elsdale, 1979; Cooke and Elsdale, 1980). Heat shocks during early gastrulation resulted in grossly abnormal somitogenesis, and heat shocks at different neurula stages resulted in abnormalities at different positions along the anterior-posterior body axis, preceded and followed by normal segmentation anterior and posterior to the effected region (Elsdale and Pearson, 1979), suggesting that the temperature sensitivity moved back through the paraxial mesoderm ahead of visible somite formation. The observed lag between the heat shock and its visible consequence was interpreted to represent the time difference between the specification of somites and the actual segmentation process. Heat shocks at early stages were thought to affect the wave, whereas shocks at later stages transiently affected the coordination between wave and oscillator, allowing the coordination process to recover, and thus allowing segmentation to proceed normally posterior to the abnormal somites (Elsdale and Pearson, 1979; Cooke and Elsdale, 1980).

Proportional regulation of somite number has also been described for *amputated* (*am*) mouse mutant embryos. Although homozygous *am* embryos are shorter than their wild-type littermates, the same number of somites with reduced size was found (Flint *et al.*, 1978), similar to the proportionate regulation in size-reduced *Xenopus* embryos (Cooke, 1975). However, these results were interpreted as not compatible with the clock and wavefront model, due to the axial growth of mouse embryos, which occurs during all stages of somitogenesis (Flint *et al.*, 1978). Segmentation in *Xenopus* commences at the late neurula stage, when all presumptive mesoderm cells have invaginated through the blastopore lip. Thus, during early somitogenesis, the mesoderm which is subsequently subdivided along the anterior-posterior axis, is already present. In contrast, in amniotes, which gastrulate by node regression, somite formation commences while mesoderm is continuously being generated, and the length of the embryo steadily increases. It is hard to imagine how a mechanism could operate that regulates somite formation according to the total size of the embryo before the total size is established. In addition, it must be questioned how representative the proportionate regulation of somite size and number is, since in quail embryos regulation of somite number was observed in the absence of reduction of somite size (Veini and Bellairs, 1983), and severely size-reduced mouse embryos did not fully regulate somite number during compensatory growth (Gregg and Snow, 1983).

Heat shock experiments with chick embryos also led to results conflicting with the clock and wavefront model. When single heat shocks were applied to primitive streak–stage embryos, repeated anomalies in the pattern of somites were observed (Primmett *et al.*, 1988, 1989; Stern *et al.*, 1988). These periodic anomalies were usually confined to one or two adjacent segments separated by six to seven normal somites. Up to four repeated anomalies were obtained after a single heat shock, suggesting that the heat shock perturbed an oscillatory process which occurs with a periodicity of six to seven somite precursors in the segmental plate (Primmett *et al.*, 1988). Based on the existence of discrete regions of partial cell cycle synchrony in the segmental plate (Stern and Bellairs, 1984a; Primmett *et al.*, 1989); a cell cycle duration of approximately 9.5 hr, which corresponds to the development of six to seven somites in chick embryos; and periodic anomalies caused also by drugs inhibiting cell cycle progression, a model was advanced that correlated the cell cycle with segmentation (Keynes and Stern, 1988; Primmett *et al.*, 1989). This model suggests that (1) cells that are destined to segment together have some degree of cell cycle synchrony; (2) the determination of cells to segment together occurs one full cell cycle before segmentation; (3) cells gated during the determination step increase their adhesion to one another shortly before segmentation, regardless of their position within the segmental plate; and (4) the increase of adhesion takes place at a fixed time point during the cell cycle. The determination phase was suggested to comprise about one-seventh of the cell cycle time including M phase. Cells reaching the end of this period emit a signal that recruits the other cells in the vicinity and in the same time window of the cell cycle to increase adhesiveness, which one cell cycle later leads to somite formation. The periodic anomalies are caused by transient arrests of the clock, altering the size of the group of cells that will segment together. Since this arrest occurs multiple times in discrete cell populations within the segmental plate, periodic disturbances become visible during subsequent development (Keynes and Stern, 1988; Primmett *et al.*, 1989). This model explains the periodic abnormalities observed after a single heat shock, which were also, albeit rarely, observed in heat-shock-treated amphibian embryos (Elsdale *et al.*, 1976), as well as the autonomous character of somite formation in segmental plate explants or after reversal of the anterior-posterior axis (Menkes and Sandor, 1977). However, it is remarkable that heat shocks caused only unilateral anomalies of segmentation in the majority of cases (85%) and bilateral anomalies at the same anterior-posterior level in only few (7.5%) cases (Primmett *et al.*, 1988). The bilateral asymmetry of anomalies was explicitly mentioned as one of the specific differences between heat shock effects in amphibian and chick embryos and has been interpreted as evidence for independent somite formation on both sides of the embryo (Veini and Bellairs, 1986). However, given the strictly synchronous formation of somites on both sides of the midline, one would expect the paraxial mesoderm on both sides to be equally sensitive to a disturbance of the mechanism of segmentation.

E. Resegmentation

During the differentiation of the somites (see later), the sclerotome cells that remain lateral to the neural tube form the vertebral arches, sclerotome cells that migrate ventromedially toward and around the notochord give rise to the vertebral bodies and intervertebral discs, and ventrolateral sclerotome in the thoracic region gives rise to the ribs. The neural arch forms in the caudal half of the lateral sclerotome; the spinal ganglion occupies the cranial half. In the mature vertebral column, however, the neural arch is attached to the vertebral body in the cranial half and the spinal ganglion is located caudal to the arch. Similarly, the mesenchymal condensations of the ribs develop in the caudal halves of segments, whereas the heads of the ribs are positioned in the region between two vertebral bodies and the ribs attach to the cranial part of the mature vertebral body (Verbout, 1986; Theiler, 1988; Christ and Ordahl, 1995). How this change in topological relations is brought about was subject of a long-standing debate that centered on the concept of "Neugliederung" (new segmentation, or resegmentation) developed by Remak in the middle of the 19th century.

Based on observations of whole chick embryos, Remak (1855) proposed that primitive vertebral bodies form initially in register with the original somites, but then the primitive vertebral bodies fuse and new divisions take place between the levels of the original fissures. According to this concept, the vertebral bodies are derived from the fusion of the caudal half of one segment with the cranial half of the adjacent segment; hence, the paraxial mesoderm becomes resegmented. Subsequently, von Ebner (1888) observed in the lateral sclerotome intrasomitic fissures, which he interpreted as the morphologic manifestation of the future boundaries between the mature vertebrae, assuming that these fissures extended into the medial region. He added the functional argument that resegmentation is essential to ensure that the segmented epaxial muscles can connect the vertebrae (von Ebner, 1892). Arguments for and against resegmentation were brought forward for almost a century. These arguments were exclusively based on histologic studies, and were very thoroughly documented and reviewed by Verbout (1976). One major argument put forward against Remak's interpretation was that the medullary mesenchyme (i.e., the sclerotome surrounding the notochord) is not segmented to begin with and hence cannot "resegment" (Baur, 1969; Verbout, 1985). This argument refers to the homogeneous histologic appearance of the early ventromedial sclerotome ("axial sclerotome"), which transiently lacks morphologically obvious signs of segment borders or segmentation. Metamerism in the axial sclerotome becomes obvious with the formation of zones of higher cell density corresponding to the posterior segment halves. This results in a metameric arrangement of regions of low and high cell density, which in mouse embryos is readily observable at day 10 of development (Theiler, 1988). A second major argument put forward against resegmentation rejected the functional necessity for the realignment of sclerotome and myotome. Since the epaxial muscles do not attach to the vertebral bodies but to the vertebral processes, which occupy only the cranial regions of the mature vertebrae, the epaxial muscles do not actually span two adjacent vertebrae, and proper muscle attachment does not require a realignment of the medial sclerotome and myotome (Baur, 1969; Verbout, 1985). It was concluded, based on anatomic and histologic grounds, that the vertebral bodies develop in their definitive positions, eliminating the need for resegmentation (Baur, 1969; Verbout, 1976, 1985). However, there is now increasing experimental evidence that the intersomitic boundaries do not correspond to borders between vertebral bodies and that cells from two adjacent somites contribute to one vertebra.

When rostral or caudal somite halves of chick embryos were replaced by equivalent somite parts from quail embryos, sclerotome cells from both somite halves contributed to the vertebrae, ribs, and intervertebral discs and were found to colonize more than one vertebra (Stern and Keynes, 1987). Similarly, when whole single somites of chick embryos were replaced by somites from an equivalent, similarly staged quail embryo, quail cells contributed to one intervertebral disc and two adjacent vertebral bodies (Bagnall et al., 1988, 1989; Huang et al., 1996), and somitocoel cells of quail embryos that were grafted into the somitocoel of a thoracic chick somite colonized the rib derived from the operated somite and the rib from the adjacent cranial segment (Huang *et al.*, 1994, 1996). This suggests that cells from one somite can contribute to at least two vertebrae. Limitations to interpretation of these experiments are that the adhesive properties of quail and chick cells differ (Saunders, 1986) and that the transplantation procedure can lead to abnormal cell displacement (Newgreen *et al.*, 1986), which may influence the distribution of grafted cells. To circumvent these problems, the fates of cells of single somites were studied after labeling with DiI. In these experiments, sclerotome cells from one somite formed a coherent group of labeled cells surrounding the notochord, did not mix with medial sclerotome cells of adjacent somites, and subsequently contributed to the caudal portion of one vertebra, the intervertebral disc, and the cranial portion of the adjacent caudal vertebra (Bagnall, 1992). The unlabeled cells in these two adjacent vertebrae were presumed to be derived from different somites. However, there was a substantial contribution of unlabeled cells also to the vertebral halves colonized by labeled cells. This was presumed to result from dilution of the label, but it does not allow these results to be interpreted unambiguously. Compelling evidence for the contribution of cells from two adjacent somites to one vertebral body stems from a study analyzing the fates of somitic cells that were genetically labeled *in situ*. When cells from a single somite were infected with a *lacZ*-expressing, replication-defective, retroviral vector, the β-galactosidase-expressing cells (descendants of the labeled cells) were most abundant in the cranial half of one vertebra and the caudal half of the adjacent rostral vertebra (Ewan and Everett, 1992).

It appears then that cells from the ventral portion of the sclerotome of each somite migrate ventromedially and form an apparently homogenous mesenchyme around the notochord along the anterior-posterior axis. During their migration, prospective axial sclerotome cells derived from one somite stay together and do not mix significantly with cells from adjacent somites. Subsequently, cells from the caudal half of one somite and cells from the cranial half of the adjacent caudal somite form one vertebral body. Cells just anterior to the caudal half of a somite (i.e., in the middle of a somite) form the intervertebral discs that separate the vertebral bodies. Therefore, cells from two somites contribute to one vertebra, and the borders between vertebral bodies do not correspond to the primary segment borders generated during somite formation.

VI. Axial Specification

A. Embryologic Studies

As already mentioned and described in detail later, the principle differentiation pattern of all of the somites is very similar and initially results in dermotomal, myotomal, and sclerotomal compartments. However, during subsequent differentiation and morphogenesis, unique anatomic structures form, depending on the position along the anterior-posterior body axis. The sclerotome gives rise to vertebrae with shapes and appendages characteristic for their position, and the myotomes form distinct muscles and fiber types which develop in a site specific-manner.

Transplantation experiments in chick embryos demonstrated that the positional specification of somite occurs early during somitogenesis and that sclerotome and myotome are specified differentially (Kieny et al., 1972; Chevallier, Kieny, and Mauger, 1977, 1978; Chevallier, Kieny, Mauger, and Sengel, 1977; Chevallier, 1979; Wachtler et al., 1982; Chada et al., 1986; Butler et al., 1988). When cervical somites were replaced with somites from the trunk region, embryos developed that had rib-like structures in the cervical vertebral column. When thoracic somites were replaced by cervical somites, embryos developed that had no ribs on the operated sides of their thoracic vertebral columns (Kieny et al., 1972). In both cases the sclerotomes of the transplanted somites developed according to their developmental history, indicating that they had been stably instructed prior to their heterotopic grafting to form structures characteristic for their original axial position. The same result was obtained when segmental plate (presomitic mesoderm) cells from the prospective cervical or thoracic regions were used for heterotopic transplantations, demonstrating that prospective sclerotome cells are already positionally specified before somite formation and morphologically visible segmentation (Kieny et al., 1972; Wachtler et al., 1982). It is noteworthy that all these experiments examined the specification of groups of cells; single cells may behave differently. It is conceivable that maintenance of positional specification after grafting requires a minimal number of cells with the same positional specification, analogous to the "community effect" that has been shown for muscle cell differentiation in amphibians (Gurdon, 1988) and the mouse (Cossu et al., 1995). In contrast to sclerotome-derived tissue, muscles that developed from heterotopic transplants of presomitic mesoderm consistently formed the muscle types that strictly corresponded to the level of the new site occupied by the transplant after grafting (Chevallier, Kieny, and Mauger, 1977, 1978; Chevallier, Kieny, Mauger, and Sengel, 1977; Chevallier, 1979; Chada et al., 1986; Butler et al., 1988), demonstrating that no stable positional specification is imposed on myogenic progenitors. Since sclerotome and myotome are not specified before somite stage III (see later discussion), these findings imply that positional information is initially imposed on the whole somite. Once somitic cells differentiate, the positional specification is fixed in the sclerotome and lost or maintained labile in the myotome.

In mice, the expression of a Hoxa-7/lacZ transgene was used to monitor the maintenance of region-specific gene expression after heterotopic transplantations of transgenic somites into wild-type mouse embryos (Beddington et al., 1992). This transgene is expressed in all somites posterior to somite 13 (which corresponds to the second thoracic vertebra). When transgenic somites from axial levels that expressed the lacZ gene were grafted into anterior regions that normally did not express *lacZ*, expression was maintained. Similarly, when trans-genic somites from anterior regions that did not express *lacZ* were transplanted to posterior regions that normally expressed *lacZ*, the transgene was not activated. In both cases the transcriptional status of the *lacZ* gene was maintained in the transplanted somites according to their developmental history, suggesting that the regionalized expression of genes in somites is stably maintained once somites have formed, consistent with the transplantation results in chick embryos. This suggests that somites are able to "remember" their position along the anterior-posterior body axis, once this position has been specified, and stably maintain position-dependent gene expression.

B. Molecular Codes

There is now ample evidence that positional specification of the paraxial meso-derm requires the function of members of the *Hox* gene family (reviewed in Krumlauf, 1994). *Hox* genes are homologues of the homeotic selector genes of the *Drosophila* antennapedia and ultra-bithorax complexes, which are essential for specifying segment identity in insects (reviewed in Lawrence and Morata, 1994). Beyond sequence homology, the clustered arrangement of *Hox* genes, as well as principal aspects of their regulation and expression, which are of func-tional significance, have been conserved between *Drosophila* and vertebrates (reviewed in Kessel and Gruss, 1990; Krumlauf, 1994). In vertebrates, approx-imately 40 *Hox* genes are grouped together into four gene complexes, the *HoxA*, *HoxB*, *HoxC*, and *HoxD* complexes or clusters, which are located on different chromosomes (for review, see McGinnis and Krumlauf, 1992). *Hox* genes are expressed in ordered arrays of spatially restricted domains along the anterior-posterior body axis in derivatives of all three germ layers. Their activation during development correlates with their ordered arrangement in the *Hox* complex, a property referred to as colinearity. Two types of colinearity, spatial and temporal, are found in vertebrate embryos. Spatial colinearity refers to the correlation between the anterior expression boundaries of a *Hox* gene and its position in a complex: the rostral boundary of expression of a *Hox* gene is (with some excep-tions) more anterior than that of a 5′ neighboring gene and more posterior than that of a 3′ neighboring gene (Gaunt *et al.*, 1988; Dollé *et al.*, 1989; reviewed in Krumlauf, 1994). Temporal colinearity refers to the correlation between the time of activation of a *Hox* gene and its position in a complex: initiation of transcrip-tion of a *Hox* gene is earlier than that of a 5′ neighboring gene and later than that of a 3′ neighboring gene (Simeone *et al.*, 1990; Izpisúa-Belmonte *et al.*, 1991; Dekker *et al.*, 1993). Thus, genes at the extreme 3′ end of a *Hox* gene cluster are activated earliest and have the most anterior boundaries of expression. In addi-tion, *Hox* genes differ in their response to retinoic acid (RA), a molecule impli-cated in the specification of axes during development (Summerbell and Maden,

1990), depending on their position in the cluster. The 3' genes have the highest sensitivity to RA, and their expression can be readily induced, whereas 5' genes have a low sensitivity or are even repressed by RA (Simeone *et al.*, 1990; Boncinelli *et al.*, 1991; reviewed in Krumlauf, 1994).

The spatial and temporal colinearity of *Hox* gene activation results in the expression of unique combinations of *Hox* genes in defined groups of somites and their derivatives along the anterior-posterior axis (Gaunt, 1988; Holland and Hogan, 1988). This led to the suggestion that a *"Hox* code," a defined set of *Hox* genes expressed in distinct vertebral segments, specifies the identity of somites and sclerotome-derived structures (Kessel and Gruss, 1991; Kessel, 1992). In this concept the sequential activation of *Hox* genes along the anterior-posterior axis defines successively more posterior axial levels. The *Hox* code has proved to be a useful conceptual framework to correlate phenotypes and altered *Hox* gene expression. However, not all observed defects can be readily explained with simple alterations of the combinatorial expression of single *Hox* genes as a result of experimental manipulation. Relative levels of different *Hox* gene products in the same segments also appear to be important (Pollock *et al.*, 1992), and *Hox* genes may have additional functions, for example in the control of cell proliferation, which may explain the deletion of axial structures that has been observed in double mutants (Condie and Capecchi, 1994). In addition, there is substantial overlap in expression domains among members of paralogous *Hox* groups, which may lead to compensation of some aspects of the loss-of-function phenotypes.

The role for *Hox* genes in positional specification has been analyzed by interfering with or altering the expression of single *Hox* genes or by simultaneously perturbing the expression of more than one *Hox* family member. Interference with the expression of more than one *Hox* family member was achieved by RA treatment of embryos in utero and by mutation of genes required for the maintenance of transcriptional states of *Hox* genes (see later discussion). Administration of RA *in vivo* resulted in homeotic transformations (Kessel and Gruss, 1991; Kessel, 1992). Depending on the time of RA administration, anterior and posterior homeotic transformations were obtained. Administration of RA early during gastrulation, at a stage when *Hox* genes that are sensitive to activation by RA begin to be expressed in paraxial mesoderm cells, resulted in posterior transformations accompanied by anterior shifts of *Hox* gene expression boundaries. These alterations were explained by precocious activation and subsequent ectopic anterior expression of *Hox* genes. Administration late during gastrulation, at a stage when *Hox* genes that do not respond to or are repressed by RA begin to be expressed, resulted exclusively in anterior transformations accompanied by posterior shifts of *Hox* gene expression, which was explained by the repression of *Hox* gene activation (Kessel and Gruss, 1991; Kessel, 1992). When RA was applied at later stages, during sclerotome differentiation and migration, complex anterior and posterior transformations were observed, indicating that positional information can be respecified by RA (Kessel, 1992). However, the observed transformation could not be correlated with qualitative changes of *Hox* gene expression, although quantitative changes of gene expression could not be ruled out in these experiments (Kessel, 1992).

In *Drosophila*, after the region-specific expression of homeotic genes is initiated, the long-term maintenance of their spatially restricted stable expression states requires the function of global transcriptional regulators of the *trithorax* (*trx*) and *Polycomb* group (*Pc*-G) genes, which are thought to maintain active and repressed states of homeotic gene expression by modulation of chromatin structure (reviewed in Kennison and Tamkun, 1992; Kennison, 1993; Bienz and Muller, 1995; Simon, 1995). Mouse homologues of *trx* and *Pc*-G genes have been isolated, and, as with application of RA, mutations in such genes simultaneously alter region-specific expression of several *Hox* genes and result in homeotic transformations. Null mutations in the *Pc*-G homologues *Mel18* (Akasaka *et al.*, 1996) and *Bmi-1* (van der Lugt *et al.*, 1994), as well as hypomorphic alleles of *eed* (Schumacher *et al.*, 1996), resulted in posterior homeotic transformations (i.e., segments acquired the identity of more posterior segments), whereas overexpression of *Bmi-1* (Alkema *et al.*, 1995) resulted in anterior homeotic transformations (i.e., segments acquired the identity of more anterior segments). Posterior transformations were accompanied by a shift of anterior expression boundaries of *Hox* genes (Akasaka *et al.*, 1996), and in the anterior transformations *Hox* gene expression was shifted posteriorly (Akasaka *et al.*, 1996), consistent with a role for theses genes in maintaining repressed states of *Hox* gene expression. Mice lacking one copy of the *trx* homologue *Mll* developed bidirectional homeotic transformations accompanied by altered *Hox* gene expression in heterozygous mutant embryos, whereas in homozygous embryos expression of several *Hox* genes was completely abolished (Yu *et al.*, 1995), consistent with a role for *Mll* in maintaining active states of *Hox* gene expression. Whereas *trx* and *Pc*-G homologues appear to regulate *Hox* genes by modulating chromatin structure, the homeobox-containing gene *Cdx1* may directly activate *Hox* genes from different clusters, as was concluded from presence of Cdx1-binding sites in the promoter regions of *Hox* genes and anterior transformations accompanied by posterior shifts of *Hox* gene expression in mice lacking *Cdx1* (Subramanian *et al.*, 1995).

The role of individual *Hox* genes was analyzed in gain-of-function experiments (ectopic or deregulated *Hox* gene expression in transgenic mice) and loss-of-function experiments (generation of putative null alleles by homologous recombination; reviewed in Krumlauf, 1994). Homeotic transformations of vertebral segments were observed in both gain-of-function experiments (Balling *et al.*, 1989; Kessel *et al.*, 1990; Jegalian and Derobertis, 1992; McLain *et al.*, 1992; Pollock *et al.*, 1992; Gérard *et al.*, 1996) and loss-of-function experiments (e.g., Le Mouellic *et al.*, 1992; Lufkin *et al.*, 1992; Condie and Capecchi, 1993; Gendron Maguire *et al.*, 1993; Ramirez Solis *et al.*, 1993; Small and Potter, 1993; Charite *et al.*, 1995; Suemori *et al.*, 1995; Saegusa *et al.*, 1996; Zhao *et al.*, 1996), clearly demonstrating that alteration of a single *Hox* gene can change the positional specification in distinct regions of the sclerotomal derivatives of the paraxial mesoderm. However, not all gain-of-function or loss-of-function experiments resulted in homeotic transformations. Frequently, additional abnormalities were observed, consistent with the earlier notion that *Hox* genes may have additional functions, and gene dosage effects and functional compensation may complicate the interpretation of the phenotypes. In many cases loss-of-function mutations led to anterior transformations and gain-of-function experiments to posterior transformations, consistent with the rules established by Lewis for homeotic genes in *Drosophila* (Lewis, 1978). Often, the observed transformations occurred in the anterior expression domains of the genes under investigation, suggesting that *Hox* gene function for positional specification of sclerotome-derived structures is most critical in the region close to the anterior expression domains and thereby implying that the more posterior *Hox* genes are functionally dominant.

In summary, a role for *Hox* genes in positional specification of the sclerotomal derivatives of the paraxial mesoderm has firmly been established. As mentioned earlier, the axial identity of sclerotome cells cannot be respecified by grafting of somites to heterotopic positions, whereas the positional specification of myotome tissue strictly depends on the new position of such grafts. *Hox* gene expression typically is not maintained in the myotome of differentiating somites (Kessel, 1992), suggesting that the absence of *Hox* gene function in myotomes allows for positional respecification.

VII. Differentiation of Somites

Epithelial somites are only transient structures. Shortly after they emerge from the cranial end of the presomitic mesoderm, they progressively reorganize and differentiate, first into the sclerotome and dermomyotome, and subsequently into the myotome underlying the dermomyotome (Fig. 4). Their derivatives give rise to all parts of the vertebral column, the ribs, all skeletal muscles of the trunk and limbs, part of the connective tissue, endothelium, and smooth muscle cells. One exception is the nuclei pulposi of the intervertebral discs, which are notochord-derived (for reviews, see Verbout, 1985; Theiler, 1988; Christ and Wilting, 1992; Christ and Ordahl, 1995; Lassar and Münsterberg, 1996). Interestingly, formation of an epithelial somite is not required for the cytodifferentiation of dermomyotome, myotome, and sclerotome; these are formed in mouse mutant *paraxis* and *Dll1* embryos that lack epithelial somites (Burgess *et al.*, 1996; Hrabě de Angelis *et al.*, 1997). In contrast to the formation of somites from the presomitic mesoderm, which appears to be largely independent of tissue interactions, the differentiation of somites into sclerotome, dermomyotome, and myotome is controlled by and depends on specific signals from the surrounding tissues, which subdivide the somites in dorsoventral and mediolateral regions with particular

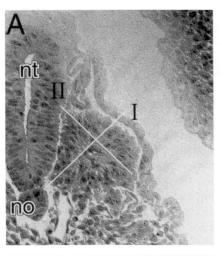

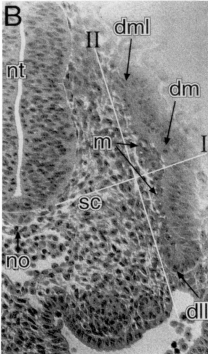

patterns of cell differentiation and cell fates (e.g., Watterson *et al.*, 1954; Avery *et al.*, 1956; Hall, 1977; Kenny Mobbs and Thorogood, 1987; Christ *et al.*, 1992; Rong *et al.*, 1992; Brand-Saberi *et al.*, 1996). However, angioblasts, the precursors of vascular endothelium, appear to develop in avian embryos throughout the somites and from somitocoel cells (Wilting *et al.*, 1995), although expression of a marker for early vascular endothelium cells appears to be restricted to the dorsolateral somite quadrant (Eichmann *et al.*, 1993). Endothelial cells derived from different parts of the somite populate correspondingly different regions of the embryo; that is, angioblasts from the dorsal and ventral somite halves colonize predominantly dorsal and ventral vessels, respectively (Wilting *et al.*, 1995), suggesting that the position in the somites determines to which vessels these cells contribute.

A. Sclerotome

The mesenchymal sclerotome forms at the ventral side of the somites (Christ and Wilting, 1992). Transplantation studies in avian embryos showed that the dorsoventral partition of somites is determined after they are formed. When the newly formed three caudal somites of quail embryos were grafted to chick embryos with reversed dorsoventral orientation, the two caudal graft somites (I and II) differentiated to normally situated sclerotomes and dermomyotomes, whereas the most cranial graft-derived somite (III) formed a dermomyotome medially and a sclerotome underlying the ectoderm (Aoyama and Asamoto, 1988). Similarly, heterotopic grafts of dorsal or ventral fragments from newly formed somites developed into muscle or cartilage, depending on the new position of the graft (Christ *et al.*, 1992, Aoyama, 1993). The formation of mesenchymal sclerotome cells is followed by their migration ventromedially and dorsomedially, and in the thoracic region also ventrolaterally. Sclerotome cells that migrate ventromedially toward and around the notochord give rise to the vertebral bodies and contribute

←——

Fig. 4   Somite compartments and differentiation. Transverse sections through a mouse day 10 embryo. A, somite at about the time of initiation of sclerotome formation in the posterior trunk region. B, differentiated somite in the upper trunk region. The epithelial dermomyotome (dm) overlies the myotome (m) dorsally; the ventral mesenchymal sclerotome (sc) fills the space between dermomyotome/myotome and neural tube. In both A and B, the dorsoventral (I) and mediolateral (II) axes are indicated by white lines. Note that the orientation of these axes corresponds to the original orientation of the somite and do not strictly correlate to the dorsoventral and mediolateral axes of the embryo at this stage, due to the changed topologic relations after neural tube formation and ventral curvature of the surface ectoderm. [In contrast, in the zebrafish, the terms dorsoventral and mediolateral are used with reference to the topologic relations in the embryo (van Eeden *et al.*, 1996)]. dml, dorsomedial lip of the dermomyotome; dll, dorsolateral lip of the dermomyotome; no, notochord; nt, neural tube.

to the intervertebral discs (Verbout, 1985; Christ and Wilting, 1992). Formation of the intervertebral discs is restricted to cells of the rostromedial sclerotome (Goldstein and Kalcheim, 1992). Sclerotome cells that migrate dorsally around the neural tube form the pedicles of the neural arches in the caudal segment halves. Removal of neural crest cells and consequent absence of spinal ganglia resulted in fused neural arches, suggesting that the spinal ganglia pattern the vertebral arches (Hall, 1977). However, the analysis of vertebral structures derived from grafts of multiple rostral-half and multiple caudal-half segments showed that only caudal sclerotome halves formed neural arches, suggesting that the cranial and caudal somite halves are committed to distinct developmental fates (Goldstein and Kalcheim, 1992). In the thoracic region, caudal sclerotome cells as well as cells from the somitocoel migrate ventrolaterally and form the ribs (Huang *et al.*, 1994, 1996).

Sclerotome formation is positively regulated by signals from the notochord and ventral spinal cord. In avian embryos, grafting of a supernumerary notochord laterally or medially to the unsegmented paraxial mesoderm suppressed myotomal differentiation on the vicinity of the graft and induced the formation of a loose mesenchyme and expansion of the expression domain of *Pax1*, a marker for sclerotomal cells (Brand Saberi *et al.*, 1993; Pourquie *et al.*, 1993). Similarly, the ventral neural tube (floor plate) induces sclerotome formation (Brand Saberi *et al.*, 1993; Pourquie *et al.*, 1993). Conversely, after experimental removal of the notochord prior to somite differentiation, or in mouse embryos carrying mutations that disrupt notochord development or integrity, expression of *Pax1* in the ventral part of the somite is reduced or lost (Dietrich *et al.*, 1993; Koseki *et al.*, 1993; Monsoro-Burq *et al.*, 1994; Ebensperger *et al.*, 1995). When the notochord was removed early, before it induced a floor plate, no sclerotome differentiation occurred at all, and myotomes fused at the midline (Monsoro-Burq *et al.*, 1994), suggesting that signals from the ventral midline tissues are essential to initiate sclerotomal differentiation. Subsequently, during vertebral development, the notochord induces chondrogenic differentiation in sclerotome cells (Grobstein and Holtzer, 1955; Avery *et al.*, 1956; Hall, 1977). However, dorsal grafts of notochords expressed chondrogenic differentiation of the processus spinosus, whereas they did not affect differentiation of the pedicles of the neural arches (Monsoro-Burq *et al.*, 1994), suggesting that cartilage differentiation of the sclerotome is differentially regulated in the dorsal and ventral parts of the vertebrae.

The ventralizing signals of the notochord and floor plate are mediated by *sonic hedgehog* (*shh*) (reviewed in Bumcrot and McMahon, 1995), a vertebrate homologue of the *Drosophila* segment polarity gene *Hedgehog*, which is expressed in the notochord and floor plate (Echelard *et al.*, 1993; Roelink *et al.*, 1994). Ectopic expression of *shh* in chick embryos was sufficient to mimic the effect of ectopic notochord or floor plate grafts on sclerotome differentiation (Johnson *et al.*, 1994). Likewise, in explants of mouse presomitic mesoderm, sclerotome differentiation was induced by *shh*-expressing cells (Fan and Tessier Lavigne,

1994), and, like the notochord, SHH repressed dermomyotome differentiation (Johnson *et al.*, 1994; Fan *et al.*, 1995). *In vitro* studies demonstrated that SHH can directly induce *Pax1* over a long range of more than 150 μm (Fan and Tessier Lavigne, 1994; Fan *et al.*, 1995). Direct contact with the surface ectoderm appears to be responsible for the inhibition of sclerotome differentiation in the dorsal somite (prospective dermomyotome) and may be required to counteract sclerotome induction throughout the whole somite by the long-range effect exerted by SHH. Consistently, tissue reconstitutions in organ culture showed that epithelium overlying somites in these cultures inhibits sclerotomal differentiation (Kenny Mobbs and Thorogood, 1987). In addition to the repression of sclerotome, contact of surface ectoderm with paraxial mesoderm induced expression of *Pax3*, *Pax7*, and *Sim1*, markers for dermomyotome, in explants of mouse paraxial mesoderm suggesting that the ectoderm is the source for inhibitory and stimulatory signals (Fan and Tessier Lavigne, 1994) (see later discussion).

Whereas SHH is sufficient to induce sclerotome *in vivo* and *in vitro*, it is not necessary for the initiation of sclerotome differentiation in mouse embryos. Embryos lacking *shh* function initially express *Pax1* in the sclerotomes but fail to maintain *Pax1* expression during subsequent development (Chiang *et al.*, 1996), suggesting that SHH is an essential maintenance signal for sclerotome and that the notochord provides additional signals that initiate ventralization of the somites. Consistent with a role of SHH as a maintenance signal, the continuous presence of SHH in somite explants is required for high levels of *Pax1* expression (Münsterberg *et al.*, 1995).

B. Dermomyotome

The differentiating sclerotome leaves behind an epithelial cap of cells, the dermomyotome. The dermomyotome is located dorsally under the surface ectoderm and gives rise to the myotome, which underlies the dermomyotome medially. The myotome gives rise to postmiotic myoblasts; the dermomyotome is the source of proliferating myoblasts and fibroblasts of the dorsal dermis (Christ *et al.*, 1983; Kaehn *et al.*, 1988). The medial part of the myotome gives rise to the epaxial intrinsic back muscles, whereas the lateral myotome contains the precursors of the hypaxial muscles of the ventral body wall and the skeletal muscles of the limbs (the latter only from somites that are at the levels of the developing limb buds) (Christ *et al.*, 1983; Ordahl and Le Douarin, 1992). The exact origin of the myotomal precursor cells from the dermomyotome is not completely clear. Histologic studies in the mouse (Ede and El-Gadi, 1986) and expression of early myogenic bHLH genes (*Myf5* and *Myod*, respectively) in the dorsomedial portion of the dermomyotome in mouse and avian embryos (Ott *et al.*, 1991; Pownall and Emerson, 1992; Smith, Kachinsky, *et al.*, 1994; Cossu *et al.*, 1996; Spörle *et al.*, 1996) suggested that cells emerging from the "dorsomedial lip" and migrating laterally on the medial surface of the dermomyotome are myogenic precursors. In mouse embryos, early expression of *Myf5* was also reported in the dorsolateral portion of the myotome. It was suggested that these cells at the lateral edge represent the precursors of the hypaxial muscles and may be specified independently from myotomal cells forming in the dorsomedial portion (Spörle *et al.*, 1996). Different specification of medial and lateral myogenic precursors is also supported by the differential activation of *Myf5* and *Myod* in the medial and lateral paraxial mesoderm by the neural tube and surface ectoderm, respectively (Cossu *et al.*, 1996). Myoblasts emanating from the dorsomedial (and dorsolateral) lip and migrating along the mediolateral axis would have to reorient, because myotome cells are arranged longitudinally. Such a reorientation has not yet been described. In contrast, histologic studies in chick embryos suggested that myotome formation starts at the craniomedial corner of the dermomyotome. Cells at the cranial edge elongate and then migrate and stretch toward the caudal somite border until they span the whole length of one segment. This process spreads laterally along the cranial edge, resulting initially in a triangular shape of growing myotomes (Kaehn *et al.*, 1988). Whereas this model conforms with the observed orientation of myotome cells, it leaves open the nature of the earliest myogenic gene–expressing cells in the medial and lateral regions of the dermomyotomes.

Like the sclerotome, dermomyotome differentiation requires cues from the surrounding tissues, as was demonstrated in numerous experiments in avian embryos. However, the interactions appear to be more complex, and conflicting experimental results for the roles of various adjacent tissues have been obtained. Dermomyotome differentiation appears to require signals from the overlying ectoderm (Gallera, 1966). Removal of the surface ectoderm prevented the expression of *gMHox*, a homeobox gene normally expressed in the dermatome (Kuratani *et al.*, 1994), and the contact of surface ectoderm with paraxial mesoderm was required in presegmental plate explants to initiate the expression of the dermomyotome markers *Pax7* and *Sim1* and to maintain or reinitiate expression of *Pax3* (Fan and Tessier Lavigne, 1994). Also, dorsal neural tube induced dermomyotomal gene expression in paraxial mesoderm explants, but, in contrast to the surface ectoderm, no direct contact between tissues was required (Fan and Tessier Lavigne, 1994). Consistent with the dermomyotome-promoting activity *in vitro*, ablation of the dorsal neural tube caused failure of dermomyotome differentiation *in vivo*. However, when the whole neural tube was removed, a central fused dermomyotome formed in the region of the operation (Spence *et al.*, 1996). This suggests that in the absence of the neural tube the surface ectoderm is sufficient to induce dermomyotome formation, whereas the presence of the ventral neural tube appears to antagonize this activity. In explant cultures, the presence of the notochord suppressed the induction of *Pax3*, *Pax7*, and *Sim1* expression by the dorsal neural tube (Fan and Tessier Lavigne, 1994). Similarly, grafting of a notochord lateral to the neural tube repressed dermomyotomal *Pax3* and *Pax7* expression in chick embryos (Goulding *et al.*, 1993, 1994), suggesting that signals from different tissues interact to regulate dermomyotome differentiation *in vivo*. However, reversal of the dorsoventral orientation of the neural tube or grafts of dorsal neural tube to the ventral somite caused dermomyotome formation in the ventral (sclerotomal) region despite the presence of the notochord (Spence *et al.*, 1996), suggesting that the dominance of dorsal or ventral signals is influenced by experimental conditions (e.g., the topologic relation of tissues after grafting) or that the antagonistic action of these signals is modified by as yet unidentified interactions.

Myogenic differentiation also requires the interaction of signals from various tissues (reviewed in Cossu *et al.*, 1996; Lassar and Münsterberg, 1996). As pointed out earlier, dorsomedial and dorsolateral myotome cells have different fates. They also appear to be specified independently and by different tissue interactions and/or signals. It is important to note that in most *in vitro* and tissue recombination experiments the induction of myogenesis in general was analyzed but it was not possible to discriminate between apaxial and hypaxial myogenic differentiation. The results from these experiments must be interpreted in the context of *in vivo* studies to allow for conclusions with respect to the distinction between epaxial and hypaxial myogenesis. Muscle-inducing activities have been found in midline structures (notochord/neural tube complex) and in the dorsal (surface) ectoderm. In cultures of paraxial mesoderm, the presence of the neural tube significantly increased myogenesis (e.g., Avery *et al.*, 1956; Vivarelli and Cossu, 1986; Kenny Mobbs and Thorogood, 1987). Similarly, the combination of notochord with somite explants (Buffinger and Stockdale, 1994; Stern and Hauschka, 1995) and dorsal ectoderm (Cossu *et al.*, 1996) was sufficient to induce myogenesis. Somites taken from different positions along the anterior-posterior axis required different interactions to initiate myogenesis, consistent with the observations in neural tube ablation experiments (see later discussion). Explanted young somites (I–III) required signals from the ventral midline (either notochord or floor plate) in conjunction with signals from the neural tube to express markers for myogenesis; more mature somites (IV–VI) required only signals from the neural tube, and somites XII and beyond did not require extrinsic signals at all (Münsterberg and Lassar, 1995; Stern and Hauschka, 1995). This suggests that transient midline signals induce a stable change in somitic cells, allowing for subsequent dorsal signals alone to promote myogenic differentiation. It must be noted that the region of the neural tube that contains myogenesis-promoting activity *in vitro* is somewhat controversial. Strong myogenic stimulatory activity was described for the ventral neural tube without the floor plate (Buffinger and Stockdale, 1995), and strong stimulatory activity (Stern *et al.*, 1995; Spence *et al.*, 1996) as well as inhibitory activity (Buffinger and Stockdale, 1995) was attributed to the dorsal neural tube, the reasons for these differences not being clear.

Signals from the midline structures appear to be essential for epaxial myogenesis, since ablation of the neural tube from chick embryos impaired the development of the epaxial myotomal muscles (Christ *et al.*, 1992; Rong *et al.*, 1992; Bober *et al.*, 1994), and epaxial but not hypaxial muscle development is severely impaired in the mouse mutation *open brain* (*obp*), most likely because of the absence of dorsal neural tube signals (Spörle *et al.*, 1996). In chick embryos, the ability to form epaxial muscle depended on the period for which paraxial mesoderm has been in contact with the neural tube prior to its ablation. Removal of the neural tube from the unsegmented paraxial mesoderm resulted in complete loss of epaxial muscles at these axial levels (Bober *et al.*, 1994), whereas somites that had been in contact with the neural tube for more than 10 hr formed epaxial myoblasts (Rong *et al.*, 1987). This suggests that once the neural tube has provided the necessary signals for myogenesis, myoblasts can be maintained for an extended period of time. However, myogenic precursor cells were found in the centrally fused dermomyotomes of embryos whose complete neural tube had been removed early, at the level of the segmental plate, suggesting either that the neural tube is not essential to initiate myogenesis but stimulates the survival or maintenance of further differentiation of myotomal cells (Bober *et al.*, 1994) or that the myogenic precursors that developed in the absence of the neural tube represented hypaxial myoblasts whose specification appears to be independent from neural tube signals (Spörle *et al.*, 1996) (see later discussion). Consistent with the ability to stimulate myogenic differentiation *in vitro*, removal of the notochord from the segmental plate resulted in impaired myogenesis in the medially fused somites that developed in the operated region. In these fused somites, *Pax3* and *Pax7* expression domains were shifted ventrally, whereas *Myod* expression was completely lost (Goulding *et al.*, 1994), indicating that signals from the notochord are required for myogenic differentiation. The lack of detectable *Myod* expression may be attributed to the ventral location of the somites, which could prevent interactions between the somitic mesoderm and surface ectoderm. Surprisingly, grafts of a supernumerary notochord between the neural tube and paraxial mesoderm repressed *Myod* expression in the paraxial mesoderm adjacent to the graft (Goulding *et al.*, 1994). This suggests that the notochord can promote and inhibit myogenic differentiation, which may depend on additional signals from or interactions with the surrounding tissues. Surgical ablation and grafting experiments in quail embryos showed that the notochord is necessary to induce myogenic bHLH genes in dorsomedial myotomal cells, and neural tube signals subsequently contribute to the maintenance of high levels of *QmyoD* and *Qmyf5* expression. Removal of the lateral plate mesoderm led to the lateral expansion of the *QmyoD* expression domain, indicating that lateral plate signals restrict myogenic differentiation to the medial region (Pownall *et al.*, 1996).

The stimulatory signal of the notochord is probably mediated by SHH, since ectopic expression of *shh* induced the expansion of the *Myod* expression domain in the chick (Johnson *et al.*, 1994) and in zebrafish embryos (Hammerschmidt, Bitgood, *et al.*, 1996), and SHH substituted for the activity of ventral midline tissues (notochord and floor plate) in explant cultures (Münsterberg *et al.*, 1995). The different activities of notochord grafts and SHH in some of the experiments could have various explanations. First, the notochord could provide additional signals or induce surrounding tissues to produce them. Second, the notochord could alter myogenesis-promoting signals of the neural tube (see later). Third, notochord grafts alter the spatial relations between signaling and responding tissues, which could result in insufficient signals reaching the target cells in the paraxial mesoderm. *Wnt1*, *Wnt3*, or *Wnt4*, all of which are expressed in the dorsal neural tube, substituted for the neural tube signals in explant cultures (Münsterberg *et al.*, 1995). This suggests that combinatorial signals from SHH and WNT proteins are also required for axial myogenesis *in vivo*. The synergistic action of SHH (from the ventral midline) and WNT (from the dorsal neural tube) could also explain an apparent paradox: in explant cultures, *Pax1* was induced at lower SHH concentrations than *Myod* (Münsterberg *et al.*, 1995), suggesting that different threshold levels may elicit differential responses of the paraxial meso- derm to SHH signals, but *in vivo* myogenic cells develop further away from the sources of SHH than the sclerotome, which would be more consistent with sclerotome induction at high concentrations of SHH close to the axial midline. Mouse embryos lacking SHH initiate myogenesis, suggesting that *shh* function is not to initiate but to enhance myogenic gene expression (Chiang *et al.*, 1996); this suggests that myogenic differentiation promoted by SHH may be attributable to the expansion of already existing myogenic precursors rather than the induc- tion of myogenic progenitor cells. A role for SHH in the expansion of myogenic precursors was also suggested in zebrafish embryos. In this teleost, different *hedgehog* family members expressed in the notochord appear to be required at different stages of myogenic specification, leading to distinct muscle lineages (Currie and Ingham, 1996).

In contrast to axial myogenesis, myogenic differentiation in the lateral myo- tome appears to be independent from axial tissues. The differentiation of hypax- ial and limb myoblasts was not affected in embryos with ablated neural tubes (Rong *et al.*, 1992) or in mouse *opb* mutant embryos (Spörle *et al.*, 1996), indicating that dorsal neural tube–derived signals are not essential for the differ- entiation of myogenic precursor cells in the lateral region of the myotome and their migration to the somatopleura and limbs. During their migration, hypaxial myogenic precursors express high levels of *Pax3*, a marker for dermomyotome and early myotome (Williams and Ordahl, 1994; Cossu *et al.*, 1996), but none of the myogenic bHLH genes (Tajbakhsh and Buckingham, 1994; Williams and Ordahl, 1994). Separation of the paraxial mesoderm from the lateral mesoderm resulted in down regulation of *Pax3* and concomitant activation of *Myod* and *Myf5* in the lateral myotome. Conversely, when lateral plate mesoderm was grafted between the neural tube and paraxial mesoderm, *Pax3* was strongly up regulated in the medial myotome, accompanied by repression of *Myf5* and *Myod* (Pourquie *et al.*, 1995). This suggests that signals from the lateral mesoderm maintain myogenic cells in the lateral myotome compartment as undifferentiated progenitors. Similarly, intermediate mesoderm inhibited myogenic differentiation when placed next to medial and lateral myotome explants *in vitro* (Gamel *et al.*, 1995). These experiments suggested that *Pax3* may mediate the repression or delay of myogenic differentiation in the lateral (hypaxial) myotome. However, subsequent studies demonstrated that *Pax3* acts upstream of *Myod*, in concert with *Myf5*, to active *Myod* expression (Maroto *et al.*, 1997; Tajbakhsh *et al.*, 1997), indicating that other factors mediate the repression or delay of lateral myogenesis. Grafts of BMP4-expressing cells mimicked the effect of lateral plate mesoderm and induced expression of *Sim1*, a bHLH gene specifically expressed in the lateral somite derivatives and migrating hypaxial myoblasts (Pourquie *et al.*, 1996), suggesting that *Sim1* is involved in the regulation of myogenesis in hypaxial muscle. The effects of the lateral plate mesoderm were counteracted by signals from the neural tube, suggesting that along the mediolateral axis somites are patterned by antagonistic signals, in this case from the neural tube and lateral plate (Pourquie *et al.*, 1996).

Signals from the neural tube are also essential for the differentiation of the dermal somite compartment, as demonstrated in chick embryos. Insertion of an impermeable membrane between neural tube and dermomyotome prevented dorsal dermis formation. Treatment of the membrane with neurotrophin-3 (NT-3) completely restored dermis differentiation, and in explant cultures NT-3 stimulated epithelial mesenchymal transition of dermatome cells. NT-3-blocking antibodies disrupted early dermis differentiated in embryos, indicating that NT-3 is a neural tube signal required for dermatome differentiation (Brill *et al.*, 1995).

In summary, there is now mounting experimental evidence that various combinatorial stimulatory and antagonistic signals regulate somite differentiation, and the molecular mechanisms underlying this regulation are emerging (Fig. 5). However, some studies report conflicting effects of tissues and signals or their precise localization that have yet to be resolved. Similarly, the signals that initiate various differentiation pathways remain to be identified.

VIII. Summary and Conclusions

We are still far from understanding "somitogenesis" as a whole, but there is an emerging picture of the tissue interactions and molecular mechanisms that underlie and govern various aspects of this essential multistep patterning process in vertebrates. The ability to form segmental units appears to be a property specific to the paraxial mesoderm (as opposed to lateral or limb mesoderm), and this ability is probably acquired during early development, when paraxial mesoderm is specified and emerges from the primitive streak. Signaling molecules expressed in the primitive streak and tail bud are prime candidates involved in specifying paraxial (as well as other mesodermal) fates. Increasing levels of

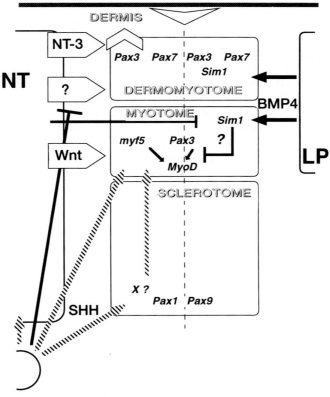

Fig. 5 Combinatorial and antagonistic signals regulate somite differentiation. Long-range signals from the ventral midline (SHH), dorsal neural tube (Wnt and NT-3), and lateral mesoderm (BMP4), as well as short-range signals from the surface ectoderm, regulate the differentiation of somitic cells according to their position along the dorsoventral and mediolateral axes. For details, see text. Arrows indicate stimulatory effects, lines ending with bars indicate inhibitory effects. NT, neural tube; LP, lateral plate.

signaling molecules may be required in posterior regions of the embryo, and combinatorial signals may be essential to specify the paraxial mesoderm along the entire anterior-posterior axis. However, most of the pivotal signals, and the ways in which they are integrated and interact, remain enigmatic. Once the paraxial mesoderm is formed, segmentation proceeds largely without the requirement for continuous interactions with surrounding tissues. Somitomeres represent a morphologic pattern in the mesenchymal presomitic mesoderm, but their significance for somite formation is unclear. Molecular patterns are established in the presomitic mesoderm and probably are of functional significance. Cell interactions within the paraxial mesoderm appear to be involved in forming segment borders and ensuring their maintenance during subsequent differentiation of somites. These interactions are, at least in part, mediated by components of the conserved Notch signaling pathway, which may have multiple functions during somitogenesis. Epithelial somites are clearly a result of segmentation, but epithelialization is not the mechanism to form segments, supporting the idea that the basic mechanisms that govern segmentation in the mesoderm of vertebrates are very similar in different species despite divergent types of resulting segments (i.e., epithelial somites versus rotated myotomes). Concomitantly with segmentation, segment polarity and positional specification are established. How these processes are linked to, and depend on, each other is unknown, as is how they are regulated and how segmentation is coordinated on both sides of the neural tube. In contrast to early patterning in the presomitic mesoderm, patterning of the mature somites during their subsequent differentiation is the result of extensive tissue interactions. Virtually all tissues in close proximity to somites provide signals that are involved in induction or inhibition of particular differentiation pathways, but how these pathways are initiated is less clear. Some of the molecules mediating inductive signals and tissue interactions are known, and a growing number of candidate genes are potentially involved in regulating various steps of somitogenesis. The roles of these genes have yet to be analyzed. In addition, the molecular genetic analysis of mutations affecting somitogenesis, which were collected in the mouse and more recently in the zebrafish (Driever *et al.*, 1996; Haffter *et al.*, 1996; van Eeden *et al.*, 1996), promises to add important new insights into this process. Much remains to be done, but the tools are at hand to provide further understanding of the molecular mechanisms underlying somitogenesis.

Acknowledgments

We thank Rosa Beddington, Tim O'Brien, Tom Gridley, Michael Kessel, Janet Rossant, and Ralf Spörle for their critical comments on the manuscript, which helped to improve it, and Yumiko Saga and Sally Dunwoodie for communicating unpublished work. This work was funded by National Science Foundation Grant IBN 9506156 to A. G.

References

Akasaka, T., Kanno, M., Balling, R., Mieza, M. A., Taniguchi, M., and Koseki, H. (1996). A role for mel-18, a Polycomb group–related vertebrate gene, during the anteroposterior specification of the axial skeleton. *Development* **122**, 1513–1522.

Alkema, M. J., van der Lugt, N. M., Bobeldijk, R. C., Berns, A., and van Lohuizen, M. (1995). Transformation of axial skeleton due to overexpression of bmi-1 in transgenic mice. *Nature* **374**, 724–727.

Ang, S. L., and Rossant, J. (1994). HNF-3β is essential for node and notochord formation in mouse development. *Cell* **78**, 561–574.

Aoyama, H. (1993). Developmental plasticity of the prospective dermatome and the prospective sclerotome region of an avian somite. *Dev. Growth Differ.* **35**, 507–519.

Aoyama, H., and Asamoto, K. (1988). Determination of somite cells: Independence of cell differentiation and morphogenesis. *Development* **104**, 15–28.

Artavanis Tsakonas, S., Matsuno, K., and Fortini, M. E. (1995). Notch signaling. *Science* **268**, 225–232.

Avery, G., Chow, M., and Holtzer, H. (1956). An experimental analysis of the development of the spinal column. V. Reactivity of chick somites. *J. Exp. Zool.* **132**, 409–423.

Bagnall, K. M. (1992). The migration and distribution of somite cells after labelling with the carbocyanine dye, DiI: The relationship of this distribution to segmentation in the vertebrate body. *Anat. Embryol. (Berl.)* **185**, 317–324.

Bagnall, K. M., Higgins, S. J., and Sanders, E. J. (1988). The contribution made by a single somite to the vertebral column: Experimental evidence in support or resegmentation using the chick–quail chimaera model. *Development* **103**, 69–85.

Bagnall, K. M., Higgins, S. J., and Sanders, E. J. (1989). The contribution made by cells from a single somite to tissues within a body segment and assessment of their integration with similar cells from adjacent segments. *Development* **107**, 931–943.

Bagnall, K. M., Sanders, E. J., and Berdan, R. C. (1992). Communication compartments in the axial mesoderm of the chick embryo. *Anat. Embryol. (Berl.)* **186**, 195–204.

Balling, R., Mutter, G., Gruss, P., and Kessel, M. (1989). Craniofacial abnormalities induced by ectopic expression of the homeobox gene Hox-1.1 in transgenic mice. *Cell* **58**, 337–347.

Baur, R. (1969). Zum Problem der Neugliederung der Wirbelsäule. *Acta Anat. (Basel)* **73**, 321–356.

Beddington, R. S. (1982). An autoradiographic analysis of tissue potency in different regions of the embryonic ectoderm during gastrulation in the mouse. *J. Embryol. Exp. Morphol.* **69**, 265–285.

Beddington, R. (1987). Isolation, culture and manipulation of post-implantation mouse embryos. In "Mammalian Development: A Practical Approach" (M. Monk, Ed.), pp. 43–69. IRL Press, Oxford.

Beddington, R. S., Puschel, A. W., and Rashbass, P. (1992). Use of chimeras to study gene function in mesodermal tissues during gastrulation and early organogenesis. *Ciba Found. Symp.* **165**, 61–74; discussion, 67–74.

Beddington, R. S., and Smith, J. C. (1993). Control of vertebrate gastrulation: Inducing signals and responding genes. *Curr. Opin. Genet. Dev.* **3**, 655–661.

Bellairs, R. (1963). The development of somites in the chick embryo. *J. Embryol. Exp. Morphol.* **11**, 697–714.

Bellairs, R. (1985). A new theory about somite formation in the chick. *Prog. Clin. Biol. Res.* **171**, 25–44.

Bellairs, R., and Sanders, E. J. (1986). Somitomeres in the chick tail bud: An SEM study. *Anat. Embryol. (Berl.)* **175**, 235–240.

Bellairs, R., and Veini, M. (1984). Experimental analysis of control mechanisms in somite segmentation in avian embryos. II. Reduction of material in the gastrula stages of the chick. *J. Embryol. Exp. Morphol.* **9**, 183–200.

Bettenhausen, B., Hrabě de Angelis, M., Simon, D., Guenet, J. L., and Gossler, A. (1995). Transient and restricted expression during mouse embryogenesis of Dll1, a murine gene closely related to *Drosophila* Delta. *Development* **121**, 2407–2418.

Bienz, M., and Muller, J. (1995). Transcriptional silencing of homeotic genes in *Drosophila*. *Bioessays* **17**, 775–784.

Bober, E., Brand Saberi, B., Ebensperger, C., Wilting, J., Balling, R., Paterson, B. M., Arnold, H. H., and Christ, B. (1994). Initial steps of myogenesis in somites are independent of influence from axial structures. *Development* **120**, 3073–3082.

Boncinelli, E., Simeone, A., Acampora, D., and Mavillio, F. (1991). *HOX* gene activation by retinoic acid. *Trends Genet.* **7**, 329–334.

Bonnerot, C., and Nicolas, J. F. (1993). Clonal analysis in the intact mouse embryo by intragenic homologous recombination. *C. R. Acad. Sci. III* **316**, 1207–1217.

Brand-Saberi, B., Ebensperger, C., Wilting, J., Balling, R., and Christ, B. (1993). The ventralizing effect of the notochord on somite differentiation in chick embryos. *Anat. Embryol. (Berl.)* **188**, 239–245.

Brand-Saberi, B., Wilting, J., Erbensperger, C., and Christ, B. (1996). The formation of somite compartments in the avarian embryo. *Int. J. Dev. Biol.* **40**, 411–420.

Brill, G., Kahane, N., Carmeli, C., von Schack, D., Barde, Y. A., and Kalcheim, C. (1995). Epithelial-mesenchymal conversion of dermatome progenitors requires neural tube–derived signals: Characterization of the role of Neurotrophin-3. *Development* **121**, 2583–2594.

Bronner Fraser, M. (1986). Analysis of the early stages of trunk neural crest migration in avian embryos using monoclonal antibody HNK-1. *Dev. Biol.* **115**, 44–55.

Brustis, J., and Gipoulou, J. (1973). Potentialites d'organisation et de differenciation du mesoderm somitique non segmente associe aux tissus axiaux (corde dorsale et tube nerveux) chez les amphibiens anoures. *C.r. hebd. Seanc. Acad. Sci. Paris* **276**, 85–88.

Buffinger, N., and Stockdale, F. E. (1994). Myogenic specification in somites: Induction by axial structures. *Development* **120**, 1443–1452.

Buffinger, N., and Stockdale, F. E. (1995). Myogenic specification of somites is mediated by diffusible factors. *Dev. Biol.* **169**, 96–108.

Bumcrot, D. A., and McMahon, A. P. (1995). Somite differentiation: Sonic signals somites. *Curr. Biol.* **5**, 612–614.

Burgess, R., Cserjesi, P., Ligon, K. L., and Olson, E. N. (1995). Paraxis: A basic helix-loop-helix protein expressed in paraxial mesoderm and developing somites. *Dev. Biol.* **168**, 296–306.

Burgess, R., Rawls, A., Brown, D., Bradley, A., and Olson, E. (1996). Requirement of the *paraxis* gene for somite formation and muscoskeletal patterning. *Nature* **384**, 570–573.

Butcher, E. O. (1929). The development of the somites in the white rat (*Mus novegicus albinus*) and the fate of the myotome, neural tube and gut in the tail. *Am. J. Anat.* **44**, 381–424.

Butler, J., Cosmos, E., and Cauwenbergs, P. (1988). Positional signals: Evidence for a possible role in muscle fibre-type patterning of the embryonic avian limb. *Development* **102**, 763–772.

Butros, J. (1967). Limited axial structures in nodeless chick blastoderms. *J. Embryol. Exp. Morphol.* **17**, 119–130.

Catala, M., Teillet, M.-A., and Le Douarin, N. M. (1995). Organization and development of the tail bud analyzed with the quail–chick chimera system. *Mech. Dev.* **51**, 51–65.

Chada, K., Magram, J., and Constantini, F. (1986). An embryonic pattern of expression of a human fetal globin gene in transgenic mice. *Nature* **319**, 685–689.

Charite, J., de Graaff, W., and Deschamps, J. (1995). Specification of multiple vertebral identities by ectopically expressed Hoxb-8. *Dev. Dyn.* **204**, 13–21.

Cheney, C. M., and Lash, J. W. (1984). An increase in cell–cell adhesion in the chick segmental plate results in a meristic pattern. *J. Embryoyl. Exp. Morphol.* **79**, 1–10.

Chevallier, A. (1979). Role of the somitic mesoderm in the development of the thorax in bird embryos. II. Origin of thoracic and appendicular musculature. *J. Embryol. Exp. Morphol.* **49**, 73–88.

Chevallier, A., Kieny, M., and Mauger, A. (1977). Limb-somite relationship: Origin of the limb musculature. *J. Embryol. Exp. Morphol.* **41**, 245–258.

Chevallier, A., Kieny, M., and Mauger, A. (1978). Limb-somite relationship: Effect of removal of somitic mesoderm on the wing musculature. *J. Embryol. Exp. Morphol.* **43**, 263–278.

Chevallier, A., Kieny, M., Mauger, A., and Sengel, P. (1977). Developmental fate of the somitic mesoderm in the chick embryo. *In* "Vertebrate Limb and Somite Morphogenesis." (D. A. Ede, J. R. Hinchliffe, and M. Balls, Eds.), pp. 421–423. Cambridge University Press, Cambridge.

Chiang, C., Litingtung, Y., Lee, E., Young, K. E., Corden, J. L., Westphal, H., and Beachy, P. A. (1996). Cyclopia and defective axial patterning in mice lacking Sonic hedgehod gene function. *Nature* **383,** 407–413.

Chitnis, A., Henrique, D., Lewis, J., Ish Horowicz, D., and Kintner, C. (1995). Primary neurogenesis in *Xenopus* embryos regulated by a homologue of the *Drosophila* neurogenic gene *delta*. *Nature* **375,** 761–766.

Christ, B. (1970). Experimente zur Lageentwicklung der Somiten. *Verh. Anat. Ges.* **64,** 555–564.

Christ, B., Brand Saberi, B., Grim, M., and Wilting, J. (1992). Local signalling in dermomyotomal cell type specification. *Anat. Embryol. (Berl.)* **186,** 505–510.

Christ, B., Jacob, H. J., and Jacob, M. (1972). Experimentelle Untersuchungen zur Somitenentstehung beim Hühnerembryo. *Z. Anat. Entwickl.-Gesch.* **138,** 82–97.

Christ, B., Jacob, H. J., and Jacob, M. (1974a). Die Somitogenese beim Huhnerembryo: Zur Determination der Segmentierungsrichtung. *Verh. Anat. Ges.* **68,** 573–579.

Christ, B., Jacob, H. J., and Jacob, M. (1974b). Ueber den Ursprung der Flugmuskulatur: Experimentelle Untersuchungen an Wachtel- und Huehnerembryonen. *Experientia* **30,** 1446–1448.

Christ, B., Jacob, M., and Jacob, H. J. (1983). On the origin and development of the ventrolateral abdominal muscles in the avian embryo: An experimental and ultrastructural study. *Anat. Embryol. (Berl.)* **166,** 87–101.

Christ, B., and Ordahl, C. P. (1995). Early stages of chick somite develoment. *Anat. Embryol. (Berl.)* **191,** 381–396.

Christ, B., and Wilting, J. (1992). From somites to vertebral column. *Ann. Anat.* **174,** 23–32.

Cockroft, D. L. (1990). Dissection and culture of postimplantation embryos. *In* "Postimplantation Mammalian Embryos: A Practical Approach" (A. J. Copp and D. L. Cockroft, Eds.), pp. 15–40. Oxford University Press, Oxford.

Condie, B. G., and Capecchi, M. R. (1993). Mice homozygous for a targeted disruption of Hoxd-3 (Hox-4.1) exhibit anterior transformations of the first and second cervical vertebrae, the atlas and the axis. *Development* **119,** 579–595.

Condie, B. G., and Capecchi, M. R. (1994). Mice with targeted disruptions in the paralogous genes hoxa-3 and hoxd-3 reveal synergistic interactions. *Nature* **370,** 304–307.

Conlon, F. L., Lyons, K. M., Takaesu, N., Barth, K. S., Kispert, A., Herrmann, B., and Robertson, E. J. (1994). A primary requirement for nodal in the formation and maintenance of the primitive streak in the mouse. *Development* **120,** 1919–1928.

Conlon, R. A., Reaume, A. G., and Rossant, J. (1995). *Notch1* is required for the coordinate segmentation of somites. *Development* **121,** 1533–1545.

Cooke, J. (1975). Control of somite number during morphogenesis of a vertebrate, *Xenopus laevis. Nature* **254,** 196–199.

Cooke, J. (1977). The control somite number during amphibian development: Models and experiments. *In* "Vertebrate Limb and Somite Morphogenesis" (D. A. Ede, J. A. Hinchliffe, and M. Balls, Eds.), pp. 434–448. Cambridge University Press, Cambridge.

Cooke, J. (1978). Somite abnormalities caused by short heat shocks to pre-neurula stages of *Xenopus laevis. J. Embryol. Exp. Morphol.* **45,** 283–294.

Cooke, J. (1981). The problem of periodic patterns in embryos. *Philos. Trans. R. Soc. Lond. B Biol. Sci.* **295,** 509–524.

Cooke, J., and Elsdale, T. (1980). Somitogenesis in amphibian embryos. III. Effects of ambient temperature and of developmental stage upon pattern abnormalities that follow short temperature shocks. *J. Embryol. Exp. Morphol.* **58,** 107–118.

Cooke, J., and Zeeman, E. C. (1976). A clock and wavefront model for control of the number of repeated structures during animal morphogenesis. *J. Theor. Biol.* **58,** 455–476.

Cossu, G., Kelly, R., Di Donna, S., Vivarelli, E., and Buckingham, M. (1995). Myoblast differentiation during mammalian somitogenesis is dependent upon a community effect. *Proc. Natl. Acad. Sci. U.S.A.* **92,** 2254–2258.

Cossu, G., Kelly, R., Tajbakhsh, S., Di Donna, S., Vivarelli, E., and Buckingham, M. (1996). Activation of different myogenic pathways: myf-5 is induced by the neural tube and MyoD by the dorsal ectoderm in mouse paraxial mesoderm. *Development* **122,** 429–437.

Cossu, G., Tajgbakhsh, S., and Buckingham, M. (1996). How is myogenesis initiated in the embryo? *Trends Genet.* **12,** 218–223.

Couly, G. F., Coltey, P. M., and Le Douarin, N. M. (1992). The development fate of the cephalic mesoderm in quail-chick chimeras. *Development* **114,** 1–15.

Crossin, K. L., Hoffman, S., Grumet, M., Thiery, J. P., and Edelman, G. M. (1986). Site-restricted expression of cytotactin during development of the chicken embryo. *J. Cell Biol.* **102,** 1917–1930.

Crossley, P. H., and Martin, G. R. (1995). The mouse Fgf8 gene encodes a family of polypeptides and is expressed in regions that direct outgrowth and patterning in the developing embryo. *Development* **121,** 439–451.

Cserjesi, P., Brown, D., Ligon, K. L., Lyons, G. E., Copeland, N. G., Gilbert, D. J., Jenkins, N. A., and Olson, E. N. (1995). Scleraxis: A basic helix-loop-helix protein that prefigures skeletal formation during mouse embryogenesis. *Development* **121,** 1099–1110.

Currie, P. D., and Ingham, P. W. (1996). Induction of a specific muscle cell type by a hedgehog-like protein in zebrafish. *Nature* **382,** 452–455.

Davies, J. A., Cook, G. M., Stern, C. D., and Keynes, R. J. (1990). Isolation chick somites of a glycoprotein fraction that causes collapse of dorsal root ganglion growth cones. *Neuron* **4,** 11–20.

de Celis, J. F., Garcia Bellido, A., and Bray, S. J. (1996). Activation and function of Notch at the dorsal-ventral boundary of the wing imaginal disc. *Development* **122,** 359–369.

Dekker, E. J., Pannese, M., Houtzager, E., Boncinelli, E., and Durston, A. (1993). Colinearity in the *Xenopus laevis* Hox-2 complex. *Mech. Dev.* **40,** 3–12.

Del Amo, F. F., Smith, D. E., Swiatek, P. J., Gendron Maguire, M., Greenspan, R. J., McMahon, A. P., and Gridley, T. (1992). Expression pattern of Motch, a mouse homolog of *Drosophila* Notch, suggests an important role in early postimplantation mouse development. *Development* **115,** 737–744.

de la Pompa, J. L., Wakeham, A., Correia, K. M., Samper, E., Brown, S., Aguilea, R. J., Nakano, T., Honjo, T., Mak, T. W., Rossant, J., and Conlon, R. A. (1997). Conservation of the Notch signalling pathway in mammalian neurogenesis. *Development* **124,** 1139–1148.

Dickinson, M. E., and McMahon, A. P. (1992). The role of Wnt genes in vertebrate development. *Curr. Opin. Genet. Dev.* **2,** 562–566.

Dietrich, S., Schubert, F. R., and Gruss, P. (1993). Altered Pax gene expression in murine notochord mutants: The notochord is required to initiate and maintain ventral identity in the somite. *Mech. Dev.* **44,** 189–207.

Doherty, D., Feger, G., Younger Shepherd, S., Jan, L. Y., and Jan, Y. N. (1996). Delta is a ventral to dorsal signal complementary to Serrate, another Notch ligand, in *Drosophila* wing formation. *Genes Dev.* **10,** 421–434.

Dollé, P., Izpisua-Belmonte, J. C., Falkenstein, H., Renucci, A., and Duboule, D. (1989). Coordinate expression of the murine Hox-5 complex homeobox-containing genes during limb pattern formation. *Nature* **342,** 767–772.

Drake, C. J., Davis, L. A., Hungerford, J. E., and Little, C. D. (1992). Perturbation of beta 1 integrin-mediated adhesions results in altered somite cell shape and behavior. *Dev. Biol.* **149,** 327–338.

Drake, C. J., and Little, C. D. (1991). Integrins play an essential role in somite adhesion to the embryonic axis. *Dev. Biol.* **143,** 418–421.

Driever, W., Solnica-Krezel, L., Schier, A. F., Neuhauss, S. C. F., Malicki, J., Stemple, D. L., Stainier, D. Y. R., Zwartkruis, F., Abdelilah, S., Rangini, Z., Belak, J., and Boggs, C. (1996). A genetic screen for mutations affecting embryogenesis in zebrafish. *Development* **123,** 37–46.

Duband, J. L., Dufour, S., Hatta, K., Takeichi, M., Edelman, G. M., and Thiery, J. P. (1987). Adhesion molecules during somitogenesis in the avian embryo. *J. Cell Biol.* **104,** 1361–1374.

Duband, J. L., and Thiery, J. P. (1987). Distribution of laminin and colleagens during avian neural crest development. *Development* **101,** 461–478.

Duband, J. L., Volberg, T., Sabanay, I., Thiery, J. P., and Geiger, B. (1988). Spatial and temporal distribution of the adherens-junction-associated adhesion molecule A-CAM during avian embryogenesis. *Development* **103,** 325–344.

Dunwoodie, S. L., Henrique, D., Harrison, S. M., and Beddington, R. S. P. (1997). Mouse *Dll3*: A novel divergent *Delta* gene which may complement the function of other *Delta* homologues during early pattern formation in the mouse embryo. *Development* **124,** 3065–3076.

Dush, M. K., and Martin, G. R. (1992). Analysis of mouse evx-genes-evx-1 displays graded expression in the primitive streak. *Dev. Biol.* **151,** 273–287.

Ebensperger, C., Wilting, J., Brand-Saberi, B., Mizutani, Y., Christ, B., Balling, R., and Koseki, H. (1995). *Pax-1*, a regulator of sclerotome development is induced by notochord and floor plate signals in avian embryos. *Anat. Embryol. (Berl.)* **191,** 297–310.

Echelard Y., Epstein, D. J., St. Jacques, B., Shen, L., Mohler, J., McMahon, J. A., and McMahon, A. P. (1993). Sonic hedgehog, a member of a family of putative signaling molecules, is implicated in the regulation of CNS polarity. *Cell* **75,** 1417–1430.

Ede, D. A., and El-Gadi, A. O. A. (1986). Genetic modifications of developmental acts in chick and mouse somite development. *In* "Somites in Developing Embryos" (R. Bellairs, D. Ede, and J. Lash, Eds.), pp. 209–224. Plenum Press, New York.

Eichmann, A., Marcelle, C., Breant, C., and Le Douarin, N. M. (1993). Two molecules related to the VEGF receptor are expressed in early endothelial cells during avian embryonic development. *Mech. Dev.* **42,** 33–48.

Elsdale, T., and Pearson, M. (1979). Somitogenesis in amphibia. II. Origins in early embryogenesis of two factors involved in somite specification. *J. Embryol. Exp. Morphol.* **53,** 245–267.

Elsdale, T., Pearson, M., and Whitehead, M. (1976). Abnormalities in somite segmentation following heat shock to *Xenopus* embryos. *J. Embryol. Exp. Morphol.* **35,** 625–635.

Ewan, K. B., and Everett, A. W. (1992). Evidence for resegmentation in the formation of the vertebral column using the novel approach of retroviral-mediated gene transfer. *Exp. Cell Res.* **198,** 315–320.

Fan, C. M., Porter, J. A., Chiang, C., Chang, D. T., Beachy, P. A., and Tessier Lavigne, M. (1995). Long-range sclerotome induction by sonic hedgehog: Direct role of the amino-terminal cleavage product and modulation by the cyclic AMP signaling pathway. *Cell* **81,** 457–465.

Fan, C. M., and Tessier Lavigne, M. (1994). Patterning of mammalian somites by surface ectoderm and notochord: Evidence for sclerotome induction by a hedgehog homolog. *Cell* **79,** 1175–1186.

Faust, C., Schumacher, A., Holdener, B., and Magnuson, T. (1995). The *eed* mutation disrupts anterior mesoderm production in mice. *Development* **121,** 273–285.

Fehon, R. G., Kooh, P. J., Rebay, I., Regan, C. L., Xu, T., Muskavitch, M. A., and Artavanis Tsakonas, S. (1990). Molecular interactions between the protein products of the neurogenic loci Notch and Delta, two EGF-homologous genes in *Drosophila*. *Cell* **61,** 523–534.

Ferguson, E. L. (1996). Conservation of dorsal-ventral patterning in arthropods and chordates. *Curr. Opin. Genet. Dev.* **6,** 424–431.

Flint, O. P., Ede, D. A., Wilby, O. K., and Proctor, J. (1978). Control of somite number in normal and *amputated* mutant mouse embryos: An experimental and a thoretical analysis. *J. Embryol. Exp. Morphol.* **45,** 189–202.

Fraser, R. (1960). Somite genesis in the chick. III. The role of induction. *J. Exp. Zool.* **145,** 151–167.

Fraser, S. (1993). Segmentation moves to the fore. *Curr. Biol.* **3,** 787–789.

Furuta, Y., Ilic, D., Kanazawa, S., Takeda, N., Yamamoto, T., and Aizawa, S. (1995). Mesodermal defect in late phase of gastrulation by a targeted mutation of focal adhesion kinase, FAK. *Oncogene* **11**, 1989–1995.

Gallera, J. (1966). Mise en évidence du rôle de l'éctoblaste dans la différenciation des somites chez les oiseaux. *Re. Suisse Zool.* **73**, 492–503.

Gamel, A. J., Brand-Saberi, B., and Christ, B. (1996). Halves of epithelial somites and segmental plate show distinct muscle differentiation behavior *in vitro* compared to entire somites and segmental plate. *Dev. Biol.* **172**, 625–639.

Gaunt, S. J. (1988). Mouse homeobox gene transcripts occupy different but overlapping domains in embryonic germ layers and organs: A comparison of *Hox-3.1* and *Hox-1.5*. *Development* **103**, 135–144.

Gaunt, S. J., Sharpe, P. T., and Duboule, D. (1988). Spatially restricted domains of homeogene transcripts in mouse embryos: Relation to a segmented body plan. *Development* **104** (Suppl.), 169–179.

Gendron Maguire, M., Mallo, M., Zhang, M., and Gridley, T. (1993). Hoxa-2 mutant mice exhibit homeotic transformation of skeletal elements derived from cranial neural crest. *Cell* **75**, 1317–1331.

George, E. L., Georges Labouesse, E. N., Patel King, R. S., Rayburn, H., and Hynes, R. O. (1993). Defects in mesoderm, neural tube and vascular development in mouse embryos lacking fibronectin. *Development* **119**, 1079–1091.

Georges Labouesse, E. N., George, E. L., Rayburn, H., and Hynes, R. O. (1996). Mesodermal development in mouse embryos mutant for fibronectin. *Dev. Dyn.* **207**, 145–156.

Gérard, M., Chen, J. Y., Gronemeyer, H., Chambon, P., Duboule, D., and Zákány, J. (1996). In vivo targeted mutagenesis of a regulatory element required for positioning the Hoxd-11 and Hoxd-10 expression boundaries. *Genes Dev.* **10**, 2326–2334.

Goldstein, R. S., and Kalcheim, C. (1992). Determination of epithelial half-somites in skeletal morphogenesis. *Development* **116**, 441–445.

Goldstein, R. S., Teillet, M. A., and Kalcheim, C. (1990). The microenvironment created by grafting rostral half-somites is mitogenic for neural crest cells. *Proc. Natl. Acad. Sci. U.S.A.* **87**, 4476–4480.

Gont, L. K., Steinbeisser, H., Blumberg, B., and de Robertis, E. M. (1993). Tail formation as a continuation of gastrulation: The multiple cell population of the *Xenopus* tailbud derive from the late blastopore lip. *Development* **119**, 991–1004.

Goulding, M. D., Lumsden, A., and Gruss, P. (1993). Signals from the notochord and floor plate regulate the region-specific expression of two Pax genes in the developing spinal cord. *Development* **117**, 1001–1016.

Goulding, M., Lumsden, A., and Paquette, A. J. (1994). Regulation of Pax-3 expression in the dermomyotome and its role in muscle development. *Development* **120**, 957–971.

Grabowski, C. (1956). The effects of the excision of Hensen's node on the development of the chick embryo. *J. Exp. Zool.* **133**, 301–344.

Greco, T. L., Takada, S., Newhouse, M. M. McMahon, J. A., McMahon, A. P., and Camper, S. A. (1996). Analysis of the *vestigial tail* mutation demonstrates that Wnt-3a gene dosage regulates mouse axial development. *Genes Dev.* **10**, 313–324.

Gregg, B. C., and Snow, M. H. (1983). Axial abnormalities following disturbed growth in mitomycin C-treated mouse embryos. *J. Embryol. Exp. Morphol.* **73**, 135–149.

Griffith, C. M., Wiley, M. J., and Sanders, E. J. (1992). The vertebrate tail bud: Three germ layers from one tissue. *Anat. Embryol. (Berl.)* **185**, 101–113.

Grobstein, C., and Holtzer, H. (1955). In vitro studies of cartilage induction in mouse somite mesoderm. *J. Exp. Zool.* **128**, 333–356.

Gurdon, J. B. (1988). A community effect in animal development. *Nature* **336**, 772–774.

Haenlin, M., Kramatschek, B., and Campos-Ortega, J. A. (1990). The pattern of transcription of the neurogenic gene Delta of *Drosophila* melanogaster. *Development* **110**, 905–914.

Haffter, P., Granato, M., Brand, M., Mullins, M. C., Hammerschmidt, M., Kane, D. A., Odenthal, J., van Eeden, F. J. M., Jiang, Y.-J., Heisenberg, C.-P., Kelsh, R. N., Furutani-Seiki, M., Vogelsang, E., Euchle, D., Schach, U., Fabian, C., and Nusslein-Volhard, C. (1996). The identification of genes with unique and essential functions in the development of the zebrafish, *Danio rerio. Development* **123**, 1–36.

Hall, B. K. (1977). Chondrogenesis of the somitic mesoderm. *Adv. Anat. Embroyol. Cell Biol.* **53**, 3–47.

Halpern, M. E., Thisse, C., Ho, R. K., Thisse, B., Riggleman, B., Trevarrow, B., Weinberg, E. S., Postlethwait, J. H., and Kimmel, C. B. (1995). Cell-autonomous shift from axial to paraxial mesodermal development in zebrafish floating head mutants. *Development* **121**, 4257–4264.

Hamilton, L. (1969). The formation of somites in *Xenopus. J. Embryol. Exp. Morphol.* **22**, 253–264.

Hammerschmidt, M., Bitgood, M. J., and McMahon, A. P. (1996). Protein kinase A is a common negative regulator of Hedgehog signaling in the vertebrate embryo. *Genes Dev.* **10**, 647–658.

Hammerschmidt, M., Serbedzija, G. N., and McMahon, A. P. (1996). Genetic analysis of dorso-ventral pattern formation in the zebrafish: Requirement of a BMP-like ventralizing activity and its dorsal repressor. *Genes Dev.* **10**, 2452–2461.

Hartenstein, A. Y., Rugendorff, A., Tepass, U., and Hartenstein, V. (1992). The function of the neurogenic genes during epithelial development in the *Drosophila* embryo. *Development* **116**, 1203–1220.

Hatada, Y., and Stern, C. D. (1994). A fate map of the epiblast of the early chick embryo. *Development* **120**, 2879–2889.

Haub, O., and Goldfarb, M. (1991). Expression of the fibroblast growth factor-5 gene in the mouse embryo. *Development* **112**, 397–406.

Hebert, J. M., Boyle, M., and Martin, G. R. (1991). mRNA localization studies suggest that murine FGF-5 plays a role in gatrulation. *Development* **112**, 407–415.

Heitzler, P., and Simpson, P. (1993). Altered epidermal growth factor-like sequences provide evidence for a role of Notch as a receptor in cell fate decisions. *Development* **117**, 1113–1123.

Herrmann, B. G., Labeit, S., Poustka, A., King, T. R., and Lehrach, H. (1990). Cloning of the T gene required in mesoderm formation in the mouse [see comments]. *Nature* **343**, 617–622.

Heston, W. E. (1951). The vestigial tail mouse. *J. Hered.* **42**, 71–74.

Hogan, B. L. M. (1996). Bone morphogenetic proteins in development. *Curr. Opin, Genet. Dev.* **6**, 432–438.

Holland, P. W. H., and Hogan, B. L. M. (1988). Expression of homeo box genes during mouse development: A review. *Genes Dev.* **2**, 773–782.

Hrabě de Angelis, M., and Kirchner, C. (1993). Fibroblast growth factor induces primitive streak formation in rabbit pre-implantation embryos in vitro. *Anat. Embryol. (Berl.)* **187**, 269–273.

Hrabě de Angelis, M., McIntyre II, J., and Gossler, A. (1997). Maintenance of somite borders in mice requires the *Delta* homologue *Dll1. Nature* **386**, 717–721.

Huang, R., Zhi, Q., Neubüser, A., Müller, T. S., Brand-Saberi, B., Christ, B., and Wilting, J. (1996). Function of somite and somitocoel cells in the formation of the vertebral motion segment in avian embryos. *Acta Anat. (Basel)* **155**, 231–241.

Huang, R., Zhi, Q., Wilting, J., and Christ, B. (1994). The fate of somitocoele cells in avian embryos. *Anat. Embryol. (Berl.)* **190**, 243–250.

Hynes, R. O. (1992). Integrins: Versatility, modulation, and signaling in cell adhesion. *Cell* **69**, 11–25.

Hynes, R. O. (1996). Targeted mutations in cell adhesion genes: What have we learned from them? *Dev. Biol.* **180**, 402–412.

Ingham, P. W., and Martinez Arias, A. (1992). Boundaries and fields in early embryos. *Cell* **68**, 221–235.

Izpisúa-Belmonte, J. C., Falkenstein, H., Dollé, P., Renucci, A., and Duboule, D. (1991). Murine genes related to the *Drosophlia* AbdB homeotic gene are sequentially expressed during development of the posterior part of the body. *EMBO J.* **10**, 2279–2289.

Jacob, M., Wachtler, F., Jacob, H. J., and Christ, B. (1986). On the problem of metamerism in the head mesenchyme of chick embryos. *In* "Somites in Developing Embryos" (R. Bellairs, D. Ede, and J. Lash, Eds.), pp. 79–90. Plenum Press, New York.

Jacobson, A. G. (1988). Somitomeres: Mesodermal segments of vertebrate embryos. *Development* **104** (Suppl.), 209–220.

Jacobson, A. G., and Meier, S. (1984). Morphogenesis of the head of a newt: Mesodermal segments, neuromeres, and distribution of neural crest. *Dev. Biol.* **106**, 181–193.

Jacobson, A., and Meier, S. (1986). Sommitomeres: The primordial body segments. *In* "Somites in Developing Embryos" (R. Bellairs, D. Ede, and J. Lash, Eds.), pp. 1–18. Plenum Press, New York.

Jacobson, A. G., and Tam, P. P. (1982). Cephalic neurulation in the mouse embryo analyzed by SEM and morphometry. *Anat. Rec.* **203**, 375–396.

Jarriault, S., Brou, C., Logeat, F., Schroeter, E. H., Kopan, R., and Israel, A. (1995). Signalling downstream of activated mammalian notch. *Nature* **377**, 355–358.

Jegalian, B. G., and Derobertis, E. M. (1992). Homeotic transformations in the mouse induced by overexpression of a human *hox3.3* transgene. *Cell* **71**, 901–910.

Jen, W. C., Wettstein, D., Turner, D., Chitnis, A., and Kintner, C. (1997). The Notch ligand, X-Delta-2, mediates segmentation of the paraxial mesoderm in *Xenopus* embryos. *Development* **124**, 1169–1178.

Johnson, D. R. (1986). "The Genetics of the Skeleton." Clarendon Press, Oxford.

Johnson, R. L., Laufer, E., Riddle, R. D., and Tabin, C. (1994). Ectopic expression of Sonic hedgehog alters dorsal-ventral patterning of somites. *Cell* **79**, 1165–1173.

Kaehn, K., Jacob, H. J., Christ, B., Hinrichsen, K., and Poelmann, R. E. (1988). The onset of myotome formation in the chick. *Anat. Embryol. (Berl.)* **177**, 191–201.

Kalcheim, C., and Teillet, M. A. (1989). Consequences of somite manipulation on the pattern of dorsal root ganglion development. *Development* **106**, 85–93.

Keller, R. E. (1975). Vital dye mapping of the gastrula and neurula of *Xenopus laevis*. I. Prospective areas and morphogenetic movements of the superficial layer. *Dev. Biol.* **42**, 222–241.

Keller, R. E. (1976). Vital dye mapping of the gastrula and neurula of *Xenopus laevis*. II. Prospective areas and morphogenetic movements of the deep layer. *Dev. Biol.* **51**, 118–137.

Kennison, J. A. (1993). Transcriptional activation of *Drosophila* homeotic genes from distant regulatory elements. *Trends Genet.* **9**, 75–79.

Kennison, J. A., and Tamkun, J. W. (1992). *Trans*-regulation of homeotic genes in *Drosophila*. *New Biol.* **4**, 91–96.

Kenny Mobbs, T., and Thorogood, P. (1987). Autonomy of differentiation in avian branchial somites and the influence of adjacent tissues. *Development* **100**, 449–462.

Kessel, M. (1992). Respecification of vertebral identities by retinoic acid. *Development* **115**, 487–501.

Kessel, M., Balling, R., and Gruss, P. (1990). Variations of cervical vertebrae after expression of a Hox-1.1 transgene in mice. *Cell* **61**, 301–308.

Kessel, M., and Gruss, P. (1990). Murine developmental control genes. *Science* **249**, 374–379.

Kessel, M., and Gruss, P. (1991). Homeotic transformations of murine vertebrae and concomitant alteration of *Hox* codes induced by retinoic acid. *Cell* **67**, 89–104.

Kessler, D. S., and Melton, D. A. (1994). Vertebrate embryonic induction: Mesodermal and neuronal patterning. *Science* **266**, 596–604.

Keynes, R. J., and Stern, C. D. (1984). Segmentation in the vertebrate nervous system. *Nature* **310**, 786–789.

Keynes, R. J., and Stern, C. D. (1988). Mechanisms of vertebrate segmentation. *Development* **103**, 413–429.

Kielbowna, L. (1981). The formation of somites and early myotomal myogenesis in *Xenopus laevis, Bombina variegata*, and *Pelobates fuscus. J. Embryol. Exp. Morphol.* **64**, 295–304.

Kieny, M., Mauger, A., and Sengel, P. (1972). Early recognition of the somite mesoderm as studied by the development of the axial skeleton of the chick embryo. *Dev. Biol.* **28**, 142.

Kimmel, C. B., Sepich, D. S., and Trevarrow, B. (1988). Development of segmentation in zebrafish. *Development* **104**, 197–207.

Kimura, Y., Matsunami, H., Inoue, T., Shimamura, K., Uchida, N., Ueno, T., Miyazaki, T., and Takeichi, M. (1995). Cadherin-11 expressed in association with mesenchymal morphogenesis in the head, somite, and limb bud of early mouse embryos. *Dev. Biol.* **169**, 347–358.

Koseki, H., Wallin, J., Wilting, J., Mizutani, Y., Kispert, A., Ebensperger, C., Herrmann, B. G., Christ, B., and Balling, R. (1993). A role for Pax-1 as a mediator of notochordal signals during the dorsoventral specification of vertebrae. *Development* **119**, 649–660.

Krumlauf, R. (1994). *Hox* genes in vertebrate development. *Cell* **78**, 191–201.

Kuratani, S., Martin, J. F., Wawersik, S., Lilly, B., Eichele, G., and Olson, E. N. (1994). The expression pattern of the chick homeobox gene gMHox suggests a role in patterning of the limbs and face and in compartmentalization of somites. *Dev. Biol.* **161**, 357–369.

Lanot, R. (1971). La formation de somites chez l'embryon d'oiseau: etude experimentale. *J. Embryol. Exp. Morphol.* **26**, 1–20.

Larue, L., Antos, C., Butz, S., Huber, O., Delmas, V., Dominis, M., and Kemler R. (1996). A role for cadherins in tissue formation. *Development* **122**, 3185–3194.

Lash, J. W., Linask, K. K., and Yamada, K. M. (1987). Synthetic peptides that mimic the adhesive recognition signal of fibronectin: Differential effects on cell–cell and cell–substratum adhesion in embryonic chick cells. *Dev. Biol.* **123**, 411–420.

Lash, J. W., Seitz, A. W., Cheney, C. M., and Ostrovsky, D. (1984). On the role of fibronectin during the compaction stage of somitogenesis in the chick embryo. *J. Exp. Zool.* **232**, 197–206.

Lassar, A. B., and Münsterberg, A. (1996). The role of positive and negative signals in somite patterning. *Curr. Opin. Neurobiol.* **6**, 57–63.

Lawrence, P. A., and Morata, G. (1994). Homeobox genes: Their function in *Drosophila* segmentation and pattern formation. *Cell* **78**, 181–189.

Lawson, K. A., Meneses, J. J., and Pedersen, R. A. (1991). Clonal analysis of epiblast fate during germ layer formation in the mouse embryo. *Development* **113**, 891–911.

Lawson, K. A., and Pedersen, R. A. (1992). Clonal analysis of cell fate during gastrulation and early neurulation in the mouse. *Ciba Found. Symp.* **165**, 3–21.

Le Mouellic, H., Lallamand, Y., and Brulet, P. (1992). Homeosis in the mouse induced by a null mutation in the Hox-3.1 gene. *Cell* **69**, 251–264.

Lewis, E. B. (1978). A gene complex controlling segmentation in *Drosophila. Nature* **276**, 565–570.

Lipton, B. H., and Jacobson, A. G. (1974a). Analysis of normal somite development. *Dev. Biol.* **38**, 73–90.

Lipton, B. H., and Jacobson, A. G. (1974b). Experimental analysis of the mechanisms of somite morphogenesis. *Dev. Biol.* **38**, 91–103.

Lufkin, T., Mark, M., Hart, C. P., Dolle, P., LeMeur, M., and Chambon, P. (1992). Homeotic transformation of the occipital bones of the skull by ectopic expression of a homeobox gene. *Nature* **359**, 835–841.

MacMurray, A., and Shin, H. S. (1988). The antimorphic nature of the *Tc* allele at the mouse T locus. *Genetics* **120,** 545–550.

Mansour, S. L., Goddard, J. M., and Capecchi, M. R. (1993). Mice homozygous for a targeted disruption of the protooncognee *int-2* have developmental defects in the tail and inner ear. *Development* **117,** 13–28.

Mansour, S. L., Thomas, K. R., and Capecchi, M. R. (1988). Disruption of the proto-oncogene *int-2* in mouse embryo-derived stem cells: A general strategy for targeting mutations to non-selectable genes. *Nature* **336,** 348–352.

Maroto, M., Reshef, R., Münsterberg, A., Koester, S., Goulding, M., and Lassar, A. B. (1997). Ectopic *Pax-3* activates *MyoD* and *Myf-5* expression in mesoderm and neural tissue. *Cell* **89,** 139–148.

Martindale, M. Q., Meier, S., and Jacobson, A. G. (1987). Mesodermal metamerism in the teleost, *Oryzias latipes* (the medaka). *J. Morphol.* **193,** 241–252.

McGinnis, W., and Krumlauf, R. (1992). Homeobox genes and axial patterning. *Cell* **68,** 283–302.

McLain, K., Schreiner, C., Yager, K. L., Stock, J. L., and Potter, S. S. (1992). Ectopic expression of *hox-2.3* induces craniofacial and skeletal malformations in transgenic mice. *Mod.* **39,** 3–16.

McMahon, A. P., Gavin, B. J., Parr, B., Bradley, A., and McMahon, J. A. (1992). The Wnt family of cell signalling molecules in postimplantation development of the mouse. *Ciba Found. Symp.* **165,** 199–212.

Meier, S. (1979). Development of the chick embryo mesoblast: Formation of the embryonic axis and establishment of the metameric pattern. *Dev. Biol.* **73,** 25–45.

Meier, S. (1981). Development of the chick embryo mesoblast: Morphogenesis of the prechordal plate and cranial segments. *Dev. Biol.* **83,** 49.

Meier, S. (1982). The development of segmentation in the cranial region of vertebrate embryos. *Scanning Electron Micros.* **3,** 1269–1282.

Meier, S., and Jacobson, A. G. (1982). Experimental studies of the origin and expression of metameric pattern in the chick embryo. *J. Exp. Zool.* **219,** 217–232.

Meier, S. and Packard, D. S., Jr. (1984). Morphogenesis of the cranial segments and distribution of neural crest in the embryos of the snapping turtle, *Chelydra serpentina. Dev. Biol.* **102,** 309–323.

Meier, S., and Tam, P. P. (1982). Metameric pattern development in the embryonic axis of the mouse. I. Differentiation of the cranial segments. *Differentiation* **21,** 95–108.

Meinhardt, H. (1986a). Hierarchical inductions of cell states: A model for segmentation in *Drosophila. J. Cell Sci. Suppl.* **4,** 357–381.

Meinhardt, H. (1986b). Models of segmentation. *In* "Somites in Developing Embryos" (R. Bellaris, D. A. Ede, and J. W. Lash, Eds.), pp. 179–189. Plenum Press, New York.

Menkes, B., and Sandor, S. (1977). Somitogenesis: Regulation potencies, sequence determination and primordial interacations. *In* "Vertebrate Limb and Somite Morphogenesis" (D. A. Ede, J. R. Hinchliffe, and M. Balls, Eds.), pp. 405–419. Cambridge University Press, Cambridge.

Mishina, Y., Suzuki, A., Ueno, N., and Behringer, R. R. (1995). *Bmpr* encodes a type I bone morphogenetic protein receptor that is essential for gastrulation during mouse embryogenesis. *Genes Dev.* **9,** 3027–3037.

Monsoro-Burq, A. H., Bontoux, M., Teillet, M. A., and Le Douarin, N. M. (1994). Heterogeneity in the development of the vertebra. *Proc. Natl. Acad. Sci. U.S.A.* **91,** 10435–10439.

Morin-Kensicki, E., and Eisen, J. (1997). Screrotome development and peripheral nervous system segmentation in the embryonc zebrafish. *Development* **124,** 159–167.

Müller, M., Weizäcker, E., and Campos-Ortega, J. A. (1996). Expression domains of a zebrafish homologue of the *Drosophila* pair-rule gene *hairy* correspond to primordia of alternating somites. *Development* **122,** 2071–2078.

Münsterberg, A. E., Kitajewski, J., Bumcrot, D. A., McMahon, A. P., and Lassar, A. B. (1995). Combinatorial signaling by Sonic hedgehog and Wnt family members induces myogenic bHLH gene expression in the somite. *Genes Dev.* **9**, 2911–2922.

Münsterberg, A. E., and Lassar, A. B. (1995). Combinatorial signals from the neural tube, floor plate and notochord induce myogenci bHLH gene expression in the somite. *Development* **121**, 651–660.

New, D. A. T., Coppola, P. T., and Terry, S. (1973). Culture of explanted rat embryos in rotating tubes. *J. Reprod. Fertil.* **35**, 135–138.

Newgreen, D. F., Scheel, M., and Kastner, V. (1986). Morphogenesis of sclerotome and neural crest in avian embryos: In vivo and in vitro studies on the role of notochordal extracellular material. *Cell Tissue Res.* **244**, 299–313.

Nicolet, G. (1970). Is the presumptive notochord responsible for somite genesis in the chick? *J. Embryol. Exp. Morphol.* **24**, 467–478.

Nicolet, G. (1971). The young notochord can induce somite genesis by means of diffusible substances in the chick. *Experientia* **27**, 938–939.

Niswander, L., and Martin, G. R. (1992). Fgf-4 expression during gastrulation, myogenesis, limb and tooth development in the mouse. *Development* **114**, 755–768.

Noden, D. M. (1988). Interactions and fates of avian craniofacial mesenchyme. *Development* **103** (Suppl.), 121–140.

Oka, C., Nakano, T., Wakeham, A., de la Pompa, J. L., Mori, C., Sakai, T., Okazaki, S., Kawaichi, M., Shiota, K., Mak, T. W., and Honjo, T. (1995). Disruption of the mouse RBP-J kappa gene results in early embryonic death. *Development* **121**, 3291–3301.

Ooi, V. E., Sanders, E. J., and Bellairs, R. (1986). The contribution of the primitive streak to the somites in the avian embryo. *J. Embryol. Exp. Morphol.* **92**, 193–206.

Ordahl, C. P., and Le Douarin, N. M. (1992). Two myogenic lineages within the developing somite. *Development* **114**, 339–353.

Orr Urtreger, A., Bedford, M. T., Burakova, T., Arman, E., Zimmer, Y, Yayon, A., Givoil, D., and Lonai, P. (1993). Developmental localization of the splicing alternatives of fibroblast growth factor receptor-2 (FGFR2). *Dev. Biol.* **158**, 475–486.

Ostrovsky, D., Cheney, C. M., Seitz, A. W., and Lash, J. W. (1983). Fibronectin distribution during somitogenesis in the chick embryo. *Cell Differ.* **13**, 217–223.

Ostrovsky, D., Sanger, J. W., and Lash, J. W. (1988). Somitogenesis in the mouse embryo. *Cell Differ.* **23**, 17–25.

Ott, M.-O., Bober, E., Lyons, G., Arnold, H., and Buckingham, M. (1991). Early expression of the myogenic regulatory gene, *myf-5* in precursor cells of skeletal muscle in the mouse embryo. *Development* **111**, 1097–1107.

Packard, D. S., Jr. (1980a). Somite formation in cultured embryos of the snapping turtle, *Chelydra serpentina*. *J. Embryol. Exp. Morphol.* **59**, 113–130.

Packard, D. S., Jr. (1980b). Somitogenesis in cultured embryos of the Japanese quail, *Coturnix coturnix japonica*. *Am. J. Anat.* **158**, 83–91.

Packard, D. S., Jr., and Jacobson, A. G. (1976). The influence of axial structures on chick somite formation. *Dev. Biol.* **53**, 36–48.

Packard, D. S., Jr., and Meier, S. (1984). Morphological and experimental studies of the somitomeric organization of the segmental plate in snapping turtle embryos. *J. Embryol. Exp. Morphol.* **84**, 35–48.

Packard, D. S., Jr., Zheng, R. Z., and Turner, D. C. (1993). Somite pattern regulation in the avian segmental plate mesoderm. *Development* **117**, 779–791.

Parameswaran, M., and Tam, P. P. L. (1995). Regionalisation of cell fate and morphogenetic movement of the mesoderm during mouse gastrulation. *Dev. Genet.* **17**, 16–28.

Parr, B. A., and McMahon, A. P. (1994). Wnt genes and vertebrate development. *Curr. Opin. Genet. Dev.* **4**, 523–528.

Parsons, J. T., Schaller, M. D., Hildebrand, J., Leu, T. H., Richardson, A., and Otey, C. (1994). Focal adhesion kinase: Structure and signalling. *J. Cell. Sci. Suppl.* **18,** 109–113.

Pearson, M., and Elsdale, T. (1979). Somitogenesis in amphibian embryos. I. Experimental evidence for an interaction between two temporal factors in the specification of somite pattern. *J. Embryol. Exp. Morphol.* **51,** 27–50.

Peterson, P. E., Pow, C. S., Wilson, D. B., and Hendrickx, A. G. (1993). Distribution of extracellular matrix components during early embryonic development in the macaque. *Acta Anat. (Basel)* **146,** 3–13.

Poelmann, R. E. (1981). The formation of the embryonic mesoderm in the early post-implantation mouse embryo. *Anat. Embryol. (Berl.)* **162,** 29–40.

Polezhaev, A. A. (1992). A mathematical model of the mechanism of vertebrate somitic segmentation. *J. Theor. Biol.* **156,** 169–181.

Pollock, R. A., Jay, G., and Bieberich, C. J. (1992). Altering the boundaries of *Hox3.1* expression: Evidence for antipodal gene regulation. *Cell* **71,** 911–923.

Pourquie, O., Coltey, M., Breant, C., and Le Douarin, N. M. (1995). Control of somite patterning by signals from the lateral plate. *Proc. Natl. Acad. Sci. U.S.A.* **92,** 3219–3223.

Pourquie, O., Coltey, M., Teillet, M. A., Odahl, C., and Le Douarin, N. M. (1993). Control of dorsoventral patterning of somitic derivatives by notochord and floor plate. *Proc. Natl. Acad. Sci. U.S.A.* **90,** 5242–5246.

Pourquie, O., Fan, C. M., Coltey, M., Hirsinger, E., Watanabe, Y., Breant, C., Francis West, P., Brickell, P., Tessier Lavigne, M., and Le Douarin, N. M. (1996). Lateral and axial signals involved in avian somite patterning: A role for BMP4. *Cell* **84,** 461–471.

Pow, C. S., and Hendrickx, A. G. (1995). Localization of integrin subunits alpha 6 and beta 1 during somitogenesis in the long-tailed macaque (*M. fascicularis*). *Cell Tissue Res.* **281,** 101–108.

Pownall, M. E., and Emerson, C. P., Jr. (1992). Sequential activation of three myogenic regulatory genes during somite morphogenesis in quail embryos. *Dev. Biol.* **151,** 67–79.

Pownall, M. E., Strunk, K. E., and Emerson, C. P., Jr. (1996). Notochord signals control the transcriptional cascade of myogenic bHLH genes in somites of quail embryos. *Development* **122,** 1475–1488.

Primmett, D. R., Norris, W. E., Carlson, G. J., Keynes, R. J., and Stern, C. D. (1989). Periodic segmental anomalies induced by heat shock in the chick embryo are associated with the cell cycle. *Development* **105,** 119–130.

Primmett, D. R., Stern, C. D., and Keynes, R. J. (1988). Heat shock causes repeated segmental anomalies in the chick embryo. *Development* **104,** 331–339.

Probstmeier, R., Bilz, A., and Schneider Schaulies, J. (1994). Expression of the neural cell adhesion molecule and polysialic acid during early mouse embryogenesis. *J. Neurosci. Res.* **37,** 324–335.

Psychoyos, D., and Stern, C. D. (1996). Fates and migratory routes of primitive streak cells in the chick embryo. *Development* **122,** 1523–1534.

Rabinowitz, J. E., Rutishauser, U., and Magnuson, T. (1996). Targeted mutation of Ncam to produce a secreted molecule results in a dominant embryonic lethality. *Proc. Natl. Acad. Sci. U.S.A.* **93,** 6421–6424.

Radice, G. L., Rayburn, H., Matsunami, H., Knudsen, K. A., Takeichi, M., and Hynes, R. O. (1997). Developmental defects in mouse embryos lacking N-cadherin. *Dev. Biol.* **181,** 64–78.

Ramirez Solis, R., Zheng, H., Whiting, J., Krumlauf, R., and Bradley, A. (1993). Hoxb-4 (Hox-2.6) mutant mice show homeotic transformation of a cervical vertebra and defects in the closure of the sternal rudiments. *Cell* **73,** 279–294.

Reaume, A. G., Conlon, R. A., Zirngibl, R., Yamaguchi, T. P., and Rossant, J. (1992). Expression analysis of a *Notch* homologue in the mouse embryo. *Dev. Biol.* **154** 377–387.

Remak, R. (1855). "Untersuchungen über die Entwicklung der Wirbelthiere." Reimer, Berlin.

Rickmann, M., Fawcett, J. W., and Keynes, R. J. (1985). The migration of neural crest cells and the growth of a motor axons through the rostral half of the chick somite. *J. Embryol. Exp. Morphol.* **90,** 437–455.

Roelink, H., Augsburger, A., Heemskerk, J., Korzh, V., Norlin, S., Ruiz i Altaba, A., Tanabe, Y., Placzek, M., Edlund, T., Jessell, T. M., et al. (1994). Floor plate and motor neuron induction by vhh-1, a vertebrate homolog of hedgehog expressed by the notochord. *Cell* **76,** 761–775.

Rong, P. M., Teillet, M. A., Ziller, C., and Ledouarin, N. M. (1992). The neural tube/notochord complex is necessary for vertebral but not limb and body wall striated-muscle differentiation. *Development* **115,** 657–672.

Rong, P. M., Ziller, C., Pena Melian, A., and Le Douarin, N. M. (1987). A monoclonal antibody specific for avian early myogenic cells and differentiated muscle. *Dev. Biol.* **122,** 338–353.

Ruoslahti, E., and Pierschbacher, M. D. (1986). Arg-Gly-Asp: A versatile cell recognition signal. *Cell* **44,** 517–518.

Saegusa, H., Takahashi, N., Noguchi, S., and Suemori, H. (1996). Targeted disruption in the mouse Hoxc-4 locus results in axial skeleton homeosis and malformation of the xiphoid process. *Dev. Biol.* **174,** 55–64.

Saga, Y., Hata, N., Kobayashi, S., Magnuson, T., Seldin, M. F., and Taketo, M. M. (1996). MesP1: A novel basic helix-loop-helix protein expressed in the nascent mesodermal cells during mouse gastrulation. *Development* **122,** 2769–2778.

Sanders, E. J. (1986). A comparison of the adhesiveness of somitic cells from chick and quail embryos. *In* "Somites in Developing Embryos" (R. Bellairs, D. Ede, and J. Lash, Eds.), pp. 191–200. Plenum Press, New York.

Sandor, S., and Fazakas Todea, I. (1980). Researches on the formation of axial organs in the chick embryo. X. Further investigations on the role of ecto- and endoderm in somitogenesis. *Morphol. Embryol. Bucur.* **26,** 29–32.

Sasaki, H., and Hogan, B. L. (1993). Differential expression of multiple fork and head related genes during gastrulation and axial pattern formation in the mouse embryo. *Development* **118,** 47–59.

Schaller, M. D., and Parsons, J. T. (1994). Focal adhesion kinase and associated proteins. *Curr. Opin. Cell Biol.* **6,** 705–710.

Schoenwolf, G. C. (1984). Histological and ultrastructural studies of secondary neurulation in mouse embryos. *Am. J. Anat.* **169,** 361–376.

Schoenwolf, G. C., Garcia Martinez, V., and Dias, M. S. (1992). Mesoderm movement and fate during avian gastrulation and neurulation. *Dev. Dyn.* **193,** 235–248.

Schumacher, A., Faust, C., and Magnuson, T. (1996). Positional cloning of a global regulator of anterior-posterior patterning in mice. *Nature* **383,** 250–253.

Selleck, M. A., and Stern, C. D. (1991). Fate mapping and cell lineage analysis of Hensen's node in the chick embryo. *Development* **112,** 615–626.

Selleck, M. A., and Stern, C. D. (1992). Commitment of mesoderm cells in Hensen's node of the chick embryo to notochord and somite. *Development* **114,** 403–415.

Simeone, A., Acampora, D., Arcioni, L., Andrews, P. W., Boncinelli, E., and Mavilio, F. (1990). Sequential activation of *HOX2* homeobox genes by retinoic acid in human embryonal carcinoma cells. *Nature* **346,** 763–766.

Simon, J. (1995). Locking in stable states of gene expression: Transcriptional control during *Drosophila* development. *Curr. Opin. Cell Biol.* **7,** 376–385.

Slack, J. M. W. (1991). "From Egg to Embryo: Regional Specification in Early Development." Cambridge University Press, Cambridge.

Slack, J. M. (1994). Inducing factors in *Xenopus* early embryos. *Curr. Biol.* **4,** 116–126.

Small, K. M., and Potter, S. S. (1993). Homeotic transformations and limb defects in Hox A11 mutant mice. *Genes Dev.* **7,** 2318–2328.

Smith, J. L., Gesteland, K. M., and Schoenwolf, G. C. (1994). Prospective fate map of the mouse primitive streak at 7.5 days of gestation. *Dev. Dyn.* **201**, 279–289.

Smith, L. J. (1964). The effects of transection and extirpation on axis formation and elongation in the young mouse embryo. *J. Embryol. Exp. Morph.* **12**, 787–803.

Smith, T. H., Kachinsky, A. M., and Miller, J. B. (1994). Somite subdomains, muscle cell origins, and the four muscle regulatory factor proteins. *J. Cell. Biol.* **127**, 95–105.

Snell, G. D., and Stevens, L. C. (1991). Early embryology. *In* "Biology of the Laboratory Mouse" (E. Green, Ed.), pp. 205–245. Dover Publications, New York.

Spence, M. S., Yip J., and Erickson, C. A. (1996). The dorsal neural tube organizes the dermamyotome and induces axial myocytes in the avian embryo. *Development* **122**, 231–241.

Spörle, R., Gunther, T., Struwe, M., and Schughart, K. (1996). Severe defects in the formation of epaxial musculature in open brain (opb) mutant mouse embryos. *Development* **122**, 79–86.

Spratt, N. T. (1955). Analysis of the organiser centre in the early chick embryo. I. Localization of prospective notochord and somite cells. *J. Exp. Zool.* **128**, 121–164.

Spratt, N. T. (1957a). Analysis of the organiser centre in the early chick embryo. II. Studies of the mechanics of notochord elongation and somite formation. *J. Exp. Zool.* **134**, 577–612.

Spratt, N. T. (1957b). Analysis of the organiser centre in the early chick embryo. III. Regulative properties of the chorda and somite centres. *J. Exp. Zool.* **135**, 319–354.

Stern, C. D., and Bellairs, R. (1984a). Mitotic activity during somite segmentation in the early chick embryo. *Anat. Embryol. (Berl.)* **169**, 97–102.

Stern, C. D., and Bellairs, R. (1984b). The roles of node regression and elongation of the area pellucida in the formation of somites in avian embryos. *J. Embryol. Exp. Morphol.* **81**, 75–92.

Stern, C. D., Fraser, S. E., Keynes, R. J., and Primmett, D. R. (1988). A cell lineage analysis of segmentation in the chick embryo. *Development* **104** (Suppl.), 231–244.

Stern, C. D., Hatada, Y., Selleck, M. A., and Storey, K. G. (1992). Relationships between mesoderm induction and the embryonic axes in chick and frog embryos. *Development* (Suppl.), 151–156.

Stern, C. D., Jaques, K. F., Lim, T. M., Fraser, S. E., and Keynes, R. J. (1991). Segmental lineage restrictions in the chick embryo spinal cord depend on the adjacent somites. *Development* **113**, 239–244.

Stern, C. D., and Keynes, R. J. (1987). Interactions between somite cells: The formation and maintenance of segment boundaries in the chick embryo. *Development* **99**, 261–272.

Stern, H. M., Brown, A. M., and Hauschka, S. D. (1995). Myogenesis in paraxial mesoderm: Preferential induction by dorsal neural tube and by cells expressing Wnt-1. *Development* **121**, 3675–3686.

Stern, H. M., and Hauschka, S. D. (1995). Neural tube and notochord promote *in vitro* myogenesis in single somite explants. *Dev. Biol.* **167**, 87–103.

Stott, D., Kispet, A., and Herrmann, B. G. (1993). Rescue of the tail defect of Brachyury mice. *Genes Dev.* **7**, 197–203.

Subramanian, V., Meyer, B. I., and Gruss, P. (1995). Disruption of the murine homeobox gene Cdx1 affects axial skeletal identities by altering the mesodermal expression domains of Hox genes. *Cell* **83**, 641–653.

Suemori, H., Takahashi, N., and Noguchi, S. (1995). Hoxc-9 mutant mice show anterior transformation of the vertebrae and malformation on the sternum and ribs. *Mech. Dev.* **51**, 265–273.

Summerbell, D., and Maden, M. (1990). Retinoic acid, a developmental signalling molecule. *Trends Neurosci.* **13**, 142–147.

Tajbakhsh, S., and Buckingham, M. E. (1994). Mouse limb muscle is determined in the absence of the earlieset myogenic factor myf-5. *Proc. Natl. Acad. Sci. U.S.A.* **91**, 747–751.

Tajbakhsh, S., Rocancourt, D., Cossu, G., and Buckingham, M. (1977). Redefining the genetic hierarchies controling skeletal myogenesis: *Pax-3* and *Myf-5* act upstream of *MyoD*. *Cell* **89**, 127–138.

Takada, S., Stark, K. L., Shea, M. J., Vassileva, G., McMahon, J. A., and McMahon, A. P. (1994). Wnt-3a regulates somite and tailbud formation in the mouse embryo. *Genes Dev.* **8**, 174–189.

Tam, P. P. (1986). A study of the pattern of prospective somites in the presomitic mesoderm of mouse embryos. *J. Embryol. Exp. Morphol.* **92**, 269–285.

Tam, P. P. (1988). The allocation of cells in the presomitic mesoderm during somite segmentation in the mouse embryo. *Development* **103**, 379–390.

Tam, P. P. L., and Beddington, R. S. P. (1986). The metameric organization of the presomitic mesoderm and somite specification in the mouse embryo. *In* "Somites in Development Embryos" (R. Bellairs, D. Ede, and J. Lash, Eds.), pp. 17–36. Plenum Press, New York.

Tam, P. P., and Beddington, R. S. (1987). The formation of mesodermal tissues in the mouse embryo during gastrulation and early organogenesis. *Development* **99**, 109–126.

Tam, P. P., and Beddington, R. S. (1992). Establishment and organization of germ layers in the gastrulating mouse embryo. *Ciba Found. Symp.* **165**, 27–41; discussion, 42–29.

Tam, P. P., and Meier, S. (1982). The establishment of a somitomeric pattern in the mesoderm of the gastrulating mouse embryo. *Am. J. Anat.* **164**, 209–225.

Tam, P. P. L., Meier, S., and Jacobson, A. G. (1982). Differentiation of the metameric pattern in the embryonic axis of the mouse. II. Somitomeric organization of the presomitic mesoderm. *Differentiation* **21**, 109–122.

Tam, P. P., and Tam, S. S. (1992). The somitogenetic potential of cells in the primitive streak and the tail bud of the organogenesis-stage mouse embryo. *Development* **115**, 703–715.

Tam, P. P., and Trainor, P. A. (1994). Specification and segmentation of the paraxial mesoderm. *Anat. Embryol. (Berl.)* **189**, 275–305.

Technau, G. M., and Campos-Ortega, J. A. (1987). Cell autonomy of expression of neurogenic genes of *Drosophila melanogaster. Proc. Natl. Acad. Sci. U.S.A.* **84**, 4500–4504.

Teillet, M. A., Kalchein, C., and Le Douarin, N. M. (1987). Formation of the dorsal root ganglia in the avian embryo: Segmental origin and migratory behavior of neural crest progenitor cells. *Dev. Biol.* **120**, 329–347.

Tepass, U., and Hartenstein, V. (1995). Neurogenic and proneural genes control cell fate specification in the *Drosophila* endoderm. *Development* **121**, 393–405.

Theiler, K. (1988). "Vertebral Malformations." Springer-Verlag, Berlin.

Thiery, J. P., Duband, J. L., and Tucker, G. C. (1985). Cell migration in the vertebrate embryo: Role of cell adhesion and tissue environment in pattern formation. *Annu. Rev. Cell Biol.* **1**, 91–113.

Tomasiewicz, H., Ono, K., Yee, D., Thompson, C., Goridis, C., Rutishauser, U., and Magnuson, T. (1993). Genetic deletion of a neural cell adhesion molecule variant (N-CAM-180) produces distinct defects in the central nervous system. *Neuron* **11**, 1163–1174.

Tosney, K. W. (1987). Proximal tissues and pattened neurite outgrowth at the lumbosacral level of the chick embryo: Deletion of the dermamyotome. *Dev. Biol.* **122**, 540–558.

Tosney, K. W. (1988). Proximal tissues and patterned neurite outgrowth at the lumbosacral level of the chick embryo: Partial and complete deletion of the somite. *Dev. Biol.* **127**, 266–286.

Trainor, P. A., and Tam, P. P. (1995). Cranial paraxial mesoderm and neural crest cells of the mouse embryo: Co-distribution in the craniofacial mesenchyme but distinct segregation in branchial arches. *Development* **121**, 2569–2582.

Trainor, P. A., Tan, S. S., and Tam, P. P. (1994). Cranial paraxial mesoderm: Regionalisation of cell fate and impact on craniofacial development in mouse embryos. *Development* **120**, 2397–2408.

Vallés, A. M., Boyer, B., and Thiery, J. P. (1991). Adhesion systems in embryonic epithelial-to-mesenchyme transformations and in cancer invasion and metastatis. *In* "Cell Motility Factors" (I. D. Goldberg, Ed.), pp. 17–34. Birkhäuser Verlag, Basel.

van der Lugt, N. M. T., Domen, J., Linders, K., van Roon, M., Robanus-Maandag, E., te Riele,

H., van der Valk, M., Deschamps, J., Sofroniew, M., van Lohuizen, M., and Berns, A. (1994). Posterior transformation, neurological abnormalities, and severe hematopoietic defects in mice with a targeted deletion of the *bmi-1* proto-oncogene. *Genes Dev.* **8,** 757–769.

van Eeden, F. J. M., Granato, M., Schach, U., Brand, M., Furutani-Seiki, M., Haffter, P., Hammerschmidt, M., Heisenberg, C.-P., Jiang, Y.-J., Kane, D. A., Kelsh, R. M., Mullins, M. C., Odenthal, J., Warga, R. M., Allende, M. L., Weinberg, E. S., and Nusslein-Volhard, C. (1996). Mutations affecting somite formation and patterning in the zebrafish, *Danio rerio. Development* **123,** 153–164.

Vässin, H., Bremer, K. A., Knust, E., and Campos-Ortega, J. A. (1987). The neurogenic gene *Delta* of *Drosophila melanogaster* is expressed in neurogenic territories and encodes a putative transmembrane protein with EGF-like repeats. *EMBO J.* **6,** 3431–3440.

Veini, M., and Bellairs, R. (1983). Experimental analysis of control mechanisms in somite segmentation in avian embryos. I. Reduction of material at the blastula state in *Coturnix coturnix japonica. J. Embryol. Exp. Morphol.* **74,** 1–14.

Veini, M., and Bellairs, R. (1986). Health shock effects in chick embryos. *In* "Somites in Developing Embryos" (R. Bellaris, D. Ede, and J. Lash, Eds.), pp. 135–146. Plenum Press, New York.

Veini, M., and Bellairs, R. (1991). Early mesoderm differentiation in the chick embryo. *Anat. Embryol. (Berl.)* **183,** 143–149.

Verbout, A. J. (1976). A critical review of the 'neugliederung' concept in relation to the development of the vertebral column. *Acta Biotheor.* **25,** 219–258.

Verbout, A. J. (1985). The development of the vertebral column. *Adv. Anat. Embryol. Cell Biol.* **90,** 1–122.

Vivarelli, E., and Cossu, G. (1986). Neural control of early myogenic differentiation in cultures of mouse somite. *Dev. Biol.* **117,** 319–325.

von Ebner, V. (1888). Urwirbel und Neugliederung der Wirbelsäule. *Sitzungsber. Akad. Wiss. Wien III* **97,** 194–206.

von Ebner, V. (1892). Über die Beziehungen der Wirbel zu den Urwirbeln. *Sitzungsber. Akad. Wiss. Wien III* **101,** 235–260.

Wachtler, F., Christ, B., and Jacob, H. J. (1982). Grafting experiments on determination and migratory behaviour of presomitic, somitic and somatopleural cells in avian embryos. *Anat. Embryol. (Berl.)* **164,** 369–378.

Watterson, R. L., Fowler, I., and Fowler, B. J. (1954). The role of the neural tube and notochord in development of the axial skeleton of the chick. *Am. J. Anat.* **95,** 337–399.

Weinstein, D. C., Ruiz i Altaba, A., Chen, W. S., Hoodless, P., Prezioso, V. R., Jessell, T. M., and Darnell, J. E., Jr. (1994). The winged-helix transcription factor HNF-3 beta is required for notochord develoment in the mouse embryo. *Cell* **78,** 575–588.

Wharton, K. A., Yedvobnick, B., Finnerty, V. G., and Artavanis Tsakonas, S. (1985). opa: A novel family of transcribed repeats shared by the Notch locus and other developmentally regulated loci in *D. melanogaster. Cell* **40,** 55–62.

Wilkinson, D. G., Bhatt, S., and Herrmann, B. G. (1990). Expression pattern of the mouse T gene and its role in mesoderm formation [see comments]. *Nature* **343,** 657–659.

Wilkinson, D. G., Peters, G., Dickson, C., and McMahon, A. P. (1988). Expression of the FGF-related proto-oncogene *int-2* during gastrulation and neurulation in the mouse. *EMBO J.* **7,** 691–695.

Williams, B. A., and Ordahl, C. P. (1994). Pax-3 expression in segmental mesoderm marks early stages in myogenic cell specification. *Development* **120,** 785–796.

Williams, R., Lendahl, U., and Lardelli, M. (1995). Complementary and combinatorial patterns of Notch gene family expression during early mouse development. *Mech. Dev.* **53,** 357–368.

Wilson, V., and Beddington, R. S. P. (1996). Cell fate and morphogenetic movement in the late mouse primitive streak. *Mech. Dev.* **55,** 79–89.

Wilting, J., Brand Saberi, B., Huang, R., Zhi, Q., Kontges, G., Ordahl, C. P., and Christ, B. (1995). Angiogenic potential of the avian somite. *Dev. Dyn.* **202,** 165–171.

Winnier, G., Blessing, M., Labosky, P. A., and Hogan, B. L. (1995). Bone morphogenetic protein-4 is required for mesoderm formation and patterning in the mouse. *Genes Dev.* **9,** 2105–2116.

Yamaguchi, T. P., Conlon, R. A., and Rossant, J. (1992). Expression of the fibroblast growth factor receptor FGFR-1/flg during gastrulation and segmentation in the mouse embryo. *Dev. Biol.* **152,** 75–88.

Yamaguchi, T. P., Harpal, K., Henkemeyer, M., and Rossant, J. (1994). fgfr-1 is required for embryonic growth and mesodermal patterning during mouse gastrulation. *Genes Dev.* **8,** 3032–3044.

Yamaguchi, T. P., and Rossant, J. (1995). Fibroblast growth factors in mammalian development. *Curr. Opin. Genet. Dev.* **5,** 485–491.

Yu, B. D., Hess, J. L., Horning, S. E., Brown, G. A., and Korsmeyer, S. J. (1995). Altered Hox expression and segmental identity in Mll-mutant mice. *Nature* **378,** 505–508.

Zhao, J. J., Lazzarini, R. A., and Pick, L. (1996). Functional dissection of the mouse Hox-a5 gene. *EMBO J.* **15,** 1313–1322.

Index

Contents of Previous Volumes

Volume 37

Meiosis and Gametogenesis

Guest edited by Mary Ann Handel